AC Electrokinetics:
colloids and nanoparticles

MICROTECHNOLOGIES AND MICROSYSTEMS SERIES

Series Editor: **Professor R. Pethig**
University of Wales, Bangor, UK

* Forthcoming

AC Electrokinetics: colloids and nanoparticles

Hywel Morgan

and

Nicolas G Green

RESEARCH STUDIES PRESS LTD.
Baldock, Hertfordshire, England

RESEARCH STUDIES PRESS LTD.

16 Coach House Cloisters, 10 Hitchin Street, Baldock, Hertfordshire, SG7 6AE, England
www.research-studies-press.co.uk

and

Institute *of* Physics PUBLISHING, Suite 929, The Public Ledger Building,
150 South Independence Mall West, Philadelphia, PA 19106, USA

Marketing:

Institute *of* Physics PUBLISHING, Dirac House, Temple Back, Bristol, BS1 6BE, England
www.bookmarkphysics.iop.org

Distribution:

NORTH AMERICA

AIDC, 50 Winter Sport Lane, PO Box 20, Williston, VT 05495-0020, USA
Tel: 1-800 632 0880 or outside USA 1-802 862 0095, Fax: 802 864 7626, E-mail: orders@aidcvt.com

UK AND THE REST OF WORLD

Marston Book Services Ltd, P.O. Box 269, Abingdon, Oxfordshire, OX14 4YN, England
Tel: + 44 (0)1235 465500 Fax: + 44 (0)1235 465555 E-mail: *direct.order@marston.co.uk*

Library of Congress Cataloguing-in-Publication Data

Morgan, Hywel, 1960-

    Electrokinetic technologies for sub-micron particles / Hywel Morgan and Nicolas G. Green.
      p. cm. -- (Microtechnologies and Microsystems series ; 2)
    Includes bibliographical references and index.
    ISBN 0-86380-255-9 (alk. paper)
    1. Nanoparticles. 2. Electrokinetics. I. Green, Nicolas G., 1972- II. Title. III. Series.

TA418.78 .M67 2001

620'.5--dc21

                                  2001031870

British Library Cataloguing in Publication Data

A catalogue record for this book is available from the British Library.

ISBN 0 86380 255 9

Printed in Great Britain by SRP Ltd., Exeter

To Heather and Amy

Editorial foreword

The first book in this series, by Michael Koch, Alan Evans and Arthur Brunnschweiler, provided a comprehensive introduction to the theory, modeling and fabrication of microfluidic systems.

In this book Hywel Morgan and Nicolas Green extend our understanding of the interaction of electric fields with particles and fluids in microsystems. The authors show how electrokinetic phenomena can be employed to manipulate and characterise particles within such microsystems. They have also taken the bold step to explore the relatively unchartered but potentially rich hunting ground of systems that operate at the submicrometric or nanometric scale.

Two important concepts are brought out in this book, namely the inter- relationship between electrokinetic and electrohydrodynamic forces for controlling the behaviour of fluids and particles, and the subtle balance that can exist between deterministic and stochastic forces. I know of no other source where these two topics, of such fundamental importance to the development of many microsystems, are developed and presented with such clarity and authority.

Ron Pethig
School of Informatics
University of Wales, Bangor

and

Aura BioSystems
Palo Alto, California

September 2002

Acknowledgements

First and foremost we would like to acknowledge the support and enthusiasm shown by the series editor, Ron Pethig, over the period of time it took to write this manuscript and for inviting us to put pen to paper in the first instance. We give special thanks to Mike Hughes, David Holmes, David Bakewell, Lili Cui, Mary Flynn, Bill Monaghan, Mary Robertson, Joel Milner and Frazer Rixon for their contributions to the many experimental images that have been reproduced in the book; to Antonio Ramos for many hours of stimulating discussion on electrohydrodynamics and for his contributions to chapters five and eight; to Antonio Castellanos and Antonio González for their enthusiasm that has led to such fruitful collaborations; and to Richard Berry for his hospitality at Oxford.

We would also like to thank Dr Lyndia Green and Mrs Amy Green for taking the time to read the manuscript and correct grammar and spelling. We gratefully acknowledge Karen Kaler for supplying figure 12.9 and J. Lyklema, J. de Coninck and S. Rovillard for supplying figure 6.9. Thanks must also go to Mike Arnold, Yuri Feldman, Jan Gimsa, Irina Ermolina and Margot Thomas for stimulating discussions; and to Heather, Amy, Ruth and Owen for their patience with the intrusion that this work made on the authors' lives.

Nicolas Green would like to acknowledge the award of a Royal Academy of Engineering Fellowship; Hywel Morgan would like to acknowledge the award of a Royal Society Leverhulme Trust Senior Research Fellowship which, together with a visiting fellowship to St Catherine's College Oxford ensured that this work could be brought to a conclusion.

Finally if there are any errors or omission, then of course the fault is entirely ours, but please inform us.

Sepetmber 2002
Hywel Morgan
Nicolas G Green

Table of contents

Chapter One

Introduction and overview

The past decade has seen an explosion of interest in microsystems for biological and chemical analysis, with particular emphasis directed towards the development of a "Lab-on-a-chip". The underpinning concept of the Lab-on-a-chip is the integration of several functional units such as separation, concentration and detection systems onto the one device. This technology should enable a diverse range of tasks to be performed on a single microchip, with applications in a wide range of fields such as screening combinatorial libraries or rapid diagnosis of complex disease.

One of the first and often most important tasks in a biochemical process is the manipulation, concentration and/or separation of different types of bioparticles or biomolecules. There are many techniques that can be used for manipulating or sorting particles, either singly or *en masse*, including optical tweezers, ultrasound, magnetic sorting (MACS), fluorescence (FACS), filtration, centrifugation and electric field-based approaches. Many of these methods have been miniaturised onto microchips and for recent developments in this field, the reader should consult the proceedings of the annual micro Total Analysis Systems (µTAS) conference (Ramsey and van den Berg 2001). A widely used technique, both for moving and separating materials is that of electrophoresis or electroosmosis. In this method, large DC voltages are used to move conducting electrolytes or charged molecules along a network of narrow channels, often etched into a glass substrate. Electric field-based manipulation and separation methods are highly suitable for integration into microdevices, and so successful has this technology been, that Lab-on-a-chip systems, based on electrophoretic separation and analysis systems, are now commercially available (for examples see the websites given in the reference list).

The motion of particles and fluids in DC fields has been the subject of study for over 100 years, but it is only during the past three decades that the behaviour of particles in AC fields, *i.e.* AC electrokinetics, has been studied in any great detail. The pioneer in this field was Herbert Pohl, who wrote the classic text on dielectrophoresis (Pohl 1978).

Owing to the ease with which microelectrodes can be fabricated into Lab-on-a-chip microdevices, there is much to be gained from exploring and exploiting the interaction of AC as well as DC fields with particles and fluids.

In this book we describe the underlying principles and theory of AC electrokinetics paying particular attention to applications for sub-micrometre and nanoparticles. The relevance of the subject to the movement of both particles **and** fluids within microsystems is discussed. AC electrokinetics is concerned with the study of the effect of AC electric fields on particles. Electric fields often cause fluid motion and in order to describe the consequence of exposing conducting electrolytes to high-strength AC fields, we introduce the science of electrohydrodynamics (EHD). An understanding of the inter-relationship between these two disciplines will lead to the development of new technologies for controlling the behaviour of particles and fluids in microdevices using AC fields.

The book begins with a general introduction to electrostatics and electrokinetic forces, paying particular attention to the issue of charge movement and accumulation at interfaces. We then present a general overview of fluid dynamics, detailing the forces on fluids, in particular the electrical forces. We examine the behaviour of colloidal particles (sub-micrometre to nanoparticles), exploring the issues surrounding deterministic *vs.* stochastic forces. The book concludes with a comprehensive overview of the current state of the art in AC electrokinetics and its application to the characterisation, manipulation and separation of particles, again emphasising the sub-micrometre and nano world.

1.1   AC electrokinetics: a beginner's guide

The application of an AC electric field to a suspension of particles can move both the particles and the fluid in different ways. In order to understand how fluids and colloidal particles move within microsystems, it is important to be able to determine the scale and range of forces that govern the behaviour under different experimental conditions.

1.1.1   Forces on particle and fluids

The particle and the fluid medium can only move through the action of an external force, which broadly speaking can be divided into two categories. The first is a random or stochastic force over which we have little control, whilst the second type of force is a deterministic force, which is almost totally under our control. The random force originates in the thermal energy or temperature of the system and is due to molecules continuously bumping into each other, or into the suspended particles, causing them to move about in a random manner. This is Brownian motion and does not lead to a net unidirectional particle movement. We have little control over this force, other than through changing the viscosity of the suspending medium or the temperature.

Deterministic forces, on the other hand, are almost completely under our control and can be exploited to move particles in well-defined ways. One obvious example of a deterministic force is gravity, which causes both particle and fluid movement. Depending on the time scale of an experiment, this force can be ignored, or indeed utilised to produce, for example, selective sedimentation of particles suspended in a fluid. For particles denser than the surrounding medium, gravity pulls the particle downwards, but for sub-micrometre or nanoparticles this force is usually small and has little or no effect during the time course of a typical experiment. As an example, consider a 500 nm diameter latex sphere which has a density slightly greater than water (1050 kg m^{-3} compared with 1000 kg m^{-3}). In the earth's gravity the particle only moves a distance the order of its diameter every minute, so that over a few minutes the particle will have moved only a few micrometres. By contrast, a biological cell has ten times this diameter and a much greater density and therefore moves up to 100 times faster, so that complete sedimentation of cells from a suspension can occur in as little as a few minutes.

For most applications involving sub-micrometre particles, gravity is of little consequence and the forces that tend to dominate in microdevices are viscous forces and electrical forces. The viscous retardation force is characterised by the Stokes drag acting on the particle. Small particles have low inertia so that during the time frame of an experiment, a change in the velocity of the particle can only arise from a change in the force acting on the particle. Thus, if we can measure the velocity of a particle as a function of time or distance, we can measure the force acting upon it. Another way of looking at this is to say that a particle will always move at a constant velocity in a constant force field (when suspended in a viscous fluid).

Electrical forces can act both on particles and on the suspending fluid. The major electrical forces acting on small particles suspended in a fluid such as water, are electrophoresis (EP) and/or dielectrophoresis (DEP). EP occurs due to the action of the electric field on the fixed, net charge of the particle, whilst DEP only occurs when there are induced charges, and only results in motion in a **non-uniform** field (this can be a DC or an AC field).

Electric fields also produce a force on the suspending medium if there are variations or gradients in the charge density of the fluid. Electroosmosis has its origin in the Coulomb force, which describes the interaction of the field with free charge (in the double layer), which causes constant fluid motion in a DC field. In an AC field, these charges move but the net-effect averaged over one cycle is zero, since charges move an equal amount in each half-cycle but in opposite directions. However, if the field is non-uniform a steady fluid motion can be produced in an AC field. In addition, if a temperature gradient is present then the AC field will also cause fluid motion due to local variations in the permittivity and/or conductivity of the fluid. Predicting the movement of particles due to the combination of these forces can be a complicated task.

The magnitudes of the three dominant forces outlined above, *viz.* Brownian motion, gravity and dielectrophoresis, depend on several variables, principally the dimensions of the system, the size of the particle, and for the electrokinetic forces, with voltage and conductivity. The changing effect of the three forces on particle behaviour can be illustrated with reference to figure 1.1. Here the displacement of a particle has been calculated over a time interval of 1 second, plotted as a function of particle size. The first point to note is the effect of Brownian motion. For large particles this is minuscule, resulting in a displacement of the order of nanometres,

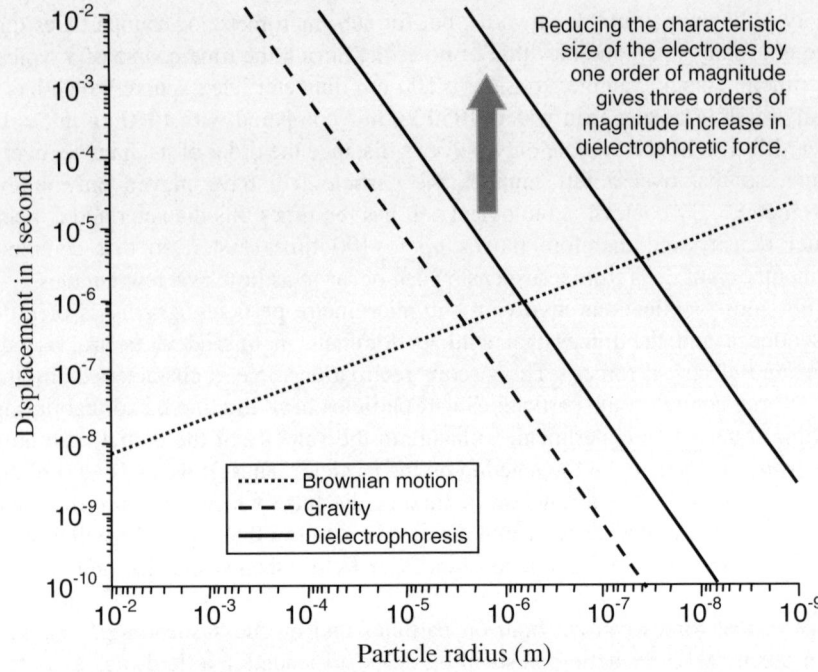

Fig. 1.1 Diagram of the relative magnitudes of displacement in one second due to three physical mechanisms: Brownian motion, gravity and dielectrophoresis as a function of particle radius from 1 centimetre down to 1 nanometre.

so that to all extents and purposes it can be ignored even for particles up to 1 μm in radius. Compare this with the influence of the gravitational force. This scales linearly with volume so that the bigger the particles the further they move (as we all know!). In a time frame of a few minutes, this force can lead to relatively large displacements of particles such as cells, but its effect is almost insignificant for sub-micrometre particles, giving rise to displacements of the order of nanometres for a 0.5 μm diameter particle.

Contrast these effects with dielectrophoretic forces. The plot shows two lines of constant dielectrophoretic force, as would be typically generated using micrometre scale electrodes. For particles larger than 1 μm, the figure shows that it is a relatively trivial matter to ensure that the DEP force dominates over both the gravitational force and Brownian motion, *i.e.* DEP is the sole deterministic force over this period of time. However, the situation is not as simple for smaller particles. Take for example a 50 nm diameter particle. The lower DEP force line shows that, although gravity can be ignored, over short time intervals we see that Brownian motion dominates particle displacement. A stronger DEP force leads to a larger deterministic force so that movement becomes possible. As we shall see

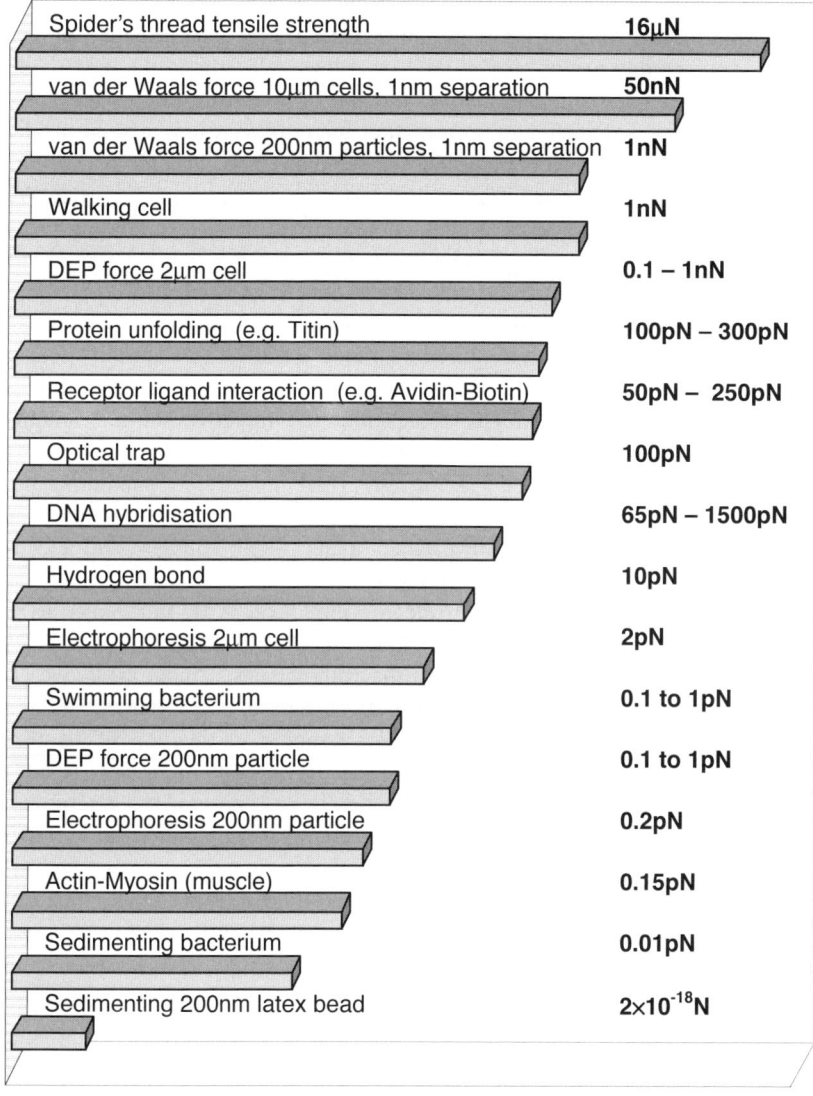

Fig. 1.2 Schematic figure of the relative sizes of some forces typical of the scales of microsystems and nanoscale particles.

later, the DEP force scales with the square of the voltage and inversely with the cube of the distance, so that decreasing the characteristic dimensions of the electrode by one order of magnitude can lead to a three orders of magnitude increase in the DEP force, as shown in the figure. This is a clear advert for using microelectrodes!

It is interesting at this point to examine these forces in a broader context by making reference to some of the wide range of forces that can be found in nature. Figure 1.2 shows a series of bars plotted on a log scale, giving the magnitude of some forces and force interactions. At the top of the scale comes the spider's thread, which has a diameter of only a few microns, is stronger than steel and requires a force of 10 μN to break. Next we see that when particles are very close together (~1 nm), short-range forces dominate the interactions (van der Waals forces). Single ligand interaction forces compare with the forces holding proteins together and these are an order of magnitude lower than the dielectrophoretic force required to manipulate cells; 0.1–1 nN. The hydrogen bond plays a crucial role in stabilising many macromolecular interactions and is an order of magnitude larger than the electrophoretic force required to move a cell. The naturally occurring motive force generated by a bacterium is around 0.1–1 pN, which lies at the upper boundary of the AC electrokinetic forces required to move sub-micrometre or nanoscale particles. For a 200 nm particle, electrophoresis requires a force of 0.2 pN; the dielectrophoretic force is an order of magnitude smaller and the sedimentation force is negligible. In these examples, the electrokinetic forces have been calculated for two sizes of particle using fixed voltages and electrode dimensions commensurate with experiments.

The figure clearly shows that AC electrokinetic forces are in the pico to nano Newton range and, as we shall see in this book, these forces allow us to manipulate particles from cells down to viruses and macromolecules. As an aside, we could ask the question, "Is there an upper limit to the size of an object that can be moved using dielectrophoresis?" Consider a spider hanging at the end of its thread. Would it be possible to pull the spider by dielectrophoresis in an electric field so that the thread breaks? Surprisingly, the answer is yes provided the electrodes are of the same dimensions as the spider and that the spider can withstand ~10 kVolts (at a suitable frequency).

Having explored the issues of forces, particularly deterministic *vs.* stochastic (a topic we will return to specifically in Chapter Nine), we now turn our attention to exploring the fundamentals of electrokinetics.

1.1.2   Charge movement in a field

The key to understanding dielectrophoresis lies in an appreciation of the charge distribution at the interface between two materials of different conductivity and/or permittivity. Conductivity is a measure of the ease with which charge can move through a material, while permittivity is a measure of the energy storage or charge accumulation (at interfaces) in a system.

In dielectrophoresis, the two materials are a small particle and an aqueous electrolyte (such as potassium chloride). In the presence of an applied electric field, charge moves and piles up at either side of the interface between the particle and the electrolyte, as shown schematically in figure 1.3. The amount of charge at the interface depends on the field strength and the electrical properties (conductivity and permittivity) of both the particle and the electrolyte.

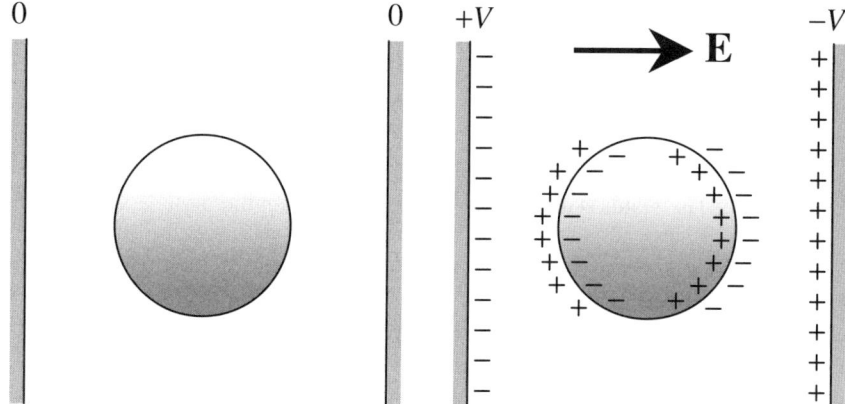

Fig. 1.3 Schematic diagram of how a dielectric particle suspended in a dielectric fluid polarises in a uniform applied electric field **E**.

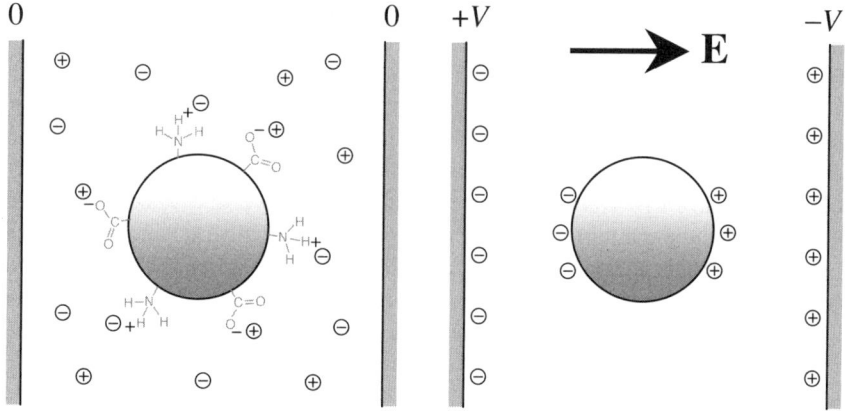

Fig. 1.4 Schematic outline of how a "real" particle polarises in an electrolyte. The chemical groups on the surface of the particle become charged in solution and ions from the electrolyte associate themselves with the surface charge. When a field is applied, these associated ions move around the surface adding to the polarisation of the particle.

In reality, the particle could be a latex sphere, a cell, a virus or a protein molecule. The surfaces of these particles generally have ionisable groups, such as carboxylic acid (COOH) or amino (NH_2) groups, as shown schematically in figure 1.4. In an electrolyte at near neutral pH, most of the groups are charged, negatively for the acid and positively for the basic groups. Changing the pH will alter the net charge on the surface. Interestingly, most cells possess more acid than basic groups and so have a characteristic net negative surface charge density. In order to maintain charge neutrality, ions of opposite charge are attracted to the surface forming a thin layer of counter charge, called the double layer. If the particle is now subjected to an external electric field, the double layer charges experience a force and move. They can either move around the particle towards the electrode of opposite polarity or radially, accumulating at the interface between the particle and the electrolyte. The result is an imbalance in the charge on the particle, as shown in the figure.

1.1.3   The induced dipole

The charge distribution around a particle can be described by introducing the concept of polarisability. Polarisability is a measure of the ability of a material to respond to a field (polarise), but it is also a measure of the ability of a material to produce charge at interfaces.

If we examine the simple picture of a spherical particle suspended in an electrolyte, as discussed in the previous section, and subjected to a uniform applied electric field, there are three cases to consider. When the polarisability of the particle is much greater than the electrolyte, there are more charges just inside the interface than outside, as shown in figure 1.5(a). This means that there is a difference in the charge density on either side of the particle (as shown in the figure), which gives rise to an effective or induced dipole across the particle, aligned with the applied field. Examples of this are a conducting particle in an insulating medium, or a particle with a high dielectric constant (permittivity) in an insulating medium with low dielectric constant. The converse of this is shown in figure 1.5(b), where the particle polarisability is much less than the electrolyte and the net dipole points in the opposite direction. This could be an insulating sphere suspended in a liquid with a high dielectric constant or a high conductivity. The third case (not shown) is where the polarisability of the particle and electrolyte are the same and there is no net dipole. Note that if the field is removed the dipole disappears, so the term "induced" is always used to describe the system. The magnitude of the dipole moment depends on the amount of charge moved and the size of the particle.

These diagrams show the charge distribution some time after the field has been applied. If we reverse the direction of the applied field and wait long enough, the opposite charge distributions will be established, with the dipoles in the same relative direction with respect to the applied field. Following the application of the field, the charges do **not** move instantaneously, typically they take a few microseconds to reach equilibrium. At low frequencies, the movement of the free charge can keep pace with the changing direction of the field. However, as the field frequency increases there comes a point where the charges no longer have sufficient time to respond. At high frequencies, free charge movement is no longer the dominant mechanism responsible for charging the interface, and instead the polarisation of the bound charges (permittivity) dominates. The difference between these two states is termed a dielectric dispersion.

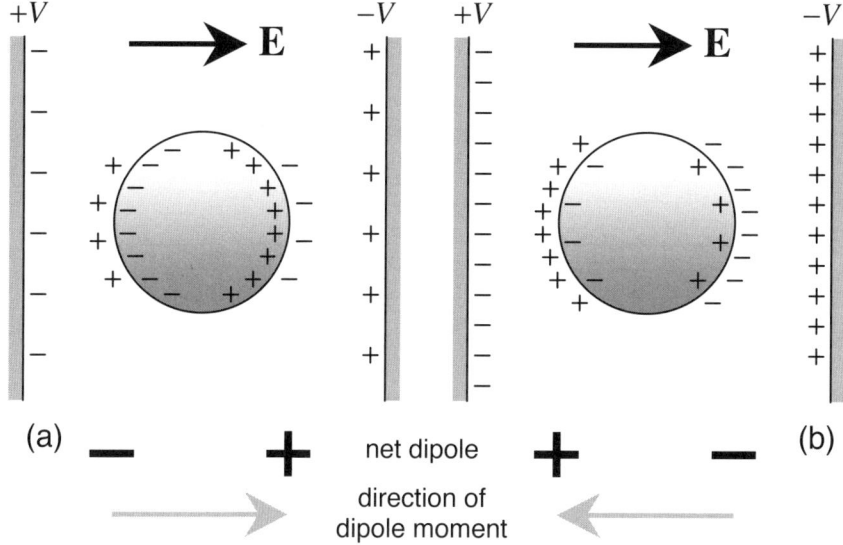

Fig. 1.5 Schematic diagram of how different dielectric particles polarise if they have a much higher (a) or much lower (b) polarisability than the suspending fluid medium. If the polarisability is higher, more charges are produced on the inside of the particle/fluid interface and there is a net dipole across the particle that is parallel to the applied field. If the polarisability is lower, more charges are produced on the outside of the interface and the net dipole points in the opposite direction, against the field.

As an example, consider a spherical particle with high conductivity and low permittivity suspended in an electrolyte of low conductivity and high permittivity (*e.g.* pure water). At low frequencies, this particle behaves in the manner shown in figure 1.5(a) since the conductivities dominate. At high frequencies, the conductivities are no longer important and the polarisation of the particle is similar to figure 1.5(b). The direction of the induced dipole changes from being in alignment with the field at low frequencies, to opposing the field at high frequencies. This serves to illustrate the frequency-dependent nature of interfacial polarisation and the induced particle dipole.

1.1.4 Dielectrophoresis
We now extend this picture and consider the same particle subjected to a non-uniform electric field, as shown in figure 1.6. This figure shows the electric field distribution around a particle for the two cases described above, for both uniform and non-uniform field geometries. Figure 1.6(a) shows a particle with a polarisability greater than the suspending medium. The electric field lines bend

towards the particle, meeting the surface at right angles as if it were a metal sphere, and the field inside the particle is nearly zero. The converse is shown in figure 1.6(b), where the particle polarisability is less than the electrolyte. The field lines now bend around the particle as if it were an insulator. The field inside is similar to that outside. When the polarisability of the particle and electrolyte are the same it is as if the particle did not exist and the field lines are parallel and continuous everywhere.

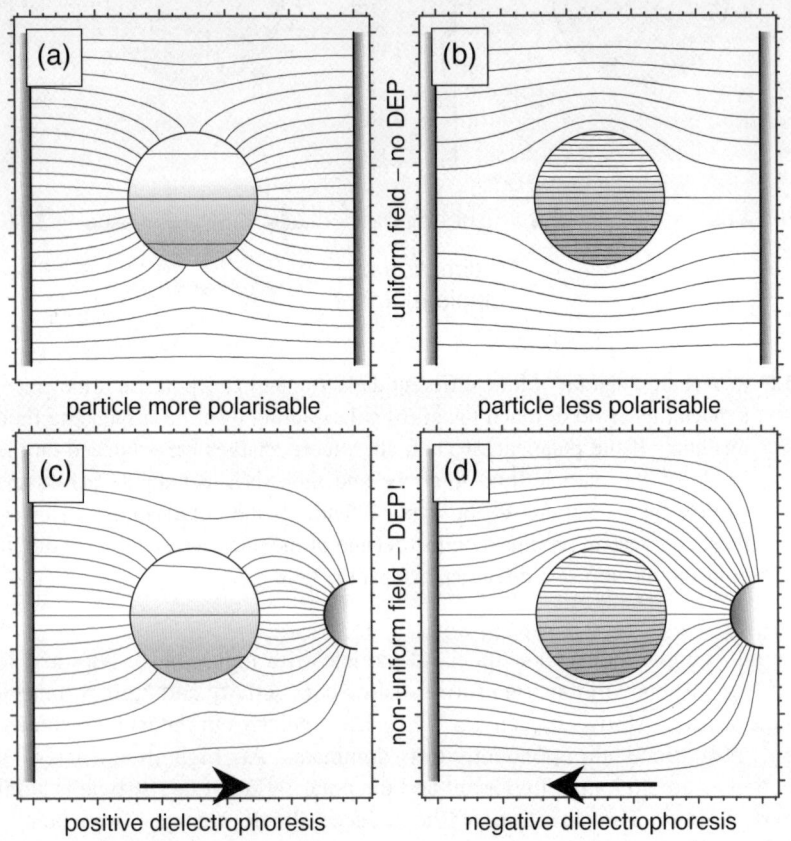

Fig. 1.6 Numerically calculated electric field lines for four different cases, defined by the particle more polarisable or less polarisable than the suspending medium, in a uniform or a non-uniform electric field. For the more polarisable particle (a) and (c), the field lines are drawn to the surface of the particle, becoming increasingly perpendicular as the polarisability increases, and the field strength inside the particle is low. For the less polarisable particle (b) and (d), the field lines are bent around the particle and the field strength inside the particle is high. The arrows show the direction of the force and movement in each case.

The field lines for the same two cases in a **non-uniform** electric field are shown in figures 1.6(c) and 1.6(d). Again the particle polarises and the field lines around the particle behave in a similar manner to the uniform field case. However, by examining the density of electric field lines, we see that the field strength on one side of the particle is greater than the other. This leads to an imbalance of forces on the induced dipole, resulting in particle movement. This effect is called dielectrophoresis. When the polarisability of the particle is greater than its surrounding, the direction of the dipole is with the field and the particle experiences a force called positive DEP; the particle moves towards the strong field region, as shown in figure 1.6(c). The opposite situation gives rise to negative DEP, figure 1.6(d), and the particle moves away from strong electric field regions. Importantly, as described previously, the induced dipole is also a function of frequency. Therefore, we see that the direction in which the particle moves is not only a function of the properties of the particle and the suspending medium but significantly also the frequency of the applied field. It is because of this effect that AC electrokinetics is such a useful tool for the manipulation and separation of particles.

Finally, let us examine the way in which the DEP force varies across a wide frequency spectrum. The force experienced by two solid homogeneous latex particles with different conductivities suspended in an electrolyte is plotted in figure 1.7 as a function of frequency. Note that the force is constant at low frequencies for both particles, since the free charge has ample time to respond to the field. At high frequencies, the force is again constant but opposite in sign, since the dielectric constant of the fluid is much higher than that of latex (80 compared with 2.5). At the intermediate frequency range there is a transition in the force spectrum so that the particle's behaviour switches from positive DEP to negative DEP. At one particular frequency, the polarisability of the particle is identical to the suspending medium, there is no induced dipole and the force is zero. The frequency at which this occurs is different for the two particles so that there is a frequency band over which one particle experiences positive DEP and the other negative, as shown by the shaded region in the figure.

This simple example serves to illustrate two of the main features of dielectrophoresis:

- the intrinsic electrical properties of a particle manifest themselves as a force which varies with the applied frequency of the AC field. Therefore, particles can be moved in a non-uniform AC field.
- two different particles can exhibit quite different force-frequency spectra. The force is not only different in magnitude but also in direction, so that the particles can move in opposite directions in the same field with the same frequency. Dynamic separation of particles is therefore possible.

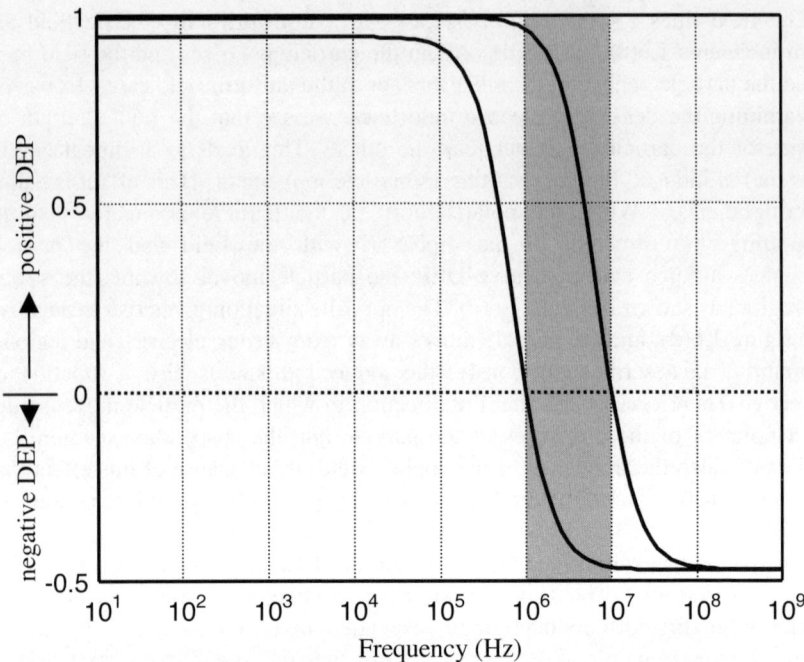

Fig. 1.7 Plot of the normalised dielectrophoretic force against frequency for two different particles. In the shaded area one particle experiences positive dielectrophoresis and the other negative dielectrophoresis, enabling separation in this frequency window.

1.2   Summary

In this brief introduction, we have described the basic underlying principles of AC electrokinetics. We have shown that for sub-micrometre and nanoparticles, thermal effects (Brownian motion and diffusion) are important, but that gravity is often of little consequence. AC electrokinetic forces can be made to dominate over thermal forces so that the deterministic movement of sub-micrometre particles is not only possible but can be used for manipulating and characterising particles. We have also briefly illustrated the value of AC electrokinetics for particle separation technologies.

In the next chapter, we examine the basic principles of the subject by looking at the physics of electrostatic interactions and the polarisation of dielectrics.

1.3   Overview

The remainder of the book falls into four categories which, broadly speaking, are: electrokinetics; fluid dynamics and electrohydrodynamics; stochastic forces and Brownian motion; and sub-micrometre particle manipulation, characterisation and separation technologies. In Chapters Two to Four, we review classical electrodynamics and derive the governing equations for the electric forces experienced by polarisable particles in electric fields.   In Chapter Five, we introduce the basic concepts of fluid dynamics, how fluids move in microsystems and how particles move in a viscous fluid.   Also, we explain classical electrohydrodynamics, the mechanism by which the electric field heats the fluid resulting in fluid motion. Chapter Six begins by outlining the fundamental physical chemistry of charge movement in an electrolyte, and how the permittivity and conductivity of the suspending medium are governed by the molecular and ionic composition. This is important since it directly influences the electrokinetic behaviour of particles, primarily by controlling the magnitude of the induced polarisation. The chapter continues by describing the electrical double layer and the effect that this has, both on the particle polarisation and the polarisation of the electrode-electrolyte interface. We then look at electroosmosis and conclude the chapter with a discussion on particle-particle interaction and colloidal stability.   In Chapter Seven, we outline the physical mechanism that governs the polarisation of a particle suspended in an electrolyte when subjected to an electric field. Both electrophoresis and dielectrophoresis are explained at the mechanistic level, particularly focussing on the interaction of the electric field with the particle charge. In Chapter Eight, we return to the subject of electrohydrodynamics and electroosmosis and show how these phenomena manifest themselves in practice. The relative magnitudes and scales of the electric field-driven fluid flows are compared with dielectrophoretic forces, by reference to experiments and numerical simulations.

Having discussed the range and types of forces present during the AC electrokinetic manipulation and characterisation of particles, Chapter Nine examines the fundamental issues surrounding the controlled manipulation and characterisation of sub-micrometre and nanoparticles. The fundamentals of Brownian motion are covered, together with the governing principles of electrokinetics in a stochastic system. We then show how to analyse the behaviour of a single particle or a large collection of particles in a force field.

The last section of the book looks at the technology of AC electrokinetics and its practice. In Chapter Ten, the design of three different electrode arrays are discussed, together with different techniques for analysing the electric fields created by these electrodes. Where possible, the analytical expression for the electric field is given so that the reader can calculate forces on particles. The book concludes with an extensive review of AC electrokinetics. Chapter Eleven shows how particles can be manipulated and their dielectric properties measured using several different techniques. For example, we show how single nanoparticles can be trapped in field cages and how the dielectrophoretic properties of different virus particles can be measured. Finally, Chapter Twelve reviews the important field of separation science and the impact that AC electrokinetics has had in this field, examining the potential for developing systems for separating sub-micrometre particles.

In writing this book our objectives have been to bring together a wide range of disciplines into a single textbook. These range, for example, from the fundamentals of colloid science to simulation of complex electric fields and from Brownian motion to experimental AC electrokinetics of sub-micrometre and nano-particles.

1.4   References

Pohl H.A. *Dielectrophoresis* Cambridge Uni. Press, Cambridge, UK (1978).

Ramsey J.M. and van den Berg A. (eds) *Micro Total Analysis Systems 2001 Proc. μTAS 2001 Symposium, Monterey, CA, USA 2001,* Kluwer Academic Publishers.

http://www.calipertech.com/technologies/index.html

http://www.chem.agilent.com/Scripts/PCol.asp?lPage=50

http://www.nanogen.com/technology/index.asp

Chapter Two

Electrostatics and Dielectrics

This chapter outlines the basic electrostatic principles required for the analysis of AC electrokinetics and electrohydrodynamics. Further comprehensive reading may be found in any number of textbooks in this area (examples include Stratton 1941; Bleaney and Bleaney 1962; Kraus 1984; Lorrain *et al.* 2000).

2.1   Charge

The AC electrokinetic force on a particle depends on the applied field, but more importantly on the amount and distribution of charge in the environment of the particle. The unit of charge is the Coulomb (C) and, for example, the electron has a charge of -1.6×10^{-19} C.

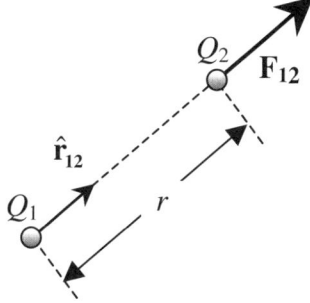

In this book we assume that the charges are stationary or have small velocities and accelerations. The movement of the charge therefore, does not change the electric field and magnetic fields can be assumed to be negligible. This is termed a quasi-electrostatic system.

Fig. 2.1 The charge Q_1 exerts a force $\mathbf{F}_{12}$ on charge Q_2.

2.1.1   Force on a charge: Coulomb's law

The force exerted by a stationary point charge Q_1 on a second stationary point charge Q_2 is described by Coulomb's law

$$\boxed{\mathbf{F}_{12} = \frac{Q_1 Q_2}{4\pi \varepsilon_o r^2} \hat{\mathbf{r}}_{12}} \qquad (2.1)$$

where r is the distance between the two charges and the unit vector $\hat{\mathbf{r}}_{12}$ points from Q_1 to Q_2. This is shown schematically in figure 2.1. Experiments show that the constant ε_o has the value 8.854×10^{-12} Farads m^{-1} in S.I. units[†]. The force is attractive if the two charges are opposite in sign and repulsive if they have the same sign. The force resulting from the action of Q_2 on Q_1 is equal and opposite.

2.1.2 Electric field, the potential and the Principle of Superposition

The *electric field strength* **E** from a charge Q, is defined as the force per unit charge that produces the Coulomb force on an arbitrary unit (1 C) positive test charge.

The electric field for a negative point charge can be found from Coulomb's Law as

$$\mathbf{E} = \frac{Q}{4\pi\varepsilon_o r^2}\hat{\mathbf{r}}_{12} \qquad (2.2)$$

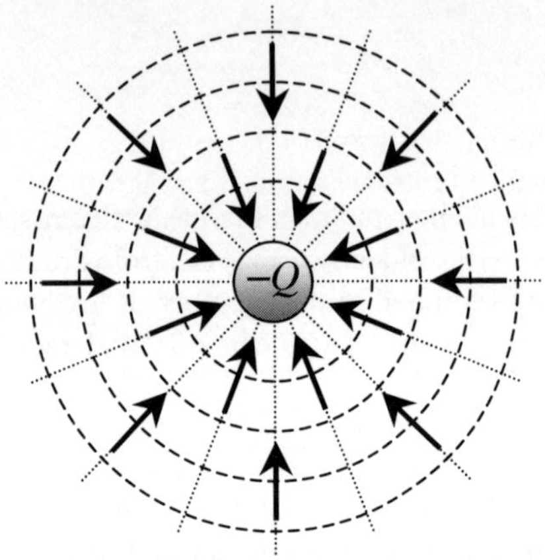

Fig. 2.2 The equipotentials (dashed), electric field lines (dotted) and electric field vectors (arrows) around a negative point charge.

The direction and magnitude of the field for a negative point charge is shown by the dotted lines and vectors in figure 2.2.

The *electric potential* is defined to be the scalar ϕ such that the electric field is given by $\mathbf{E} = -\nabla\phi$. The unit of electric potential is Volts ($V = J C^{-1}$) and the unit of **E** is therefore V m^{-1}. The lines of constant potential for a point charge are also shown in figure 2.2.

The *Principle of Superposition* states that the net effect of several charges is the linear sum of the effects of the individual charges acting alone. In the case of the electric field, the total field at an arbitrary point is the *vector* sum of the electric fields due to each charge at that point. The principle similarly applies to the potential, with the potential at a point given by the sum of the individual potentials.

2.1.3 Gauss's law and Poisson's equation

Gauss's law and Poisson's equation are important for a number of reasons, primarily because these equations are solved in order to obtain analytical or numerical representations of the electric potential or field in a system (see Chapter Ten). Gauss's Law relates the vector **E** to the volume electric charge density ρ :

[†] The permittivity of free space follows from the the value of the speed of light and the defined value of the permeability of free space, where $c^2 = (\mu_o\varepsilon_o)^{-1}$, with c the speed of light and μ_o the permeability of free space, defined to be $4\pi \times 10^{-7}$ V s (A m)$^{-1}$.

$$\nabla \cdot \mathbf{E} = \frac{\rho}{\varepsilon_o} \qquad (2.3)$$

Substituting for the potential gives

$$\nabla^2 \phi = -\frac{\rho}{\varepsilon_o} \qquad (2.4)$$

This is one form of Poisson's equation, which will be discussed in more detail later. In most cases, the charge density is zero, giving Laplace's equation

$$\nabla^2 \phi = 0 \qquad (2.5)$$

2.1.4 Electrical potential energy

The energy stored in a continuous charge distribution is given by the integral of the field over any volume υ, enclosing all the charge

$$U = \frac{1}{2} \int_\upsilon \phi \rho d\upsilon = \frac{1}{2} \int_\upsilon \varepsilon_o |\mathbf{E}|^2 d\upsilon \qquad (2.6)$$

The second form requires that the volume υ encompass all regions where $\mathbf{E}$ is present.

2.1.5 Ohm's law

The current which flows in response to an applied electric field is in most situations proportional to the field. This relationship is described by a general form of Ohm's law

$$\mathbf{J} = \sigma \mathbf{E} \qquad (2.7)$$

where $\mathbf{J}$ is the *electric current density* (Am^{-2}) and σ is the *electrical conductivity* (Sm^{-1}). The current density can be written as the product of the mean charge drift velocity $\mathbf{v}_c$ and the volume charge density *i.e.*

$$\mathbf{J} = \rho \mathbf{v}_c \qquad (2.8)$$

and substituting this into equation (2.7), we have

$$\sigma = \rho \frac{|\mathbf{v}_c|}{|\mathbf{E}|} = \rho \mu = (nq)\mu \qquad (2.9)$$

where q is the charge on the electron, μ is the mobility of the charge in the electric field (in units of $m^2\ V^{-1}\ s^{-1}$), and n is the number density of charge (m^{-3}).

2.1.6   The charge conservation equation

The *charge conservation equation* relates the rate of change with respect to time of the volume charge density ρ to the current density **J**

$$\nabla \cdot \mathbf{J} = -\frac{\partial \rho}{\partial t} \qquad (2.10)$$

In the steady state, $\dfrac{\partial \rho}{\partial t} = 0$ and therefore $\nabla \cdot \mathbf{J} = 0$.

2.2   Dipoles

The electrical dipole is formed from a simple distribution of charges and is fundamental to many aspects of electromagnetics, including AC electrokinetics. A dipole consists of two charges of the same magnitude Q and opposite sign, separated by a distance d, as shown in figure 2.3.

As described in Chapter One, in particle AC electrokinetics a dipole moment forms due to the action of a field on a polarisable particle. In a non-uniform electric field the force imbalance on the dipole moves the particle.

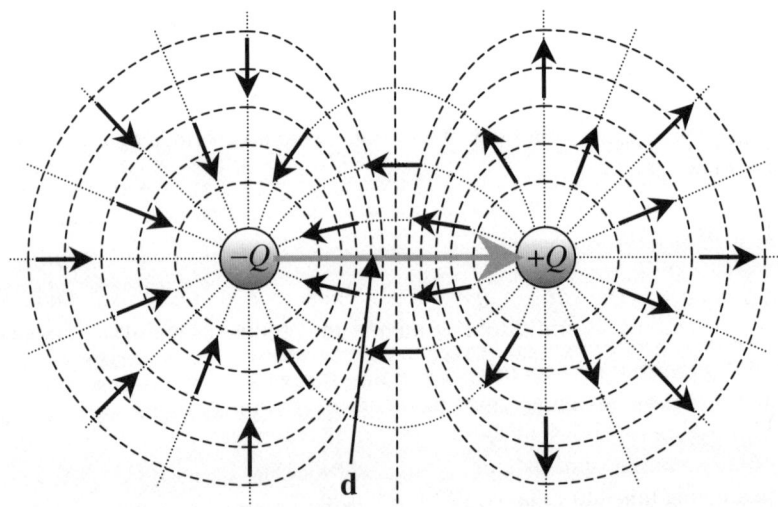

Fig. 2.3 The equipotentials (dashed), electric field lines (dotted) and electric field vectors (black arrows) around a dipole consisting of a negative and a positive charge. The grey arrow between the two charges indicates the vector **d** and the direction of the dipole moment.

2.2.1   The dipole moment

The *dipole moment* is the vector **p** directed from the negative to the positive charge, as shown in figure 2.3, such that

$$\mathbf{p} = Q\mathbf{d} \tag{2.11}$$

where the vector **d** is shown in the figure. The dipole moment has units of Coulomb-metre or Debye (where 1 Debye = 3.33×10^{-30} C m).

2.2.2   The dipole field and potential

In spherical polar co-ordinates, the potential of a dipole at distances $r \gg d$ is

$$\phi = \frac{\mathbf{p} \cdot \hat{\mathbf{r}}}{4\pi\varepsilon_o r^2} = \frac{|\mathbf{p}|\cos\theta}{4\pi\varepsilon_o r^2} \tag{2.12}$$

and the electric field is

$$\mathbf{E} = \frac{|\mathbf{p}|}{4\pi\varepsilon_o r^3}(2\cos\theta\hat{\mathbf{r}} + \sin\theta\hat{\boldsymbol{\theta}}) \tag{2.13}$$

For large distances ($r \square\ d$), the dipole field falls off as the inverse third power of r. Close to the dipole, the field expression contains higher order terms. The field and potential around a dipole are shown in figure 2.3.

Many molecules possess permanent dipole moments (Pethig 1979; Grant *et al.* 1978). These can range in size from that of a small molecules such as water (|**p**| = 1.8 Debye) to large globular protein molecules (*e.g.* 380 Debye for Serum Albumin and 170 Debye for Myoglobin). The structure of a water molecule and the corresponding dipole moment is shown in figure 2.4. The two hydrogen atoms in the molecule are on the same side of the oxygen atom as shown, making a bond angle of 106°. In this molecule, the oxygen is slightly electronegative and the two protons slightly electropositive so that the molecule has

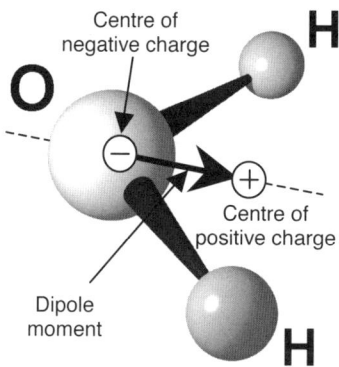

Fig. 2.4   Schematic diagram of the dipole moment of a water molecule, showing the positions of the oxygen (O) and the hydrogen (H) atoms.   The difference between the positions of the centres of positive and negative charge gives the dipole moment.

a dipole moment pointing from the oxygen atom to the point midway between the two hydrogen atoms. This is an example of a molecular fixed dipole. Importantly, an applied electric field can also *induce* a dipole in particles where none existed previously. We shall return to this later in Chapter Three.

2.2.3 The potential energy of a dipole in a field

The potential energy of a dipole in a uniform field $\mathbf{E}$ depends on the magnitude of the dipole and the angle it makes with the field. This is given by

$$U = -|\mathbf{E}|Qd\cos\theta = -\mathbf{E}\cdot\mathbf{p} \tag{2.14}$$

where θ is the angle between the dipole and the field. The energy is zero when the dipole is perpendicular to the field and maximum when the dipole opposes the field. This equation applies to an ideal dipole in a uniform time invariant field; for a non-uniform or time-dependent field the equation becomes more complicated.

2.2.4 The force on a dipole

The force on a dipole in a field is given by the following expression

$$\boxed{\mathbf{F} = (\mathbf{p}\cdot\nabla)\mathbf{E}} \tag{2.15}$$

This expression, which will be derived in Chapter Three, shows that the force on a dipole is only non-zero in a non-uniform electric field. Again, for time-dependent fields, this expression is different.

2.2.5 The torque on a dipole

When a dipole sits in a uniform electric field, each charge on the dipole experiences a force equal and opposite, tending to align the dipole parallel to the field. Figure 2.5 shows a schematic diagram of a dipole in a field, demonstrating how the two charges experience equal but opposite forces which give rise to a torque that acts to align the dipole with the field.

Therefore, if a particle contains no net charge but does contain a dipole, it will not experience a force but will experience a torque given by

$$\boxed{\mathbf{\Gamma} = \mathbf{p}\times\mathbf{E}} \tag{2.16}$$

This torque always acts to align the dipole with the field. However, small dipoles, such as molecules, will not completely align with the field since they are subject to the randomising forces of Brownian motion, a complication that will be discussed in Chapter Nine. In typical dielectric spectroscopy measurements, the electric fields are only of sufficient magnitude to align individual molecules by fractions of degrees, but the total effect of the alignments of many molecules gives rise to a large net or average alignment.

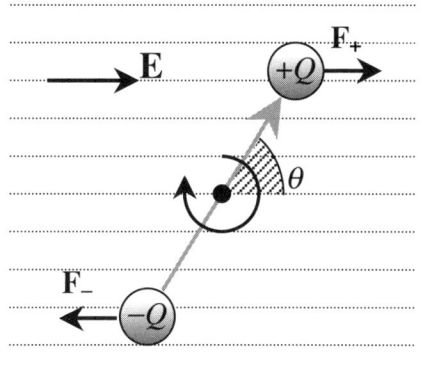

Fig. 2.5 In a uniform field **E** (indicated by the vector and the dotted field lines), the two charges experience equal and opposite forces resulting in a torque about the centre point of the dipole.

2.3   Dielectrics

A dielectric material is a material that contains charges which *polarise* under the influence of an applied electric field. These charges are bound within the material and can only move short distances when the field is applied, the negative and positive charges moving in opposite directions to form induced dipoles. Some materials also consist of molecules with permanent dipoles which polarise by orienting with the field, an example of which is pure water.

The average dipole moment of the molecules of this material is proportional to the magnitude of the field, *i.e.*

$$\mathbf{p}_{av} = \alpha \mathbf{E}' \tag{2.17}$$

where $\mathbf{E}'$ is the local electric field in the vicinity of the dipole. The proportionality α is called the polarisability (average dipole moment per unit field strength) and is a measure of the response of the dielectric to the electric field. It has units of $(C\ V^{-1}\ m^2)$ or $(F\ m^2)$.

2.3.1   The polarisation

In a dielectric consisting of n molecules per cubic metre, the *polarisation* **P** (the dipole moment per unit volume) is

$$\mathbf{P} = n\mathbf{p}_{av} = n\alpha \mathbf{E}' \tag{2.18}$$

The polarisation and the displacement of the charge gives rise to a net charge at points in the dielectric or at the surface. These charges are referred to as bound or polarisation charges. The bound volume charge density ρ_b is given by

$$\rho_b = -\nabla \cdot \mathbf{P} \tag{2.19}$$

2.3.2   Polar and non-polar dielectrics

Dielectric materials can be classified as polar and non-polar. Polar dielectrics contain molecules that possess a permanent dipole moment independent of the application of an electric field. When a field is applied, the dipoles experience a torque aligning them with the field, as discussed above. The average alignment of the molecules polarises the dielectric. Non-polar dielectrics are materials whose molecules do not possess any permanent dipole moment. An applied electric field causes slight displacement of the individual charges within or around the molecules, giving rise to an induced dipole.

2.3.3   Polarisation mechanisms

There are three basic molecular polarisation mechanisms that can occur when an electric field is applied to a dielectric: *electronic*, *atomic* and *orientational* (or dipolar). In addition, there is a long-range polarisation which is due to accumulation of charge carriers at interfaces in the dielectric. This is *interfacial polarisation*.

(i) Electronic α_e

The principle of electronic polarisation is shown in figure 2.6. In an electric field, the centre of charge of the electron cloud in an atom moves slightly with respect to the centre of charge of the nucleic charges. For an applied field of $10^6 - 10^7$ V m^{-1} (compared to the internal field of the atom which is of the order of 10^{11} V m^{-1}), the displacement is of the order of 10^{-8} times the diameter of an atom.

The electronic polarisability of a simple atom can be derived by considering the force on the charges under the influence of an external field. When the atom is polarised by this field, the centre of the negative charge is a distance d from the positive nucleus. In the steady state, the average electrostatic force on the negative charge from the external field is balanced by the force from the positive nucleus. Therefore, the local field is given by the field due to a point charge (equation (2.2))

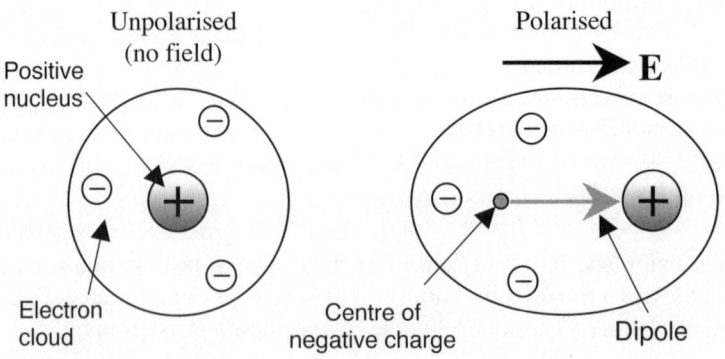

Fig. 2.6 Electronic polarisation of an atom.

at a radius of d. From equation (2.17) and equation (2.11), the polarisability of the atom (in a vacuum) is

$$\alpha = 4\pi\varepsilon_o d^3 \tag{2.20}$$

(ii) *Atomic* α_a
In a crystalline solid such as potassium chloride, the ions of different sign move in different directions when subjected to an electric field. The small displacement of the charges causes atomic polarisability.

(iii) *Orientational* α_d
As described earlier in this chapter, many molecules possess permanent dipole moments. Orientational polarisation is the polarisation arising from the alignment of these permanent dipoles in polar dielectrics.

(iv) *Counterion or Interfacial polarisation* α_i
In addition to local molecular processes, long-range charge transport causes polarisation of dielectrics. Charges become trapped at places where the dielectric is inhomogeneous (internal interfaces) or at the surface, causing macroscopic distortion of the field. Similarly, relatively long-range transport of ions occurs along and around the surface of polyelectrolytes. This manifests itself as an increase in the charge storage capacity of the dielectric or an increase in the permittivity of the dielectric. This is the basis of interfacial polarisation. It has an important role to play in AC electrokinetics, since it is the origin of the induced dipole on particles. This will be discussed in greater detail in the next chapter.

Assuming that the polarisability mechanisms act independently, then the total polarisability of a dielectric is the sum of the polarisabilities

$$\alpha_T = \alpha_e + \alpha_a + \alpha_o + \alpha_i \tag{2.21}$$

Each polarisability has its own characteristic frequency response, which will be discussed in more detail in Chapter Three.

2.3.4 The electric flux density
Considering the free ρ_f and bound ρ_b charge densities separately, Gauss's law (equation 2.3) can be written as

$$\nabla \cdot \mathbf{E} = \frac{\rho_f + \rho_b}{\varepsilon_o} \tag{2.22}$$

Using equation (2.19) for the bound charge density, this can be re-written as

$$\nabla \cdot (\varepsilon_o \mathbf{E} + \mathbf{P}) = \rho_f \qquad (2.23)$$

The vector $\mathbf{D} = \varepsilon_o \mathbf{E} + \mathbf{P}$ is called the *electric flux density* (in C m^{-2}) and its divergence is equal to the free volume charge density

$$\nabla \cdot \mathbf{D} = \rho_f \qquad (2.24)$$

2.3.5 The electric susceptibility and the relative permittivity

Most dielectrics are linear and isotropic, with $\mathbf{P}$ proportional to $\mathbf{E}$ such that

$$\mathbf{P} = \varepsilon_o \chi_{ae} \mathbf{E} \qquad (2.25)$$

where χ_{ae} is the *electric susceptibility* of the dielectric. Therefore, for an *ideal* dielectric (linear and isotropic, with zero conductivity)

$$\mathbf{D} = \varepsilon_o (1 + \chi_{ae}) \mathbf{E} = \varepsilon_o \varepsilon_r \mathbf{E} \qquad (2.26)$$

where $\varepsilon_r = 1 + \chi_{ae}$ is a dimensionless number referred to as the *relative permittivity* of the dielectric. The *permittivity* of the dielectric $\varepsilon = \varepsilon_o \varepsilon_r$ is the constant of proportionality between $\mathbf{D}$ and $\mathbf{E}$.

2.3.6 AC fields, dielectrics and complex permittivity

The movement of charge in, and the polarisation of, non-ideal dielectrics are more complicated. When AC fields are used, the situation is even more so. This section examines the polarisation of a dielectric with a permittivity ε and conductivity σ. It introduces the concept of complex permittivity, which describes the frequency dependent response of the dielectric to the field. Also introduced is the idea of using *equivalent circuits*, consisting of electrical components, to represent a physical system.

In order to examine the polarisation of a real dielectric in an AC field, consider the parallel plate capacitor shown in figure 2.7, which contains a homogeneous dielectric. The plates of the capacitor have area A and separation d, and a potential V of angular frequency ω is applied between them. With a loss-free dielectric of

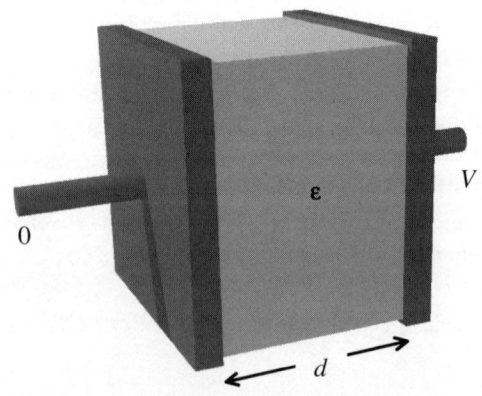

Fig. 2.7   Dielectric-filled parallel plate capacitor.

permittivity ε filling the gap, the impedance is

$$Z = \frac{1}{i\omega C} \qquad (2.27)$$

where $i^2 = -1$ and the capacitance $C = \varepsilon(A/d)$. Some dielectrics are almost loss-free and can be considered to have a constant permittivity, but this is not true of most dielectrics. In the case of lossy dielectrics, the polarisation of the medium depends on the frequency of the applied field. If the dielectric shown in figure 2.7 is lossy and has both a permittivity ε and conductivity σ, the current in the circuit is the same as if the circuit were replaced with a loss-free capacitor (capacitance C), in parallel with a resistor (resistance $R = (1/\sigma)(d/A)$). The total impedance of the circuit is then

$$Z = \frac{1}{1/R + i\omega C} = \frac{R}{1 + i\omega RC} \qquad (2.28)$$

Putting the expressions for R and C into this equation, it can be written in the form of equation (2.27) with a capacitance now given by

$$\tilde{\varepsilon}\frac{A}{d} \qquad (2.29)$$

where $\tilde{\varepsilon}$ is the *complex permittivity* given by:

$$\boxed{\tilde{\varepsilon} = \varepsilon_o \varepsilon_r - i\frac{\sigma}{\omega}} \qquad (2.30)$$

A more mathematical definition of this quantity is straightforward and can be done in one of two ways. The first is to consider Gauss's law, using the charge conservation equation (2.10) and the fact that in a harmonic field of angular frequency ω, the operator $\partial/\partial t = i\omega$. If the free charge in equation (2.24) is solely the charge responsible for the conductivity, then from equation (2.10) and (2.7), the free charge ρ_f is related to the conductivity by

$$i\omega\rho_f = -\nabla\cdot(\sigma\mathbf{E}) \qquad (2.31)$$

This can be substituted into equation (2.24) as follows

$$\nabla\cdot\mathbf{D} = \rho_f \Rightarrow \nabla\cdot(\varepsilon\mathbf{E}) = -\nabla\cdot\left(\frac{\sigma}{i\omega}\mathbf{E}\right) \Rightarrow \nabla\cdot\left(\varepsilon + \frac{\sigma}{i\omega}\mathbf{E}\right) = 0$$

to give

$$\nabla \cdot (\tilde{\varepsilon} \mathbf{E}) = 0 \tag{2.32}$$

with the same definition of the complex permittivity $\tilde{\varepsilon}$. Another way of considering this problem is to calculate the total current in the dielectric. There are two current densities, the *conduction* $\mathbf{J}_c$ (equation (2.7)) and the *displacement* $\mathbf{J}_d = \partial D / \partial t = i\omega\varepsilon\mathbf{E}$. The total current density is given by the sum of the two

$$\mathbf{J} = (\sigma + i\omega\varepsilon)\mathbf{E} = \tilde{\sigma}\,\mathbf{E} \tag{2.33}$$

where $\tilde{\sigma}$ is the complex conductivity, an expression which is equivalent to equation (2.30) with $\tilde{\sigma} = i\omega\tilde{\varepsilon}$. In many situations, a complex permittivity is written as $\tilde{\varepsilon} = \varepsilon' - i\varepsilon''$, where the real and imaginary parts may contain several terms, as will be discussed in the section 2.4.

Returning to the definition of the polarisation $\mathbf{P}$, it can be seen that in reality, both conduction currents and displacement currents flow. This results in a phase shift between the driving field and the induced polarisation, with the polarisability now a complex quantity

$$\mathbf{P} = \varepsilon_{o}\left(\frac{\tilde{\varepsilon}}{\varepsilon_{o}} - 1\right)\mathbf{E} = n\tilde{\alpha}\mathbf{E} \tag{2.34}$$

2.3.7　Energy in a dielectric

In a lossy dielectric the total current is given by the sum of $\mathbf{J}_c = \sigma\mathbf{E}$ and $\mathbf{J}_d = i\omega\varepsilon\mathbf{E}$. A useful expression is the loss tangent given by the ratio of the two

$$\tan\delta = \frac{\sigma}{\omega\varepsilon} \tag{2.35}$$

where δ is the angle between the conduction and displacement currents, referred to as the *loss angle*. In a perfectly insulating (non-lossy) dielectric, the loss angle is zero. If the complex permittivity is written as $\tilde{\varepsilon} = \varepsilon' - i\varepsilon''$, the loss angle can also be written as $\tan\delta = \varepsilon'' / \varepsilon'$. In general, the imaginary part ε'' pertains to the dissipation of energy as heat and the real part ε' to the storage of energy. The energy stored in a non-ideal dielectric is given by equation (2.6), with complex permittivity

$$U = \frac{1}{2}\int_{v} \tilde{\varepsilon}|\mathbf{E}|^{2}\,dv = \frac{1}{2}\int_{v} (\mathbf{D} \cdot \mathbf{E})dv \tag{2.36}$$

2.4   Dielectric relaxations

The polarisation of a dielectric involves the movement of charge to create dipoles. The rate of this movement is finite and as a result each polarisation mechanism has a different characteristic time associated with it. This time is the period required to achieve maximum polarisation. At low applied AC field frequencies, the dipoles have sufficient time to align with the field. As the frequency increases, the period of time that the dipole moment has to 'relax' in and then follow the field decreases.

At the frequency for which these two periods are the same, maximum energy is dissipated by the system. Below this frequency maximum polarisation and energy storage occurs. Above this frequency the polarisation no longer reaches its maximum and at very high frequencies, the mechanism does not respond to the field and no polarisation occurs. For example water, which has a relative permittivity of ~80 below 10^8 Hz, has a relaxation frequency of 2×10^{10} Hz. At frequencies greater than 10^{11} Hz, the permanent dipoles are not able to orient with the field and the permittivity of water drops to ~2. The fall in polarisability results in a decrease in energy storage (and permittivity) and is referred to as a dielectric relaxation.

2.4.1   Orientational relaxation

Of the basic polarisation mechanisms outlined in section 2.3.3, orientational polarisation has the longest relaxation time. Atomic and electronic polarisation will align with the field up to frequencies of the order of 10^{14} Hz and we can consider them to be constant. The polarisation due to these mechanisms is of the form of equation (2.25)

$$\mathbf{P}_{ae} = \varepsilon_o \chi_{ae} \mathbf{E} \tag{2.37}$$

The orientational polarisation has a characteristic *relaxation time* τ_{or} associated with the time taken for the permanent dipoles to re-orient with the field and is given by

$$\mathbf{P}_{or} = \frac{\varepsilon_o \chi_{or}}{1+i\omega\tau_{or}} \mathbf{E} \tag{2.38}$$

where χ_{or} is the low frequency limit for the orientational susceptibility. The total frequency dependent polarisation is then

$$\mathbf{P}_{tot} = \varepsilon_o \left(\chi_{ae} + \frac{\chi_{or}}{1+i\omega\tau_{or}} \right) \mathbf{E} \tag{2.39}$$

At the low frequency limit $\chi = \chi_{ae} + \chi_{or} = \varepsilon_s - 1$, where ε_s is the relative permittivity measured in a static electric field. At the high frequency limit $\chi = \chi_{ae} = \varepsilon_\infty - 1$, where ε_∞ is the relative permittivity at sufficiently high frequency that no orientational polarisation occurs. As a result, $\chi_{or} = \varepsilon_s - \varepsilon_\infty$ and equation (2.39) can be written as

$$\mathbf{P}_{tot} = \varepsilon_o (\tilde{\varepsilon}_d - 1)\mathbf{E} \tag{2.40}$$

where $\tilde{\varepsilon}_d$ is a complex term given by

$$\tilde{\varepsilon}_d = \varepsilon_\infty + \frac{\varepsilon_s - \varepsilon_\infty}{1 + i\omega\tau_{or}} \tag{2.41}$$

The total complex permittivity of the dielectric is therefore

$$\tilde{\varepsilon} = \varepsilon_o \left(\varepsilon_\infty + \frac{\varepsilon_s - \varepsilon_\infty}{1 + i\omega\tau_{or}} \right) - i\frac{\sigma}{\omega} \tag{2.42}$$

Writing this in the form $\tilde{\varepsilon} = \varepsilon' - i\varepsilon''$, ε' and ε'' are given by the Debye relations

$$\boxed{\varepsilon' = \varepsilon_o \left(\varepsilon_\infty + \frac{\varepsilon_s - \varepsilon_\infty}{1 + \omega^2\tau_{or}^2} \right)} \tag{2.43}$$

$$\boxed{\varepsilon'' = \varepsilon_o \left(\frac{(\varepsilon_s - \varepsilon_\infty)\omega\tau_{or}}{1 + \omega^2\tau_{or}^2} \right) + \frac{\sigma}{\omega}} \tag{2.44}$$

In general, a non-zero ε'' implies a phase lag between $\mathbf{D}$ and $\mathbf{E}$. This has important ramifications for energy dissipation in the system. When the loss part of the complex permittivity is zero, $\mathbf{D}$ and $\mathbf{E}$ are in phase and any energy required to produce polarisation of the material in one half-cycle of the field is given back to the driving source in the second half-cycle. In other words, the amount of energy lost through the loss mechanism per cycle is negligible.

When ε'' is non-zero, energy is dissipated in the system by two mechanisms. Firstly, if the dielectric is non-ideal and has a finite conductivity, then energy is lost through Joule heating, the second term on the r.h.s. of equation (2.44). Secondly, energy is lost due to the electrical conductivity that arises from the relaxation mechanism: the first term on the r.h.s. of equation (2.44). The electric field orients the dipoles against the randomising effects of Brownian motion. At high frequencies no energy is lost since there is insufficient time for the dipoles to orient with the field and no energy is stored. At low frequencies the dipoles are oriented and work is done in moving the dipoles in a viscous medium. The power lost per cycle of AC field is low but this increases with frequency, since the number of cycles per second increases. It reaches a maximum at the characteristic angular frequency $\omega_{or} = 1/\tau_{or}$, when the time required for maximum orientation of the dipoles is exactly equal to one half-cycle of the field.   Maximum energy dissipation from this mechanism occurs at this frequency.

The variation of ε' and ε'' with ω gives a dispersion in the relative permittivity as shown by the solid line in figure 2.8.   It can be seen that the width of the loss peak extends over approximately four decades in frequency and that for a single Debye-type relaxation the width of the peak at half height is 1.14 decades in frequency.

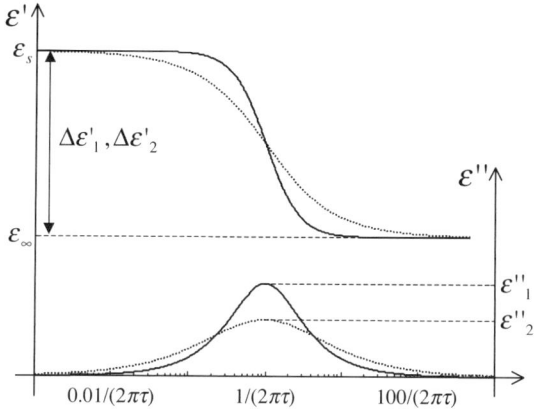

Fig. 2.8 The real and imaginary components of the complex permittivity versus frequency. The solid line indicates the dispersion for a single relaxation and the dotted line a spread of relaxations.

Such dielectric loss peaks can be analysed to obtain a great deal of information about the molecular material under investigation. For example the width of the loss peak is often wider than the 1.14 decades characteristic of the ideal Debye behaviour, where there is only a single relaxation time.   Cole and Cole (1941) modified equation (2.42) to account for a distribution in relaxation times more representative of experimental samples as follows

$$\tilde{\varepsilon} = \varepsilon_o \left(\varepsilon_\infty + \frac{\varepsilon_s - \varepsilon_\infty}{1 + i(\omega \tau_{or})^{(1-\beta)}} \right) - i \frac{\sigma}{\omega} \tag{2.45}$$

where β equals 0 for a single relaxation time and tends to 1 for an infinite number of relaxation times. A distribution in relaxation times means that the dispersion occurs at a frequency equal to the average of the relaxation times, but is broader than for a single relaxation time. This is shown by the dotted line in figure 2.8.

A useful method of analysing this type of data is by using a Cole-Cole plot, where the imaginary part ε'' is plotted against the real part ε'. For a single relaxation, the plot is a semicircle with its centre on the horizontal axis, as shown by the solid line in figure 2.9. If the plot forms an arc of a circle with the centre below the axis, then there is a distribution of relaxation times as shown by the dotted line in the figure for $\beta = 0.1$. The relationship between the angle θ and the parameter β is shown in the figure.

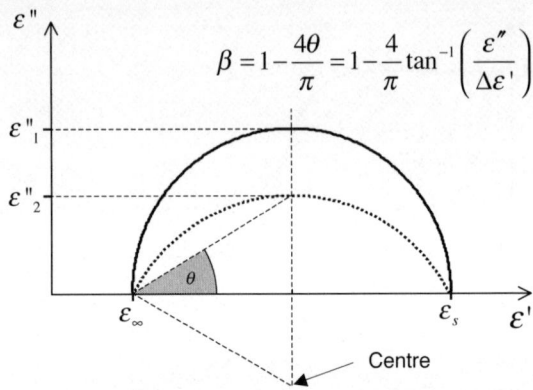

$$\beta = 1 - \frac{4\theta}{\pi} = 1 - \frac{4}{\pi}\tan^{-1}\left(\frac{\varepsilon''}{\Delta\varepsilon'}\right)$$

Fig. 2.9 A Cole-Cole plot of the imaginary versus the real part of the permittivity for a single relaxation (solid line) and a relaxation with a spread of times (dotted line). Also shown is the relationship between the angle θ and the parameter β.

The interrelationship between the real and imaginary components of the complex permittivity can be quantitatively described by the Kramers-Krönig relationships

$$\varepsilon'(\omega) - \varepsilon_\infty = \frac{2}{\pi}\int_0^\infty \frac{\varpi\varepsilon''(\varpi)}{\varpi^2 - \omega^2}d\varpi \qquad \varepsilon''(\omega) = \frac{2\omega}{\pi}\int_0^\infty \frac{\varepsilon'(\varpi) - \varepsilon_\infty}{\varpi^2 - \omega^2}d\varpi \qquad (2.46)$$

where ϖ is a dummy frequency variable over which ε'' has to be integrated to find ε' or *vice versa*. The Kramers-Krönig relations show that if either the real or the imaginary component of the complex permittivity is known, then the other can be derived from the above equations. In terms of AC electrokinetics, the Kramers-Krönig relationships can be used to equate the real and imaginary components of the polarisability of a system, or alternately to draw comparisons between electrorotation and dielectrophoresis spectra (Wang *et al.* 1992).

2.5   Summary

In this chapter we have described the fundamentals of electrostatics: the behaviour of charges in quasi-electrostatic fields. We have detailed the behaviour of a dipole in an electric field, the physical mechanism that underpins the theories of AC electrokinetics. We have also looked at the behaviour of collections of dipoles and the macroscopic properties of dielectrics formed from a collection of molecular dipoles. In the next chapter, we will take these concepts further to explain the behaviour of dielectric particles in dielectric fluids subjected to an electric field.

2.6   References

Bleaney B.I. and Bleaney B. *Electricity and Magnetism*   Clarendon Press, Oxford (1962).

Cole K.S. and Cole R.H.   *Dispersion and Absorption in Dielectrics I. Alternating Current Characteristics*   J. Chem. Phys. **9**   341-351 (1941).

Grant E.H., Sheppard R.J. and South G. *Dielectric Behaviour of Molecules in Solution* Clarendon Press, Oxford (1978).

Kraus J.D. *Electromagnetics* McGraw-Hill, New York (1984).

Lorrain P., Corson D.R. and Lorrain F.   *Fundamentals of Electromagnetic Phenomena* W.H. Freeman (2000).

Pethig R. *Dielectric and Electronic Properties of Biological Materials*   John Wiley & Sons, Chichester (1979).

Stratton J.A. *Electromagnetic Theory* McGraw Hill, New York (1941).

Wang X-B., Pethig R. and Jones T.B.   *Relationship of dielectrophoretic and electrorotational behaviour exhibited by polarized particles* J. Phys. D: Appl. Phys. **25** 905-912 (1992).

Chapter Three

Interfacial polarisation and the effective dipole of particles

Real systems often consist of a number of different dielectrics each with different electrical properties. In terms of AC electrokinetics, the system is a suspension of dielectric particles in a dielectric fluid (generally electrolytes). When an electric field is applied to the system, surface charge accumulates at discontinuities (or interfaces) between the dielectrics due to the differences in electrical properties. Since the polarisabilities of each dielectric are frequency dependent, the magnitude of the surface charge is also frequency dependent and the total complex permittivity of the system exhibits dispersions solely due to the polarisation of the interfaces. This is referred to as the *Maxwell-Wagner interfacial* polarisation.

If the system consists of a single dielectric particle and a suspending fluid, the distribution of the surface charge density around the closed surface of the particle gives rise to an induced dipole moment. The action of an electric field on this dipole gives rise to ponderomotive forces and torques; the basis of AC electrokinetics. In this chapter, we discuss interfacial polarisation, the resulting induced dipole moment of particles and the effect of this dipole moment on the electrical properties of suspensions of particles.

3.1 Interfacial polarisation

We illustrate the mechanism of interfacial polarisation using the parallel plate capacitor shown in figure 3.1, which contains slabs of two lossy dielectrics with

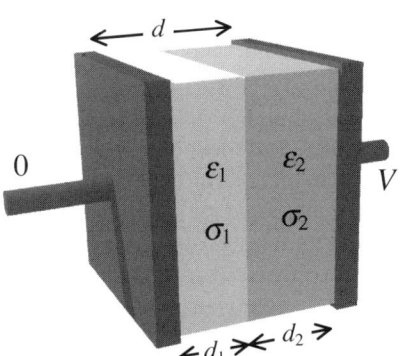

Fig. 3.1 A parallel plate capacitor, containing layers of two dielectrics with different electrical properties.

different electrical properties. In the last chapter, we showed how a parallel plate capacitor containing a lossy dielectric could be represented using an equivalent circuit. We can extend this model to analyse the capacitor shown in the figure. The two dielectrics have permittivities ε_1 and ε_2 and conductivities σ_1 and σ_2, respectively. The system can be thought of as a series combination of the equivalent circuits for each of the two lossy dielectrics. The impedance of the whole circuit is

$$Z = Z_1 + Z_2 = \frac{R_1}{1+i\omega R_1 C_1} + \frac{R_1}{1+i\omega R_2 C_2} \tag{3.1}$$

As described in section 2.3.6, this can be re-written as a single capacitance with a complex permittivity. The resulting complex permittivity contains a dispersion arising solely from the difference in the permittivities and conductivities of the two dielectrics. Using the Debye formulations (equations (2.43) and (2.44)), this complex permittivity is

$$\tilde{\varepsilon} = \varepsilon' - i\varepsilon'' = \varepsilon_o \left[\varepsilon_{hf} + \frac{\varepsilon_{lf} - \varepsilon_{hf}}{1+\omega^2\tau^2} \right] - i\varepsilon_o \left[\frac{(\varepsilon_{lf} - \varepsilon_{hf})\omega\tau_{or}}{1+\omega^2\tau^2} + \frac{\sigma}{\varepsilon_o\omega} \right] \tag{3.2}$$

where ε_{hf} is the high frequency permittivity, ε_{lf} the low frequency permittivity, τ the relaxation time and σ the system conductivity, each given by

$$\varepsilon_{hf} = \frac{d\varepsilon_1\varepsilon_2}{d_1\varepsilon_2 + d_2\varepsilon_1} \qquad \varepsilon_{lf} = \frac{d(d_1\varepsilon_1\sigma_2^2 + d_2\varepsilon_2\sigma_1^2)}{(d_1\sigma_2 + d_2\sigma_1)^2}$$

$$\tag{3.3}$$

$$\tau = \varepsilon_o \frac{d_1\varepsilon_2 + d_2\varepsilon_1}{d_1\sigma_2 + d_2\sigma_1} \qquad \sigma = \frac{d\sigma_1\sigma_2}{d_1\sigma_2 + d_2\sigma_1}$$

In general, relaxations due to interfacial polarisation occur at frequencies below those for orientational and electronic polarisation. A plot of the real and imaginary parts of the complex permittivity of an arbitrary system, taking into account all the different polarisation mechanisms, would be expected to resemble figure 3.2.

3.2   The induced *effective* dipole moment of a particle

We can see from the simple model that charge builds up at the interface between two dielectrics of different permittivity and/or conductivity. This is equally true irrespective of the geometry of the system. For a dielectric particle suspended in a dielectric fluid, when the electric field is applied, charge of opposite sign accumulates at either side of the particle. This gives rise to a dipole around the particle. This *induced* or *effective dipole moment* depends on the properties of both the particle and the suspending medium (or electrolyte), and on the frequency of

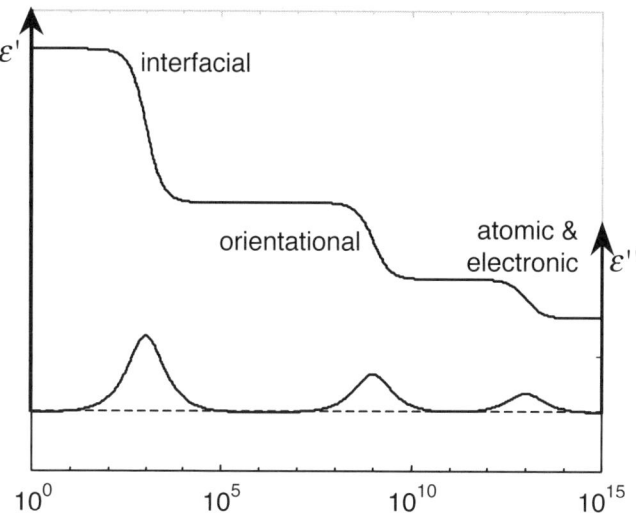

Fig. 3.2 The frequency variation of the complex permittivity of a dielectric, taking into account the typical relaxation mechanisms.

the applied field. In this section we derive expressions for the induced dipole moment of solid spherical particles, ellipsoids and more complicated shelled particles.

3.2.1 The effective dipole moment of a spherical particle

The simplest case is that of a homogeneous solid dielectric sphere of radius a suspended in a homogenous dielectric medium, shown schematically in figure 3.3. The applied electric field far from the origin is taken to be uniform and anti-parallel to the z-axis $i.e.$ $\mathbf{E} = -E\hat{\mathbf{z}}$. Without loss of generality, the sphere can be assumed to have its centre at the origin, making the problem axially symmetric and two-dimensional in spherical polar co-ordinates.

For ideal dielectrics, the Laplace equation for the electrical potential can be solved using Legendre polynomials, an exercise that can be found in electromagnetics textbooks ($e.g.$ Lorrain et al 2000). Gauss's law (equation (2.24)) gives the boundary condition for the potential at the surface of the sphere ($r = a$)

$$\varepsilon_m \mathbf{E}_m \cdot \hat{\mathbf{n}} - \varepsilon_p \mathbf{E}_p \cdot \hat{\mathbf{n}} = 0 \Rightarrow \varepsilon_m \frac{\partial \phi_m}{\partial r}\bigg|_{r=a} - \varepsilon_p \frac{\partial \phi_p}{\partial r}\bigg|_{r=a} = 0 \qquad (3.4)$$

where the subscripts p and m refer to particle and suspending medium respectively.

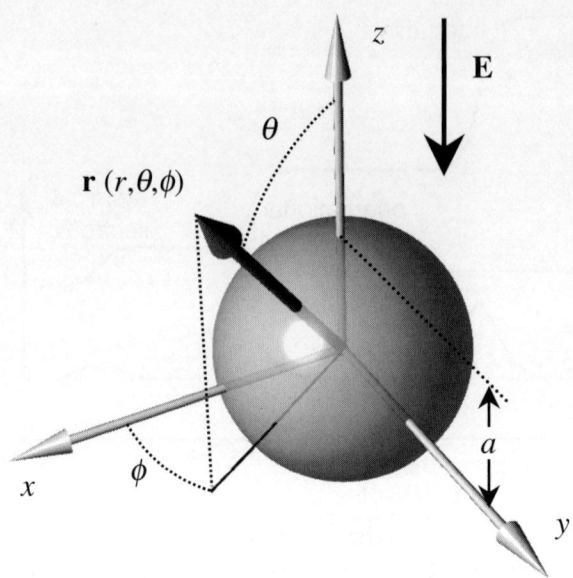

Fig. 3.3 Sphere of radius a centred on the origin, with an applied electric field anti-parallel to the z-axis. The spherical polar co-ordinates are defined as shown, making the problem two dimensional in r and θ.

If the permittivities of the dielectrics are complex and the applied field is harmonic of single frequency ω, equation (2.32) gives the same boundary condition but with the permittivities replaced by complex values. By extrapolation from Lorrain *et al.* (2000), the solutions for the potential inside and outside the dielectric sphere are

$$\phi_m = \left[\left(\frac{\tilde{\varepsilon}_p - \tilde{\varepsilon}_m}{\tilde{\varepsilon}_p + 2\tilde{\varepsilon}_m} \right) \frac{a^3}{r^3} - 1 \right] Er\cos\theta \qquad \phi_p = -\left(\frac{3\tilde{\varepsilon}_m}{\tilde{\varepsilon}_p + 2\tilde{\varepsilon}_m} \right) Er\cos\theta \qquad (3.5)$$

The external potential in the medium can be re-written as

$$\phi_m = Ea^3 \left(\frac{\tilde{\varepsilon}_p - \tilde{\varepsilon}_m}{\tilde{\varepsilon}_p + 2\tilde{\varepsilon}_m} \right) \frac{\cos\theta}{r^2} - Er\cos\theta \qquad (3.6)$$

This is the scalar sum of the potential due to the applied field (second term on the right hand side) and the potential of a dipole moment (first term, right hand side), c.f. equation (2.12). By inspection, the dipole moment is

$$\mathbf{p} = 4\pi\varepsilon_m \left(\frac{\tilde{\varepsilon}_p - \tilde{\varepsilon}_m}{\tilde{\varepsilon}_p + 2\tilde{\varepsilon}_m} \right) a^3 \mathbf{E} \tag{3.7}$$

This is the *effective dipole moment* of the sphere. It is sometimes rewritten in terms of the volume of the sphere v and a complex effective polarisability $\tilde{\alpha}$ where

$$\mathbf{p} = v\tilde{\alpha}\mathbf{E} \tag{3.8}$$

Using this equation, the magnitude of the dipole for a particle can be calculated. Taking for example a field of 10^6 V m^{-1}, the dipole moment of a particle such as a cell is of the order of 10^{-19} C m. This is equivalent to a charge of 10^{-14} C located at the poles of the cell, which equates to approximately 0.5% of the fixed charge on the cell. Equation (3.8) shows that the dipole moment scales with particle volume, so that smaller particles have smaller dipole moments and the percentage of the surface charge forming the dipole is much smaller.

From equations (3.7) and (3.8) the *effective polarisability* is therefore

$$\tilde{\alpha} = 3\varepsilon_m \left(\frac{\tilde{\varepsilon}_p - \tilde{\varepsilon}_m}{\tilde{\varepsilon}_p + 2\tilde{\varepsilon}_m} \right) = 3\varepsilon_m \tilde{f}_{CM} \tag{3.9}$$

The magnitude of the polarisability, and therefore the effective dipole moment of the particle, is frequency dependent. This dependence is described by the factor

$$\tilde{f}_{CM}(\tilde{\varepsilon}_p, \tilde{\varepsilon}_m) = \frac{\tilde{\varepsilon}_p - \tilde{\varepsilon}_m}{\tilde{\varepsilon}_p + 2\tilde{\varepsilon}_m} \tag{3.10}$$

This is referred to as the *Clausius-Mossotti factor*. It is complex, describing a relaxation in the *effective* permittivity or polarisability of the particle with a relaxation time of

$$\tau_{MW} = \frac{\varepsilon_p + 2\varepsilon_m}{\sigma_p + 2\sigma_m} \tag{3.11}$$

The angular frequency $\omega_{MW} = 2\pi f_{MW} = 1/\tau_{MW}$ is often referred to as the Maxwell-Wagner relaxation frequency since the dispersion in the dipole moment is caused by interfacial polarisation. The real part of the Clausius-Mossotti factor reaches a

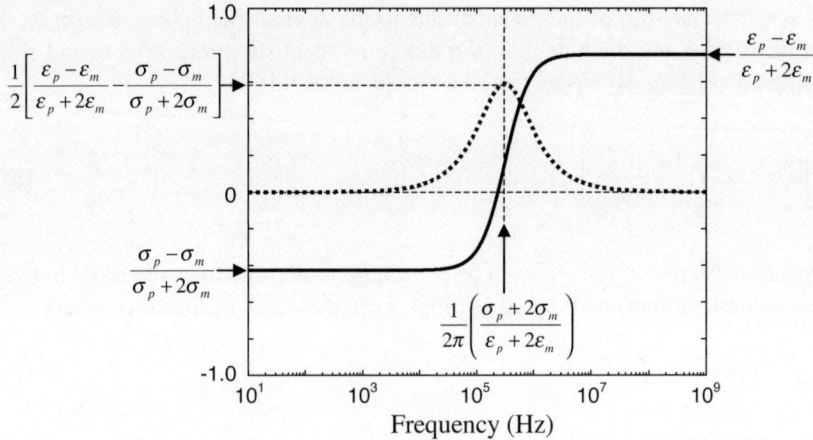

Fig. 3.4 Plot of the variation of the real (solid line) and imaginary (dotted line) parts of the Clausius-Mossotti factor with frequency. The high and low frequency limiting values of the real part are shown, as well as the value of the imaginary part at the relaxation frequency.

low frequency limiting value of $(\sigma_p - \sigma_m)/(\sigma_p + 2\sigma_m)$, *i.e.* it depends solely on the conductivity of the particle and suspending medium. Conversely, the high frequency limiting value is $(\varepsilon_p - \varepsilon_m)/(\varepsilon_p + 2\varepsilon_m)$ and the polarisation is dominated by the permittivity of the particle and suspending medium. The imaginary part is zero at high and low frequencies, and at the relaxation frequency f_{MW} has a value

$$\frac{1}{2}\left(\frac{\varepsilon_p - \varepsilon_m}{\varepsilon_p + 2\varepsilon_m} - \frac{\sigma_p - \sigma_m}{\sigma_p + 2\sigma_m}\right) \tag{3.12}$$

An example showing the frequency variation of the real and imaginary parts of the Clausius-Mossotti factor is shown in figure 3.4. Figure 3.5 shows the specific example of a latex particle in water, $\sigma_p \gg \sigma_m$ and $\varepsilon_p \ll \varepsilon_m$, giving a real part of 1 at low frequencies and $-\frac{1}{2}$ at high frequencies; the value of the imaginary part at f_{MW} is $-\frac{3}{4}$. In general, the real part of the Clausius-Mossotti factor is bounded by 1 and $-\frac{1}{2}$, and the imaginary part bounded by $+\frac{3}{4}$ and $-\frac{3}{4}$.

In AC electrokinetics, *non-uniform* fields are used. The assumption that the uniform field solution of the dipole moment can still be used in this situation is referred to as the dipole moment approximation. This approximation is valid if the size of the sphere is small compared with a characteristic distance associated with the non-uniformity of the electric field *i.e.* if the electric field is more or less uniform across the particle. The case where this approximation is not valid is discussed further in section 3.2.4.

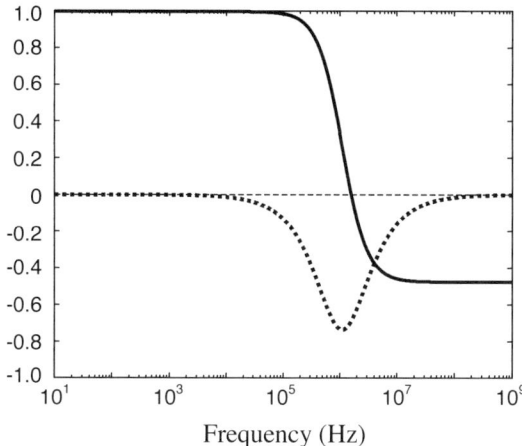

Fig. 3.5 The frequency variation of the real (solid line) and imaginary (dotted line) parts of the Clausius-Mossotti factor for a latex sphere, with $\sigma_p \gg \sigma_m$ and $\varepsilon_p \ll \varepsilon_m$.

3.2.2   The effective dipole moment of an ellipsoidal particle

The more general case of an ellipsoidal particle is also relevant since many particles are not spherical. Once again, a full calculation of the electrical potential around the particle and the corresponding dipole moment can be found in a number of textbooks (*e.g.* Jones 1995; Stratton 1941).

Consider a homogeneous, ellipsoidal, dielectric particle as shown in figure 3.6 where the half lengths of the major axes are a_1, a_2 and a_3. The effective polarisability, $\tilde{\alpha}_n = 3\varepsilon_m \tilde{K}_n$, is different for each principal axis n (where n = 1, 2, 3). $\tilde{K}_n$ is a frequency dependent factor equivalent to the Clausius-Mossotti factor, given by

$$\tilde{K}_n = \frac{\tilde{\varepsilon}_p - \tilde{\varepsilon}_m}{3(A_n(\tilde{\varepsilon}_p - \tilde{\varepsilon}_m) + \tilde{\varepsilon}_m)} \qquad (3.13)$$

A_n is called the *depolarising factor* for the axis n given by

$$A_n = \frac{1}{2} a_1 a_2 a_3 \int_0^\infty \frac{ds}{(s + a_n^2)B} \qquad (3.14)$$

where $B = \sqrt{(s + a_1^2) + (s + a_2^2) + (s + a_3^2)}$ and s is an arbitrary distance for integration.

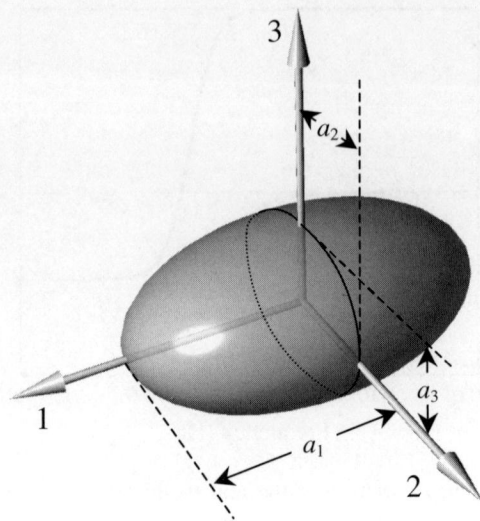

Fig. 3.6 An ellipsoid with principle axes 1, 2, 3 and half-
lengths along the axes a_1, a_2, a_3 .

The dipole moment of the ellipsoidal particle can then be written as

$$\mathbf{p}_n = \frac{4\pi}{3} a_1 a_2 a_3 \, \varepsilon_m \left[\frac{\tilde{\varepsilon}_p - \tilde{\varepsilon}_m}{\tilde{\varepsilon}_p + A_n (\tilde{\varepsilon}_p - \tilde{\varepsilon}_m)} \right] \mathbf{E}_n \qquad (3.15)$$

where $\mathbf{E}_n$ is the component of the electric field in the $n(1, 2$ or $3)$ direction. Each axis of the particle also has an associated relaxation time, given by

$$\tau_n = \frac{A_n \varepsilon_p + (1 - A_n)\varepsilon_m}{A_n \sigma_p + (1 - A_n)\sigma_m} \qquad (3.16)$$

The dipole moment for the particle therefore depends on its orientation with respect to the electric field. In fact, in a uniform AC electric field, an ellipsoidal particle experiences a frequency dependent alignment torque, which tends to orient one of the principal axes with the field. Which particular axis of the particle aligns depends on the frequency. This is discussed in greater detail in Jones (1995). For the particle described above, the time-averaged torque around axis 1 is given by

$$\langle \Gamma_1 \rangle = \frac{2}{3} \pi a_1 a_2 a_3 \varepsilon_m \left(A_3 - A_1 \right) E_2 E_3 \, \mathrm{Re} \left[\tilde{K}_2 \tilde{K}_3 \right] \qquad (3.17)$$

Here E_2 and E_3 are the components of the electric field along axes 2 and 3 respectively. The torque expressions for the other axes are found by substituting the axis subscripts according to the convention for a right-handed co-ordinate system: $1\rightarrow2\rightarrow3\rightarrow1$. One important conclusion regarding alignment torques is that lossy particles do not always align with their longest axis parallel to the field. There can exist particular frequency bands where the orientation is either parallel or perpendicular to the field. Study of the frequency-dependent orientation has been used to characterise the dielectric properties of particles such as cells (*e,g,* Teixeira-Pinto *et al.* 1960; Schwarz *et al.* 1965; Griffin 1970; Fomchemkov and Gavrilyuk 1978; Miller and Jones 1993).

Special cases of ellipsoids
There are a number of special cases of ellipsoid which are interesting to consider. For a sphere, $a_1 = a_2 = a_3$ and the depolarising factors are $A_1 = A_2 = A_3 = 1/3$. Substituting the values for the factors into equation (3.13), we see that the effective polarisability for each axis is the same as equation (3.9).

The second case is the *prolate ellipsoid,* which has the condition $a_1 \gg a_2 = a_3$. Assuming that the applied field **E** is parallel to the *x*-axis (and the major axis a_1) then the integral from equation (3.14) is

$$\int_0^\infty \frac{ds}{(s+a_2^{\,2})(s+a_1^{\,2})^{3/2}} = -\frac{1}{a_1^{\,3}e^3}\left[2e - \ln\left(\frac{1+e}{1-e}\right)\right] \tag{3.18}$$

where $e = \sqrt{1-a_2^{\,2}/a_1^{\,2}}$ is the eccentricity. In the case of some sub-micrometre particles, such as Tobacco Mosaic Virus (discussed in Chapter Eleven) which is a long thin rod, the shape can be approximated to a prolate ellipsoid, with the depolarising factor along the major axis a_1 tending to zero and to 1 along axes a_2 and a_3. Therefore, with the major axis aligned with the field, the effective polarisability of the particle simplifies to

$$\tilde{\alpha}_1 = \varepsilon_m \frac{\tilde{\varepsilon}_p - \tilde{\varepsilon}_m}{\tilde{\varepsilon}_m} \tag{3.19}$$

along the major axis. However, there is also a frequency dependent alignment torque (Jones 1995). This complicates the analysis, since equation (3.19) is not applicable if the major axis is not parallel with the field.

3.2.3   The shell model for biological particles
Biological particles, such as cells and some types of viruses, have a complicated internal structure. The common approach to theoretically modelling cells is to use a concentric multi-shell model (Irimajiri *et al.* 1979; Gimsa *et al.* 1991; Huang *et al.* 1992). The simplest case is that of a single shelled particle (*e.g.* a red blood cell), as shown schematically in figure 3.7(a).

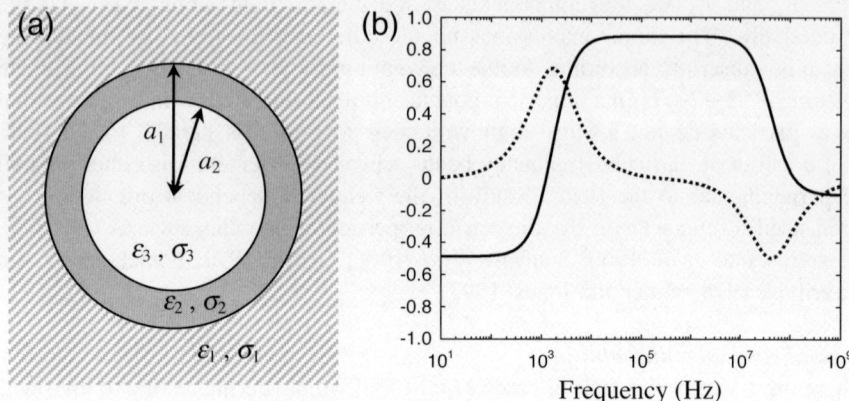

Fig. 3.7 (a) Schematic of a spherical particle with a single shell.
(b) The frequency variation of the equivalent Clausius-Mossotti factor with:
$a_1 = 2.01 \times 10^{-6}\,\mathrm{m}$, $a_2 = 2.0 \times 10^{-6}\,\mathrm{m}$, $\varepsilon_1 = 78.5\varepsilon_o$, $\varepsilon_2 = 10\varepsilon_o$, $\varepsilon_3 = 60\varepsilon_o$, $\sigma_1 = 10^{-4}\,\mathrm{S\,m^{-1}}$, $\sigma_2 = 10^{-8}\,\mathrm{S\,m^{-1}}$, $\sigma_3 = 0.5\,\mathrm{S\,m^{-1}}$.

This system has *two* intrinsic relaxation frequencies, one for each of the two interfaces. In this case the effective polarisability and the dipole moment are given by

$$\tilde{\alpha} = 3\varepsilon_1 \tilde{f}_{CM,23} = 3\varepsilon_m \left(\frac{\tilde{\varepsilon}_{23} - \tilde{\varepsilon}_1}{\tilde{\varepsilon}_{23} + 2\tilde{\varepsilon}_1} \right) \qquad \mathbf{p} = 4\pi\varepsilon_1 \tilde{f}_{CM,23} a_1^3 \mathbf{E} \qquad (3.20)$$

and the particle complex permittivity $\tilde{\varepsilon}_{23}$ in the Clausius-Mossotti factor $\tilde{f}_{CM,23}$ is given by

$$\tilde{\varepsilon}_{23} = \tilde{\varepsilon}_2 \left[\gamma_{12}^3 + 2 \left(\frac{\tilde{\varepsilon}_3 - \tilde{\varepsilon}_2}{\tilde{\varepsilon}_3 + 2\tilde{\varepsilon}_2} \right) \right] \Bigg/ \left[\gamma_{12}^3 - \left(\frac{\tilde{\varepsilon}_3 - \tilde{\varepsilon}_2}{\tilde{\varepsilon}_3 + 2\tilde{\varepsilon}_2} \right) \right] \qquad (3.21)$$

where the factor $\gamma_{12} = a_1/a_2$. This representation is equivalent to a homogeneous spherical particle of radius a_1 and permittivity $\tilde{\varepsilon}_{23}$, given by equation (3.21). The frequency variation of effective polarisability and magnitude of the dipole moment is again described by the Clausius-Mossotti factor, in this case $\tilde{f}_{CM,23}$. The real and imaginary parts of $\tilde{f}_{CM,23}$ are plotted as a function of frequency in figure 3.7(b); values of permittivities and conductivities for the plot, are shown in the figure legend.

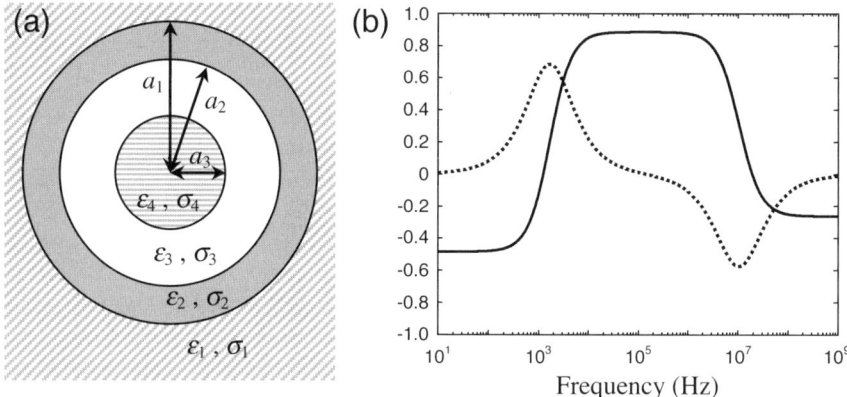

Fig. 3.8 (a) Schematic of a spherical particle with two shells.
(b) The frequency variation of the equivalent Clausius-Mossotti factor with:
$a_1 = 2.01 \times 10^{-6}\,\text{m}$, $a_2 = 2.0 \times 10^{-6}\,\text{m}$, $a_3 = 0.5 \times 10^{-6}\,\text{m}$,
$\varepsilon_1 = 78.5\varepsilon_o$, $\varepsilon_2 = 10\varepsilon_o$, $\varepsilon_3 = 60\varepsilon_o$, $\varepsilon_4 = 30\varepsilon_o$,
$\sigma_1 = 10^{-4}\,\text{S}\,\text{m}^{-1}$, $\sigma_2 = 10^{-8}\,\text{S}\,\text{m}^{-1}$, $\sigma_3 = 0.5\,\text{S}\,\text{m}^{-1}$, $\sigma_4 = 0.1\,\text{S}\,\text{m}^{-1}$.

The model can be extended to multiple shells by using equation (3.21) for each shell. Figure 3.8 shows an example with two shells. In this case the inner permittivity value $\tilde{\varepsilon}_3$ in equation (3.21) is replaced with the equivalent value $\tilde{\varepsilon}_{34}$ given by

$$\tilde{\varepsilon}_{34} = \tilde{\varepsilon}_3 \left[\gamma_{23}^3 + 2\left(\frac{\tilde{\varepsilon}_4 - \tilde{\varepsilon}_3}{\tilde{\varepsilon}_4 + 2\tilde{\varepsilon}_3} \right) \right] \Big/ \left[\gamma_{23}^3 - \left(\frac{\tilde{\varepsilon}_4 - \tilde{\varepsilon}_3}{\tilde{\varepsilon}_4 + 2\tilde{\varepsilon}_3} \right) \right] \tag{3.22}$$

where the factor $\gamma_{23} = a_2/a_3$. Figure 3.8(b) shows the variation of the real and imaginary parts of $\tilde{f}_{CM,23}$ for this particular case. Shell models are used routinely to model the complex properties of biological particles and interpret dielectric spectroscopy and electrokinetic experiments. As an example, the model illustrated in figure 3.7 can be used to represent a cell where the outer layer is a thin lipid membrane and the innermost sphere the cell cytoplasm. Figure 3.8 would represent the more detailed case of a cell with cytoplasm and nucleus (but no nuclear membrane). These models have been found to be extremely useful in the interpretation of experimental results. Values for the conductivity and permittivity of each layer can be obtained and the characteristics of the particle or cell determined. Subtle changes in the physical structure of the cell, for example transformation to a malignant state or invasion by pathogens (viruses), can be followed experimentally and corresponding changes in the dielectric properties of single cells determined.

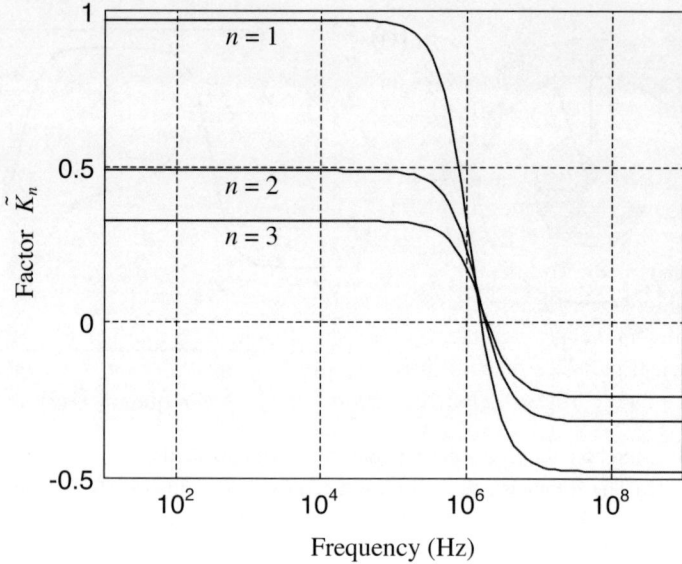

Fig. 3.9 Plot of the factor $\tilde{K}_n$ as a function of frequency for the dipole $(n = 1)$, quadrupole $(n = 2)$ and octopole $(n = 3)$ terms.

3.2.4   Particle multipoles

As discussed previously, for most experimental cases the dipole approximation is valid. However, in situations where the electric field non-uniformity is substantial or, more importantly from the experimental point of view, where there is a field null, higher order moments must be considered. The latter can occur in a negative dielectrophoretic trap, and if the particle is exactly in the centre, then the induced dipole on the particle will be zero; only higher order moments are responsible for the force on the particle.

As discussed by Jones and Washizu (1994), calculation of the dielectrophoretic force in this case requires that higher orders be examined (see Chapter Four). The calculation of the moment tensors is involved and will not be covered here (for details see Jones 1995). However, for purposes of illustration, the frequency dependence of the nth moment is described by the following expression

$$\tilde{K}_n = \frac{\tilde{\varepsilon}_p - \tilde{\varepsilon}_m}{n\tilde{\varepsilon}_p + (n+1)\tilde{\varepsilon}_m} \tag{3.23}$$

which, for the dipole moment $(n = 1)$, is equal to the Clausius-Mossotti factor (equation (3.10)). Figure 3.9 shows the frequency dependence of the factor $\tilde{K}_n$ for a spherical particle and $n = 1$ (dipole), 2 (quadrupole), 3 (octopole). The figure shows that, for example, in a field null where the dipole is zero and the trapping force is due to the higher moments, $\tilde{K}_n$ is approximately one half of the value of the Clausius-Mossotti factor. Consideration of the higher order moments is required in only a few situations.

3.3   Mixture theory and suspensions of particles

A common method of measuring the dielectric properties of a collection of similar particles is to use dielectric spectroscopy (Grant *et al.* 1978; Pethig 1979; Takashima 1989). The theory linking the response of the particle mixture to that of the individual particle and the suspending solution was first derived by Maxwell (1954), and an elegant derivation of the mixture formula is given by Jones (1995). For spherical particles dispersed in a suspending medium at a volume fraction φ, and with a dipole moment given by equation (3.7), Maxwell's mixture formula is

$$\tilde{\varepsilon}_{mix} = \tilde{\varepsilon}_m \frac{1+2\varphi \tilde{f}_{CM}}{1-\varphi \tilde{f}_{CM}} = \tilde{\varepsilon}_m \frac{1+2\varphi \left(\dfrac{\tilde{\varepsilon}_p - \tilde{\varepsilon}_m}{\tilde{\varepsilon}_p + 2\tilde{\varepsilon}_m} \right)}{1-\varphi \left(\dfrac{\tilde{\varepsilon}_p - \tilde{\varepsilon}_m}{\tilde{\varepsilon}_p + 2\tilde{\varepsilon}_m} \right)} \tag{3.24}$$

This equation gives the dielectric increment due to the particles and this is added to the permittivity of the suspending medium to give the total complex permittivity of the suspension. This is valid under the condition that the volume fraction is small (*i.e.* $\varphi \ll 1$). Further developments have been proposed to take into account the effect of high particle concentrations (Hanai *et al.* 1975; 1976; 1979; Zhang *et al.* 1983).

Dielectric spectroscopy measurements of the variation in the complex permittivity of a suspension of dielectric particles with frequency have been the accepted method for determining the effective polarisability and the induced effective dipole moment of individual particles. Whilst this technique gives the average dielectric properties of the particles, AC electrokinetic techniques give the unique properties of single particles (for a full treatment of the inter-relationship between dielectrophoresis, electrorotation and dielectric spectroscopy see Wang *et al.* 1992; 1993).

3.4   Summary

In this chapter we have discussed the induced polarisation of particles in AC electric fields. We have derived expressions for the dipole moment of particles of different geometries and internal structures. In the following chapter we turn our attention to deriving AC electrokinetic forces, *viz* dielectrophoresis, electrorotation and travelling wave dielectrophoresis. These forces arise from the interaction of an induced dipole moment and the electric field.

3.5 References

Fomchemkov V.M. and Gavrilyuk B.K. *The study of dielectrophoresis of cells using the optical technique of measuring* J. Biol. Phys. **6** 29-68 (1978).

Gimsa J., Marszalek P., Löwe U. and Tsong T.Y. *Dielectrophoresis and electrorotation of neurospora slime and murine myeloma cells* Biophys. J. **60** 749-760 (1991).

Grant E.H., Sheppard R.J and South G. *Dielectric Behaviour of Molecules in Solution* Clarendon Press, Oxford (1978).

Griffin J.L. *Orientation of human and avian erythrocyte in radio-frequency fields* Exp. Cell Res. **61** 113-120 (1970).

Hanai T., Koizumi N. and Irimajiri A. *A method for determining the dielectric constant and the conductivity of membrane-bounded particles of biological relevance* Biophys. Struct. Mechanisms **1** 285-294 (1975).

Hanai T. and Koizumi N. *Numerical estimation in a theory of interfacial polarization developed for dipserse systems in higher concentration* Bull. Inst. Chem. Res. Kyoto Univ. **54** 248-254 (1976).

Hanai T., Asami K. and Koizumi N. *Dielectric theory of concentrated suspensions of shell-spheres in particular reference to the analysis of biological cell suspensions* Bull. Inst. Chem. Res., Kyoto Univ. **57** 297-305 (1979).

Huang Y., Hölzel R., Pethig R. and Wang X-B. *Differences in the AC electrodynamics of viable and non-viable yeast cells determined through combined dielectrophoresis and electrorotation studies.* Phys. Med. Biol. **37** 1499-1517 (1992).

Irimajiri A., Hanai T. and Inouye A. *A dielectric theory of "multi-stratified shell" model with its application to a lymphoma cell* J. Theor. Biol. **78** 251-269 (1979).

Jones T.B. and Washizu M. *Multipolar dielectrophoretic force calculations* J. Electrostatics **33** 187-198 (1994).

Jones T.B. *Electromechanics of Particles* Cambridge Uni. Press, Cambridge (1995).

Lorrain P., Corson D.R. and Lorrain F. *Fundamentals of Electromagnetic Phenomena* W.H. Freeman (2000).

Maxwell J.C. *A treatise on Electricity and Magnetism* Dover Press New York 1954.

Miller R.D. and Jones T.B. *Electro-orientation of ellipsoidal erythrocytes: Theory and experiment* Biophys. J. **64** 1588-1595 (1993).

Pethig R. *Dielectric and electronic properties of biological materials* John Wiley & Sons Chichester (1979).

Schwarz G., Saito M. and Schwan H.P. *On the orientation of non-spherical particles in an alternating electric field* J. Chem. Phys. **43** 3562-3569 (1965).

Stratton J.A. *Electromagnetic theory* McGraw-Hill, New York (1941).

Takashima S. *Electrical Properties of Biopolymers and Membranes* Adam Hilger, Philadelphia (1989).

Teixeira-Pinto A.A., Nejelski Jr L.L. Cutler J.L. and Heller J.H. *The behaviour of unicellular organisms in an electromagnetic field* Exp. Cell Res. **20** 548-564 (1960).

Wang X-B., Pethig R. and Jones T.B. *Relationship of dielectrophoretic and electrorotational behaviour exhibited by polarized particles* J. Phys. D: Appl. Phys **25** 905-912 (1992).

Wang X-B., Huang Y., Hölzel, R., Burt J.P.H. and Pethig R.   *Theoretical and experimental investigations of the interdependence of the dielectric, dielectrophoretic and electrorotational behaviour of colloidal particles*   J. Phys. D: Appl. Phys.   **26** 312-322   (1993).

Zhang H.Z., Sekine K., Hanai T. and Koizumi N.   *Dielectric observations on polystyrene microcapsules and the theoretical analysis with reference to interfacial polarisation*   Coll. Polymer Sci.   **261**   381-389   (1983).

Chapter Four

Electrical forces on particles

Electrokinetics is a general term used to describe the movement of particles due to the action of electrical fields. The forces discussed in this chapter are those produced by the interaction of AC electric fields with the induced moments of the particle. There are forces that arise in uniform electric fields, such as electrophoresis and those that can only occur in a non-uniform electric field. A field with a spatially varying magnitude gives rise to dielectrophoresis and a field with rotating field vector gives rise to a torque on a particle, inducing electrorotation. Also, a field with a spatially varying phase (rather than magnitude) gives rise to travelling wave dielectrophoresis. In the following sections, we derive expressions for the different forces and torques and describe the resulting movement of the particles. We also show how these forces exhibit a frequency dependent behaviour that is determined by the combination of the dielectric properties of the particles and the suspending medium.

Electrical forces also act on the fluid and indirectly cause particle movement through the Stokes force. These effects will be discussed in more detail in Chapters Five and Eight.

4.1 Electrophoresis

Electrophoresis is the movement of a particle with a non-zero net charge produced by the Coulomb force. Biological particles generally have a finite surface charge density (usually negative, due to the presence of acid groups on the surface) and observation of the movement of these particles in a uniform electric field is used both to characterise and also to separate particles.

The Coulomb force on a particle is given by the product of the electric field and the charge on the particle (sections 2.1.1 and 2.1.2)

$$\mathbf{F}_{EP} = Q\mathbf{E} = \int_S \sigma_q dS \; \mathbf{E} \qquad (4.1)$$

where Q is the total charge on the particle which, if the particle has a surface charge density σ_q, is given by the integral of this charge density over the closed surface S of the particle. In an AC electric field, the movement due to this force is oscillatory with zero time-average.

Many particles, including biological particles, when suspended in an electrolyte, are surrounded by an electrical double layer (discussed further in Chapter Six) consisting of counter-ionic charges. This screens the particle surface charge, so that the force on the particle is not described exactly by equation (4.1). In addition, the retarding action of viscous friction from the fluid (Chapter Five) must be taken into account. A fuller analysis of electrophoresis, taking these effects into consideration, is presented in Chapter Seven.

4.2   Force on an induced dipole: Dielectrophoresis (DEP)

Dielectrophoresis is the motion of a particle produced by the interaction of a non-uniform electric field with the induced effective dipole moment of the particle (Pohl 1978). If the field is uniform, the force on each of the two poles of the dipole is equal and opposite and there is no movement. If the field is non-uniform, however, the two forces are not equal and the particle moves, as illustrated by figure 1.6 in Chapter One.

4.2.1   Translational force on a dipole in a non-uniform field

The expression for the force exerted on an induced dipole in an electric field **E** was given in Chapter Two (equation (2.15)). An excellent derivation of this expression can be found in Jones (1995). Figure 4.1 shows a dipole $\mathbf{p} = Q\mathbf{d}$ in a non-uniform electric field **E**. As shown by the figure, the two charges each experience a different value of electric field so that the dipole will experience a net force given by

$$F = QE(\mathbf{r}+\mathbf{d}) - QE(\mathbf{r}) \qquad (4.2)$$

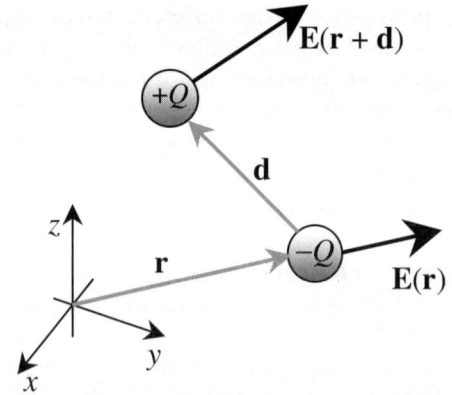

where the vector **r** is the position of the negative charge, as shown in the figure. If the length of the vector **d** is small compared with a typical dimension of the non-uniformity of the electric field, **E** can be expanded around **r** using a vector Taylor series [†] and the force on the dipole becomes

Fig. 4.1 The charges of the dipole $\mathbf{p} = Q\mathbf{d}$ in a non-uniform field **E** experience a different value of the electric field and as a result a different Coulomb force.

$$\mathbf{F} = Q\mathbf{E}(\mathbf{r}) + Q(\mathbf{d}.\nabla)\mathbf{E} + \text{higher order terms} - Q\mathbf{E}(\mathbf{r}) \qquad (4.3)$$

[†] $\mathbf{A}(\mathbf{x}+\mathbf{dx}) = \mathbf{A}(\mathbf{x}) + \mathbf{dx} \cdot \nabla \mathbf{A} +$ higher order terms.

or

$$\mathbf{F} = Q\mathbf{E}(\mathbf{r}) + Q\left(d_x\frac{\partial}{\partial x} + d_y\frac{\partial}{\partial y} + d_z\frac{\partial}{\partial z}\right)\mathbf{E} + \text{higher order terms} - Q\mathbf{E}(\mathbf{r})$$

Substituting for the dipole moment and neglecting the higher order terms gives the well known expression for the force on the dipole

$$\boxed{\mathbf{F}_{DEP} = (\mathbf{p}\cdot\nabla)\mathbf{E}}$$
(4.4)

This equation tells us that the force is zero if the electric field is uniform, *i.e.* only in a non-uniform field can dielectrophoresis occur.

Equation (4.4) gives the total force on the particle only if the length of the dipole $|\mathbf{d}|$ is smaller than a typical dimension of the field non-uniformity, an assumption referred to as the dipole approximation. In other words, if the magnitude of the electric field does not vary significantly across the dimensions of the dipole, this expression for the force is correct. If this assumption does not hold, then multipole force terms must be considered. An alternative and more exact method is to use the Maxwell stress tensor. This requires an exact solution for the potential in the system and the integration of the stress tensor around the particle to give the force (Shlögl and Sauer 1985; Jones 1995; Wang *et al.* 1997).

4.2.2   Dielectrophoresis in an AC field
We will derive the force on the particle in two stages, first considering the case of an electric field with a spatially varying field magnitude and constant phase, which is the accepted method of deriving the dielectrophoretic force. Secondly, in section 4.3 we will consider the more general case of an electric field with a spatially varying magnitude and phase, which gives rise to travelling wave dielectrophoresis.

Assuming an applied potential of a single frequency ω, then the time dependent values in the system can be represented using phasors. An arbitrary, harmonic potential can be defined as $\phi(\mathbf{x},t) = \mathrm{Re}[\tilde{\phi}(x)e^{i\omega t}]$ where $i^2 = -1$, $\mathbf{x}$ is the position, the tilde indicates the complex phasor ($\tilde{\phi} = \phi_R + i\phi_I$) and Re[...] indicates the real part of. The electric field is then given by $\mathbf{E}(\mathbf{x},t) = \mathrm{Re}[\tilde{\mathbf{E}}(\mathbf{x})e^{i\omega t}]$ where the vector $\tilde{\mathbf{E}} = -\nabla\tilde{\phi} = -(\nabla\phi_R + i\nabla\phi_I)$ is the corresponding phasor. If the phase is constant across the system, the field phasor can be assumed to be real without loss of generality (*i.e.* $\tilde{\mathbf{E}} = \mathbf{E} = -\nabla\phi_R$).

The dipole moment of the particle is therefore $\tilde{\mathbf{p}} = \upsilon\tilde{\alpha}\mathbf{E}e^{i\omega t}$ and the time-averaged force on the particle is (from equation (4.4))

$$\langle\mathbf{F}_{DEP}\rangle = \frac{1}{2}\mathrm{Re}[(\tilde{\mathbf{p}}\cdot\nabla)\mathbf{E}^*]$$
(4.5)

where * indicates a complex conjugate. Since the phasors in this case are real, the time-averaged force is

$$\langle \mathbf{F}_{DEP} \rangle = \frac{1}{2} \upsilon \operatorname{Re}[\tilde{\alpha}](\mathbf{E} \cdot \nabla)\mathbf{E} \qquad (4.6)$$

Using a vector identity [t2] and the fact that the field is irrotational ($\nabla \times \mathbf{E} = 0$) gives

$$\langle \mathbf{F}_{DEP} \rangle = \frac{1}{4} \upsilon \operatorname{Re}[\tilde{\alpha}]\nabla(\mathbf{E} \cdot \mathbf{E}) \qquad (4.7)$$

This is the dielectrophoretic force, which is commonly written as

$$\boxed{\langle \mathbf{F}_{DEP} \rangle = \frac{1}{4} \upsilon \operatorname{Re}[\tilde{\alpha}]\nabla |\mathbf{E}|^2} \qquad (4.8)$$

Often the field is quoted using *root-mean-square* (*rms*) values so that equation (4.8) becomes

$$\langle \mathbf{F}_{DEP} \rangle = \frac{1}{2} \upsilon \operatorname{Re}[\tilde{\alpha}]\nabla |\mathbf{E}_{rms}|^2$$

This is also the expression for the force in a DC field. In this book, *rms* designations will not be used, since they are unnecessary and sometimes confusing. Henceforth equation (4.8) will be used as the expression for the DEP force, where **E** is the *amplitude* of the electric field.

Inspection of this equation shows that the dielectrophoretic force depends on the volume of the particle and the gradient of the field magnitude squared (proportional to the energy density in the field). These two parameters can vary greatly in typical experimental conditions. The force also depends on the real part of the effective polarisability and therefore, on the permittivity and conductivity of both the particle and the suspending medium, as well as the frequency of the applied electric field.

For a spherical particle, the variation in the magnitude of the force with frequency is given by the real part of the Clausius-Mossotti factor (equation (3.10)). Substituting for the effective polarisability of a sphere (equation (3.9)), the full expression for the time-averaged DEP force is

$$\boxed{\langle \mathbf{F}_{DEP} \rangle = \pi \varepsilon_m a^3 \operatorname{Re}\left[\frac{\tilde{\varepsilon}_p - \tilde{\varepsilon}_m}{\tilde{\varepsilon}_p + 2\tilde{\varepsilon}_m} \right] \nabla |\mathbf{E}|^2} \qquad (4.9)$$

[t2] $\nabla(\mathbf{A} \cdot \mathbf{B}) = (\mathbf{A} \cdot \nabla)\mathbf{B} + (\mathbf{B} \cdot \nabla)\mathbf{A} + \mathbf{B} \times (\nabla \times \mathbf{A}) + \mathbf{A} \times (\nabla \times \mathbf{B})$

where the parameters in the equation have been defined in Chapter Three. Examination of equation (4.8) or (4.9) shows that the magnitude of the force depends on the particle volume (a^3), the permittivity of the suspending medium and the gradient of the field strength squared.

The real part of the Clausius-Mossotti factor defines the frequency dependence and direction of the force. Expanding on the description given in Chapter One, positive DEP occurs if the polarisability of the particle is greater than the suspending medium ($\mathrm{Re}[\tilde{f}_{CM}] > 1$) and the particle moves towards regions of high electric field strength. Negative DEP occurs if the polarisability of the particle is less than the suspending medium ($\mathrm{Re}[\tilde{f}_{CM}] < 1$) and the particles are repelled from regions of high field strength.

4.2.3   Multipoles

In certain circumstances, the dipole approximation is invalid. This occurs when the particle is close to a field null or in a region where the field strength varies greatly over particle dimensions. In these circumstances, higher order moments (quadrupole, octopole, *etc.*) and the corresponding terms in the force expression become important, as described in Chapter Three. A full derivation of the higher order terms is an involved process and has been performed exhaustively by Jones (1985; 1995), Jones and Washizu (1994; 1996) and also Washizu and Jones (1996). Wang *et al.* (1997) have also presented a detailed derivation of the force on a particle using the Maxwell stress tensor method. The reader should consult these references for further details.

4.3   Dielectrophoresis in a field with a spatially dependent phase

For a general AC field, such as that generated by the application of multiple potentials of different phase, the derivation of the dielectrophoretic force is more involved. The electric field in this case is $\mathbf{E}(\mathbf{x},t) = \mathrm{Re}[\tilde{\mathbf{E}}(\mathbf{x})e^{i\omega t}]$, where the vector $\tilde{\mathbf{E}} = -\nabla\tilde{\phi} = -(\nabla\phi_R + i\nabla\phi_I)$ is the corresponding *complex* phasor. The expression for the time-averaged force on the particle can then be derived from equation (4.5) noting that the vectors now consist of complex components. The equation for the force is

$$\langle \mathbf{F}_{DEP} \rangle = \frac{1}{2}\mathrm{Re}[(\tilde{\mathbf{p}} \cdot \nabla)\tilde{\mathbf{E}}^*] = \frac{1}{2}\upsilon\,\mathrm{Re}[\tilde{\alpha}(\tilde{\mathbf{E}} \cdot \nabla)\tilde{\mathbf{E}}^*] \qquad (4.10)$$

Using vector identities[†3] and the facts that the electric field is irrotational ($\nabla \times \tilde{\mathbf{E}} = 0$) and has zero divergence (Gauss's law) in a homogeneous dielectric, the force expression becomes

---

[†3] $\nabla(\mathbf{A} \cdot \mathbf{B}) = (\mathbf{A} \cdot \nabla)\mathbf{B} + (\mathbf{B} \cdot \nabla)\mathbf{A} + \mathbf{B} \times (\nabla \times \mathbf{A}) + \mathbf{A} \times (\nabla \times \mathbf{B})$

and $\nabla \times (\mathbf{A} \times \mathbf{B}) = (\mathbf{B} \cdot \nabla)\mathbf{A} - (\mathbf{A} \cdot \nabla)\mathbf{B} + (\nabla \cdot \mathbf{B})\mathbf{A} - (\nabla \cdot \mathbf{A})\mathbf{B}$

$$\langle \mathbf{F}_{DEP} \rangle = \frac{1}{4} \upsilon \operatorname{Re}[\tilde{\alpha}] \nabla (\tilde{\mathbf{E}} \cdot \tilde{\mathbf{E}}^*) - \frac{1}{2} \upsilon \operatorname{Im}[\tilde{\alpha}] (\nabla \times (\tilde{\mathbf{E}} \times \tilde{\mathbf{E}}^*)) \qquad (4.11)$$

where Im [...] is the imaginary part of the function. This can be re-written as

$$\langle \mathbf{F}_{DEP} \rangle = \frac{1}{4} \upsilon \operatorname{Re}[\tilde{\alpha}] \nabla \left| \tilde{\mathbf{E}} \right|^2 - \frac{1}{2} \upsilon \operatorname{Im}[\tilde{\alpha}] (\nabla \times (\operatorname{Re}[\tilde{\mathbf{E}}] \times \operatorname{Im}[\tilde{\mathbf{E}}])) \qquad (4.12)$$

where $\left| \tilde{\mathbf{E}} \right|^2 = \left| \operatorname{Re}[\tilde{\mathbf{E}}] \right|^2 + \left| \operatorname{Im}[\tilde{\mathbf{E}}] \right|^2$. If there is no spatially varying phase, the phasor of the electric field can be taken to be real (*i.e.* $\operatorname{Im}[\tilde{\mathbf{E}}] = 0$), the second term on the right hand side of equation (4.12) is then zero and the expression for the force becomes that of equation (4.8). However, if there is a spatially varying phase, as in the case of travelling wave dielectrophoresis, the complete force expression must be used. The first term in the force expression depends on the frequency in the same manner as equation (4.8) and has a similar form. The second term in the force equation depends, however, on the imaginary part of the effective polarisability, or rather the imaginary part of the Clausius-Mossotti factor. This force is zero at high and low frequencies, rising to a maximum value at the Maxwell-Wagner interfacial relaxation frequency.

Note that the form of equation (4.12) given here is different from that generally given in the literature (*e.g.* Wang *et al.* 1994) but is equivalent and lends itself readily to numerical simulation (see Chapter Ten). Another point to note is that in the presence of electrode polarisation (see Chapter Six), the potential across the medium is no longer real, it becomes complex and has an imaginary component which is not parallel with the real part. Therefore, equation (4.8) cannot be used to determine the dielectrophoretic force and equation (4.12) should be used instead.

4.4   Torque on an induced dipole: Electrorotation (ROT)

As we saw in Chapter One, the action of an externally applied electric field on a polarisable particle results in the formation of an induced dipole moment. When a dipole sits in a uniform electric field, each charge on the dipole experiences an equal and opposite force tending to align the dipole parallel to the field, *i.e.* it experiences a torque. Immediately after applying the electric field, it takes a finite amount of time for the dipole moment to become aligned with the field vector; in other words a time delay exists between the establishment of the field and the formation of the dipole. Therefore, if the field vector changes direction, the induced dipole moment vector must realign itself with the electric field vector, causing particle rotation.

Electrorotation will occur in any electric field with a spatially dependent phase. The time-averaged first order torque on the particle is

$$\mathbf{\Gamma}_{ROT} = \frac{1}{2} \operatorname{Re}[\mathbf{p} \times \mathbf{E}^*] \qquad (4.13)$$

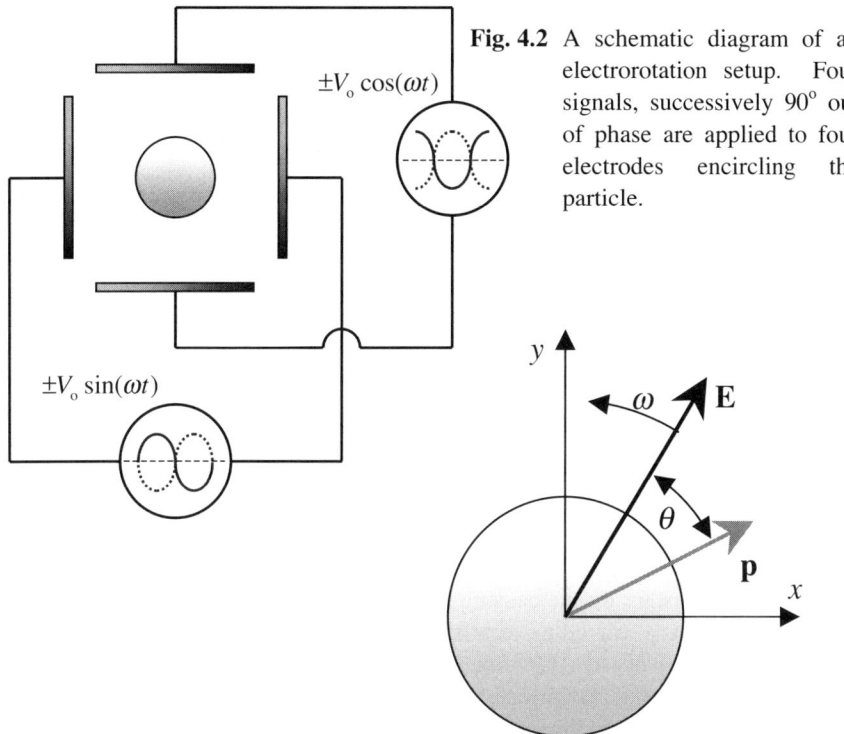

Fig. 4.2 A schematic diagram of an electrorotation setup. Four signals, successively 90° out of phase are applied to four electrodes encircling the particle.

Following the definitions given previously for the electric field and the induced effective dipole moment, the torque can be written as

Fig. 4.3 Schematic diagram showing how the induced dipole moment of a particle lags behind a rotating applied electric field.

$$\Gamma_{ROT} = \frac{1}{2}\upsilon\,\mathrm{Re}[\tilde{\alpha}(\tilde{\mathbf{E}}\times\tilde{\mathbf{E}}^*)] = -\upsilon\,\mathrm{Im}[\tilde{\alpha}](\mathrm{Re}[\tilde{\mathbf{E}}]\times\mathrm{Im}[\tilde{\mathbf{E}}]) \qquad (4.14)$$

Consider the case of a rotating or circularly polarised electric field created by the superposition of two phase-shifted AC signals as shown in figure 4.2. The electric field induces a dipole which continuously tries to align itself with the rotating electric field vector. As a result, the particle experiences a constant torque and rotates asynchronously around its axis. The torque is zero when the phase angle between the particle's polarisation vector and the applied field is zero and maximum when the phase angle is ± 90°, as shown in figure 4.3. If the induced dipole moment lags behind the field, then the direction of rotation is with the field and *vice versa* for a moment that leads the field. This can be written as

$$\Gamma_{ROT} = -v\,\mathrm{Im}[\tilde{\alpha}]\,|\mathbf{E}|^2 \tag{4.15}$$

and substituting for the particle volume we arrive at

$$\Gamma_{ROT} = -4\pi\varepsilon_m a^3\,\mathrm{Im}\left[\frac{\tilde{\varepsilon}_p - \tilde{\varepsilon}_m}{\tilde{\varepsilon}_p + 2\tilde{\varepsilon}_m}\right]|\mathbf{E}|^2 \tag{4.16}$$

As previously shown in figure 3.4, the imaginary part of the Clausius-Mossotti factor peaks at the Maxwell-Wagner interfacial frequency so that a single electrorotational peak is found for each interface, although these are not always easy to identify in practice.

This is the expression for the first order term. Expressions for the electrorotational torque incorporating higher order terms have been covered both by Jones and Washizu (1996) and Washizu and Jones (1996).

4.5   Travelling wave Dielectrophoresis (twDEP)

We have seen that when a polarisable particle is subjected to a rotating field it experiences a constant torque causing it to rotate asynchronously with the electric field vector. In a travelling electric field, as illustrated in figure 4.4, the sequentially phase-shifted AC voltages generate an electric field with a spatially dependent phase. When a particle is exposed to the travelling field, it experiences a force propelling it along the interdigitated electrode array.

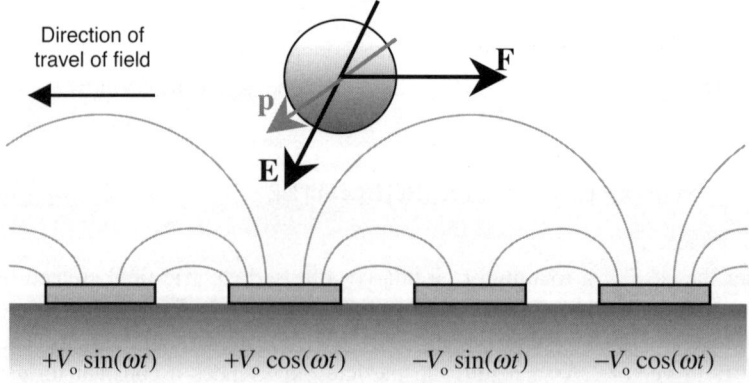

Fig. 4.4 Schematic diagram of a linear travelling wave dielectrophoresis array and the consecutive phase-shifted signals required to generate the travelling electric field. Also shown are the approximate field lines for time $t = 0$, the electric field and the dipole moment induced in the particle together with the force on the particle.

From equation (4.12) we saw that the complete expression for the force consists of two components, one of which is the DEP force and another that depends on the phase of the field. In a travelling electric field, both of these components exist, so that the full DEP force is given by

$$\langle \mathbf{F}_{DEP} \rangle = \underbrace{\frac{1}{4} \upsilon \, \mathrm{Re}[\tilde{\alpha}] \nabla |\tilde{\mathbf{E}}|^2}_{\text{DEP}} - \underbrace{\frac{1}{2} \upsilon \, \mathrm{Im}[\tilde{\alpha}] (\nabla \times (\mathrm{Re}[\tilde{\mathbf{E}}] \times \mathrm{Im}[\tilde{\mathbf{E}}]))}_{\text{twDEP}}$$

The two components can be considered separately; a dielectrophoretic force, given by the first part of the r.h.s. and an additional twDEP force which propels the particle in the opposite direction to the moving field vector, given by the second half of the r.h.s. of this equation. Generally, for twDEP to be effective, the frequency and conductivity must be chosen to fulfil two criteria: (a) the particle is

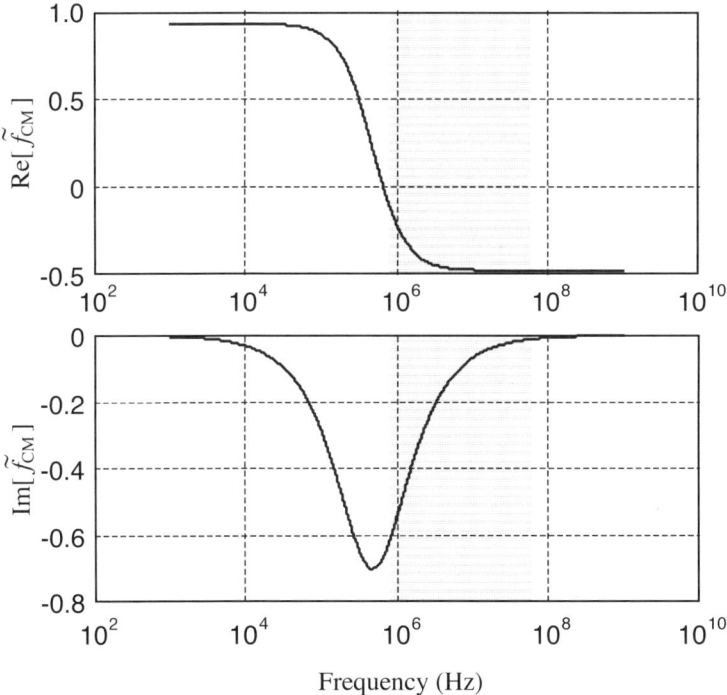

Fig. 4.5 Plot of the real and imaginary parts of the Clausius-Mossotti factor for a solid homogeneous particle plotted as a function of frequency. In the frequency range shown by the grey area, particles in a travelling wave array will experience sufficient dielectrophoretic repulsion to levitate, and a large enough imaginary part to produce a finite translational force along the array.

levitated above the electrodes by a negative DEP force and (b) the imaginary part of the Clausius-Mossotti factor is non-zero so that the particle moves along the track. These conditions are illustrated in figure 4.5, which shows a plot of the real and imaginary parts of the Clausius-Mossotti factor calculated for a latex particle suspended in a very low conductivity medium. Examination of the frequency dependence of both the real and imaginary parts shows three distinct regions, indicated by the shaded parts of the plot. At low frequencies, particles are pulled onto the electrode array by positive DEP forces, so that although the imaginary part is non-zero, twDEP does not occur in practice. At very high frequencies, particles are levitated above the electrodes, but now the imaginary part is zero so that no twDEP occurs. Only in the mid frequency range, where both the real part is substantially high **and** the imaginary part is non-zero, can twDEP occur.

4.6   Particle-particle interaction

So far, in deriving the AC electrokinetic forces we have assumed that the particles are isolated entities, or at least are sufficiently separated from each other that they do not interact. However, in many situations this is not true and particle-particle interactions must be considered. In this section we describe the phenomenology of particle-particle interaction as observed during the dielectrophoretic manipulation of particles.

From basic electrostatics we know that when two or more particles with the same sign of charge are suspended in an insulating medium, then from Coulomb's law we know that they will always repel each other. However, the situation in an electrolyte is a little more complicated, since there is free charge which moves in the applied electric field and screens the particle's charge. This effect will be discussed in greater detail in Chapter Six. The consequence is that the field produced by the particle's charge rapidly decays with distance into the suspending medium, so that any long-range electrostatic interaction with other particles does not occur. The only interaction that can occur is when the particles are very close; in this case both electrostatic interactions and van der Waals forces contribute to the total interaction force between the particles. These effects are important in determining the stability of colloids and are discussed further in Chapter Seven.

4.6.1   Two fixed dipoles

As a qualitative example, consider two fixed dipoles suspended in a uniform electric field. Both dipoles align with the field in the same direction, as shown schematically in figure 4.6. If the dipoles are close enough, each of the charges in one dipole experiences a different force from the two charges in the other dipole. The positive charge of one dipole "feels" an attractive force from the negative charge on the second dipole and *vice versa*. The action of this attractive force is to pull the dipoles together to form a chain. The rapid decay of the dipole field with distance (Chapter Two) ensures that the force imbalance rapidly goes to zero with increasing separation of the dipoles.

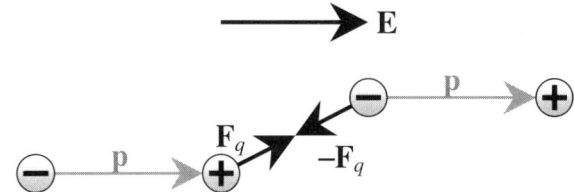

Fig. 4.6 Schematic diagram of the attractive force between two fixed dipoles aligned by an applied uniform electric field.

4.6.2 Induced dipoles

If we now consider a similar scenario, but in this case there are polarisable particles rather than fixed dipoles, then similar effects occur. We will discuss the interaction of the two induced dipoles, together with the resulting electric field and dielectrophoretic force, from a purely qualitative perspective. A more quantitative approach to the issue of particle-particle interaction can be found, for example, in Jones (1995) and Giner *et al.* (1999).

As shown in Chapter One, the behaviour of a particle in an electric field can be predicted from a phenomenological perspective by examining the electric field distribution around the particle. In order to illustrate some phenomena of particle-particle interaction we have calculated a two-dimensional field distribution for different particle configurations. This approach does not immediately lead to a quantitative value for the forces on the particle but does illustrate some interesting effects.

We begin by considering two particles aligned along the direction of a uniform electric field. The distribution in the field magnitude around the particles is shown in figure 4.7. There are two interesting cases to consider. The first is shown in figure 4.7(a), where both particles are much more polarisable than the surrounding medium (*e.g.* conducting particles in a relatively low conductivity electrolyte at low frequencies). The field strength inside the particle is low and there are low field strength regions on either side of the particles. However, there are high field strength regions at the poles of the particles, along the axis of the applied field. This is particularly obvious in the region between the two particles. Since both particles experience positive dielectrophoresis, they will always move towards high field regions and, as this figure shows, they will therefore move towards each other. The second case is when both particles are much less polarisable than the suspending medium, as shown in figure 4.7(b). The strong field regions are at the sides of the particles and the field strength outside the particles at the poles is low. Since both particles experience negative dielectrophoresis, they should move away from high field regions and therefore move towards each other. In both cases the particles will line up along the direction of the electric field and, if there is a large number of particles, chains will form.

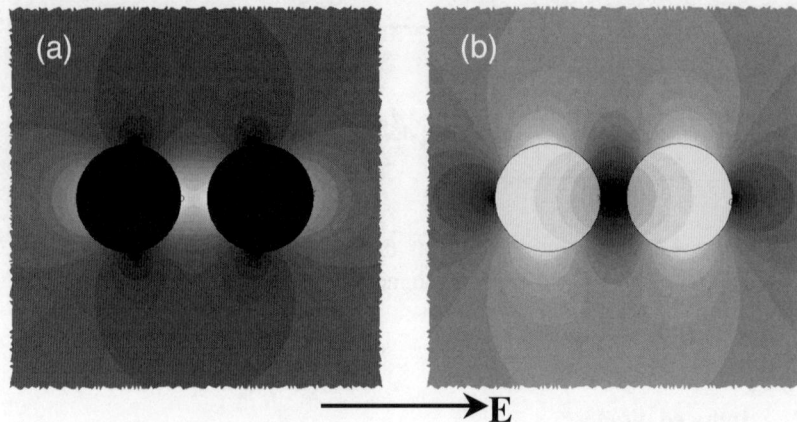

Fig. 4.7 The magnitude of the electric field around two particles aligned along a
horizontal applied electric field. In (a) both particles are much more
polarisable than the surrounding medium *i.e.* they experience positive
DEP. In (b) both particles are much less polarisable than the medium and
experience negative DEP.

Now, if we look at the same two particles aligned perpendicular to the applied
field, we can see that this arrangement is unstable. Figure 4.8(a) shows the electric
field magnitude around the more polarisable particles. The regions of strong
electric field are again at the poles of the particles and along the direction of the

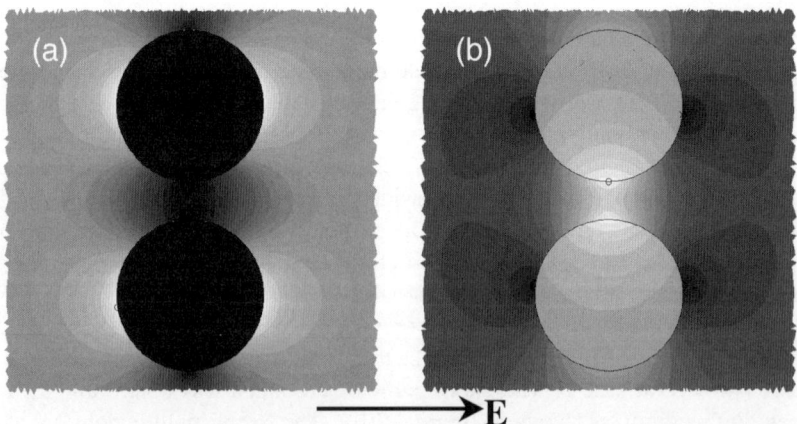

Fig. 4.8 The magnitude of the electric field around two particles aligned
perpendicular to a horizontal applied electric field. In (a), both particles
are much more polarisable than the surrounding medium and experience.
positive DEP. In (b) both particles are much less polarisable than the
suspending medium and experience negative DEP.

field. This time there is a significant region of low field strength between the two particles. Since the particles experience positive dielectrophoresis, they will be pushed away from the low field region; in other words they repel each other. This arrangement is therefore unstable; the particles move into the arrangement shown in figure 4.7(a). The same situation occurs for the less polarisable particles shown in figure 4.8(b). These particles experience negative dielectrophoresis and are attracted to the low field regions, which in this case are at the poles along the field axis. They are also pushed away from the high field region, which is between the particles. The stable configuration is therefore that shown in figure 4.7(b).

There is a third case that can be imagined: two different particles, one more polarisable and the second less polarisable than the electrolyte. In this case, we expect one particle to experience positive dielectrophoresis and the other negative dielectrophoresis. The electric field magnitude is shown in figure 4.9, for two particles aligned along the field or perpendicular to the applied field. When the particles are aligned along the field, figure 4.9(a), the more polarisable particle is attracted towards the high field points around the other particle, which are located at the sides of the particle along the axis perpendicular to the applied field. It is also repelled from the low field region between the two particles. The less polarisable particle on the right is likewise attracted to the sides of the particle along the perpendicular axis, since the low field regions around the more polarisable particle are there. The dielectrophoretic movement suggests that alignment along the field is unstable in this situation and that alignment across the field is preferred. If we then examine figure 4.9(b), which shows the two particles aligned perpendicular to the field, we see that the more polarisable (lower) particle

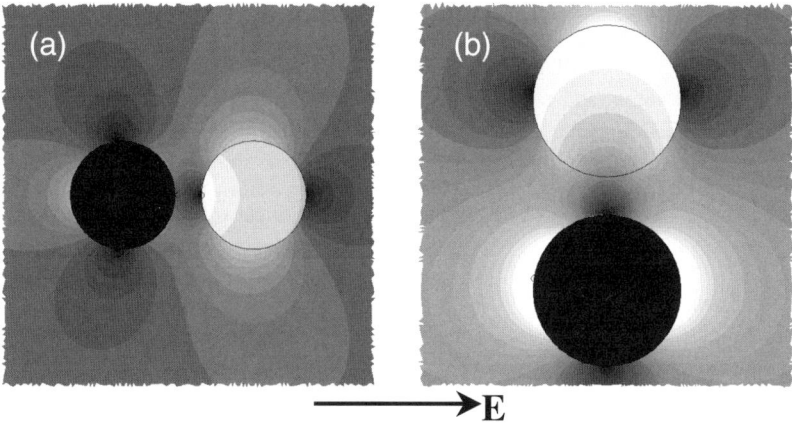

Fig. 4.9 The magnitude of the electric field around two different particles, one more polarisable and the other less polarisable than the medium, in a horizontal applied electric field. Two arrangements are shown, (a) the particles aligned parallel to the field and (b) the two particles aligned perpendicular to the field.

would be attracted towards the less polarisable (upper) particle as there are high field regions between the two. The less polarisable particle is also attracted to the more polarisable particle because of the low field region between the particles. The stable arrangement is, therefore, alignment along the axis perpendicular to the applied field.

To summarise, we see that for two or more particles of similar nature, the arrangement is the formation of long chains aligned along the applied field lines. If, however, there is a mixture of particle types, some of which experience positive and some negative dielectrophoresis, chains of like particles would form along the field lines. In addition, alternate unlike particles would align perpendicular to the field lines. This behaviour is illustrated in figure 4.10, showing the different chains comprising different particle types, as has been observed experimentally (Giner *et al.* 1999).

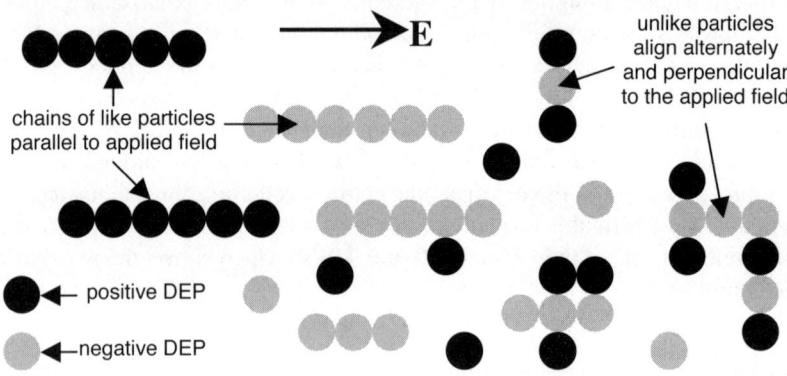

Fig. 4.10   Schematic diagram of how a mixture of particles experiencing both positive and negative dielectrophoresis align in an applied electric field **E**. Like particles form pearl chains along the field lines and unlike particles align perpendicular to the applied field and alternately according to the type of DEP the particle is experiencing.

4.7   Summary

In this chapter we have discussed the fundamental electrical force experienced by particles suspended in a conducting medium. In the next chapters we extend these concepts further by comparing the range of forces that sub-micrometre particles are subjected to when suspended in a fluid medium. In addition to this we examine the behaviour of the fluid when it is exposed to an electric field.

4.8 References

Giner V., Sancho M., Lee R.S., Martinez G. and Pethig R. *Transverse dipolar chaining in binary suspensions induced by rf fields* J. Phys D: Appl Phys **32** 1182-1186 (1999).

Jones T.B. *Multipole corrections to dielectrophoretic force* IEEE Trans. Ind. Appls. **IA-21** 930-934 (1985).

Jones T.B. *Electromechanics of Particles* Cambridge University Press, Cambridge (1995).

Jones T.B. and Washizu M. *Multipolar dielectrophoretic force calculation* J. Electrostatics **33** 187-198 (1994).

Jones T.B. and Washizu M. *Multipolar dielectrophoretic and electrorotation theory* J. Electrostatics **37** 121-134 (1996).

Pohl H.A. *Dielectrophoresis* Cambridge University Press, Cambridge UK (1978).

Schlögl R.W. and Sauer F.A. Interactions between electromagnetic fields and cells Plenum, New York (ed. Chiabrera A.) 203-251 (1985).

Wang X-B., Huang Y., Becker F.F. and Gascoyne P.R.C. *A unified theory of dielectrophoresis and travelling wave dielectrophoresis* J. Phys. D: Appl. Phys **27** 1571-1574 (1994).

Wang X-J., Wang X-B. and Gascoyne P.R.C. *General expressions for dielectrophoretic force and electrorotational torque derived using the Maxwell stress tensor method* J. Electrostatics **39** 277-295 (1997).

Washizu M. and Jones T.B. *Generalized multipolar dielectrophoretic force and electrorotational torque calculation* J. Electrostatics **38** 199-211 (1996).

Chapter Five

Fluid dynamics

Practical application of AC electrokinetics requires the use of a fluid as a suspending medium for the particles. An understanding of how the fluid behaves in microdevices is therefore essential for analysis of experimental data. The physics of basic fluid dynamics can be found in numerous textbooks (Tritton 1988; Acheson 1990; Castellanos 1998) and the application to microsystems was the subject of the previous book in this series (Koch *et al.* 2000).

In this chapter, we summarise the principal governing equations of fluid dynamics so that we can analyse the behaviour of fluids in microsystems. The movement of a particle in a fluid medium is also discussed, paying particular attention to the observed behaviour of sub-micrometre particles. Finally, the physical mechanism of electrothermal fluid flow, which is responsible for fluid flow in microsystems, is outlined. Experimental observations and numerical simulations of this effect will be discussed further in Chapter Eight.

We begin with a short summary of fluid dynamics in microsystems. When a particle is suspended in a fluid and a force is applied to the particle, the resulting movement of the particle is influenced by the fluid. When the particle moves, the fluid exerts a viscous drag force on the particle opposing the movement. The correct derivation of the relationship between particle properties and observed motion must therefore include this effect.

In addition, if the fluid moves the drag force pulls the particles along with the fluid. In microsystems, the fluid flow can be driven externally *e.g.* with a pump, or the fluid can also move under the influence of an electric field. The force producing this type of fluid movement arises from the interaction of space charge with the field or the interaction of the field with gradients in conductivity and permittivity, generated as a result of external or Joule heating. These phenomena are described by the field of electrohydrodynamics (EHD) (Castellanos 1998).

5.1   Fluid flow

A complete derivation of the equations governing the dynamics of fluid flow is an involved application of tensor mathematics, which can be found in numerous textbooks on the subject (Landau and Lifshitz 1959; Castellanos 1998). Here, we restrict ourselves to stating the governing equations, together with some of the restrictions that apply to microsystems. In our discussion, the fluid is assumed to be a continuous medium.

5.1.1   The continuity or conservation of mass equation

The local density of the fluid obeys the conservation of mass equation, similar to the charge conservation equation discussed in section 2.1.6. This states that the rate of change of mass in an arbitrary volume, is equal to the flux of mass through the surface enclosing the volume. The continuity equation can be written as

$$\frac{d\rho_m}{dt} + \rho_m \nabla \cdot \mathbf{u} = 0 \qquad\qquad (5.1)$$

where ρ_m is the mass density and $\mathbf{u}$ is the velocity of the fluid. As discussed by Castellanos (1998), the order of magnitude of the relative pressure variation is $\Delta\rho_m / \rho_m \sim (u_o / u_s)^2$, where u_o is the magnitude of a typical velocity and u_s is the speed of sound in the fluid. Typical fluid velocities in microsystems can be up to 1 mm s^{-1} and the velocity of sound in water is 1400 m s^{-1}. Therefore, pressure changes are negligible and the fluid can be considered to be incompressible, so that the continuity equation becomes

$$\nabla \cdot \mathbf{u} = 0 \qquad\qquad (5.2)$$

5.1.2   The Navier-Stokes equation

The Navier-Stokes equation is the equation of motion for the fluid and is derived from conservation of momentum arguments. For an incompressible, Newtonian fluid, the Navier-Stokes equation is

$$\rho_m \frac{\partial \mathbf{u}}{\partial t} + \rho_m (\mathbf{u} \cdot \nabla)\mathbf{u} = -\nabla p + \eta \nabla^2 \mathbf{u} + f \qquad\qquad (5.3)$$

where p is the pressure, η is the viscosity and f is the total applied body force. This equation can be simplified further by examining the scales of the problem and non-dimensionalising the Navier-Stokes equation. Each variable is scaled by a typical magnitude, e.g. $\mathbf{u} = u_o \mathbf{u}'$ where u_o is typically $10^{-4} - 10^{-3}$ ms^{-1} for microsystems and $\mathbf{u}'$ is the dimensionless velocity. The other magnitudes in the Navier-Stokes equation are time t_o, pressure p_o and a length scale l_o. Equation (5.3) then takes the form of dimensionless variables, with dimensionless scaling coefficients, the relative size of which can be examined in order to determine which terms in the equation can be neglected. This non-dimensional equation can be written as

$$\frac{\partial \mathbf{u}'}{\partial t'} + \frac{t_o u_o}{l_o}(\mathbf{u}' \cdot \nabla')\mathbf{u}' = -\frac{t_o p_o}{\rho_m u_o l_o}\nabla' p' + \frac{\eta t_o}{\rho_m l_o^2}\nabla'^2 \mathbf{u}' + \frac{t_o}{\rho_m u_o}f \qquad (5.4)$$

The ratio of the inertial term: $\rho_m(\mathbf{u} \cdot \nabla)\mathbf{u}$ to the viscous term: $\eta \nabla^2 \mathbf{u}$ can be determined from the coefficients in equation (5.4) giving a factor which is referred to as the *Reynold's number*

$$Re = \frac{\rho_m u_o l_o}{\eta} \qquad\qquad (5.5)$$

For low values of Reynold's number ($\ll$ 1), the viscous term dominates the dynamics of the fluid and for high values ($\gg$ 1) the inertial term dominates. For microsystems, typical values are $l_o \sim 10^{-4}$ m and $u_o \sim 10^{-2} - 10^{-3}$ m s^{-1}, with $\rho_m = 10^3$ kg m^{-3} and $\eta = 10^{-3}$ kg m^{-1} s^{-1}. The Reynold's number is therefore of the order 0.1 to 0.01 and the fluid flow is said to be in the viscous fluid limit. This means that if the fluid is moving under an applied pressure or force, which is suddenly removed, the fluid stops immediately. Equation (5.3) then becomes

$$\rho_m \frac{\partial \mathbf{u}}{\partial t} = -\nabla p + \eta \nabla^2 \mathbf{u} + f \qquad\qquad (5.6)$$

The solution of this equation, with the appropriate boundary conditions, gives the fluid velocity for a given body force f.

5.2 Fluid flow in microsystems
A number of generalisations can be made about flow in microsystems in order to simplify what is a complicated problem, aid understanding and assist in design. For example, the fact that the Reynold's number is low means that the Navier-Stokes equation can be used in the form of equation (5.6). Several simple pictures can be used to illustrate general aspects of fluid flow in microchannels and microsystems.

5.2.1 Laminar flow (Poiseuille flow)
In microsystems, microfabricated channels and chambers are used to guide the fluid through the device. Typical dimensions for these channels vary between ~10 μm and 1 mm. The flow in such channels is laminar *i.e.* the fluid flow follows streamlines and is free of turbulence.

As an illustration, consider the problem of steady two-dimensional fluid flow through a chamber of height $2d$. Figure 5.1 shows the problem space, with the upper and lower parallel walls of the chamber at $y = \pm d$ respectively. The chamber has a width (z-direction) much greater than the height $2d$, so that the system can be considered two-dimensional. The chamber has length l_o and a pressure p_o is applied at the start of the chamber ($x = 0$). The flow is plane and parallel, of the form $\mathbf{u} = (u_x(y,t), 0, 0)$. This automatically satisfies the continuity equation (5.2). Equation (5.6) becomes

Fig. 5.1 Diagram of the simple channel, height $2d$ and length l_o, with the axes marked. The top and bottom of the channel are at $+d$ and $-d$.

$$\rho_m \frac{\partial u_x}{\partial t} = -\frac{\partial p}{\partial x} + \eta \frac{\partial^2 u_x}{\partial y^2} \tag{5.7}$$

$$\frac{\partial p}{\partial y} = \frac{\partial p}{\partial z} = 0 \tag{5.8}$$

The second equation implies that the pressure is a function only of x and time t, but the first equation implies that $\partial p/\partial x$ is a function of y and t. As a result, the gradient term $\partial p/\partial x$ (pressure drop) is only a function of time, and for a steady flow, is constant along the chamber, with value $-p_o/l_o$. In the steady state, (5.7) becomes

$$\frac{\partial^2 u_x}{\partial y^2} = \frac{1}{\eta}\frac{\partial p}{\partial x} = -\frac{p_o}{\eta l_o} \tag{5.9}$$

Integrating this expression twice with respect to y gives the general solution

$$u_x = -\frac{p_o}{2\eta l_o} y^2 + C_1 y + C_2 \tag{5.10}$$

where C_1 and C_2 are integration constants. Applying the boundary conditions $u_x = 0$ at $y = \pm d$ gives values for the constants of $C_2 = p_o d^2/2\eta l_o$ and $C_1 = 0$. The solution for the velocity in this chamber is then

$$u_x = \boxed{\frac{1}{2\eta} \frac{p_o}{l_o} (d^2 - y^2)}$$ (5.11)

This is the equation for an inverted parabola with a maximum of

$$u_{max} = \frac{p_o d^2}{2\eta l_o}$$ (5.12)

in the centre of the chamber ($y = 0$) and an average value in the chamber of

$$u_{av} = \frac{d^2}{3\eta} \frac{p_o}{l_o}$$ (5.13)

The volume flow rate per unit width of chamber is given by

$$Q = \int_{y=-d}^{y=+d} u_x dy = \frac{2d^3}{3\eta} \frac{p_o}{l_o}$$ (5.14)

Figure 5.2 shows the flow profile for a variety of channel heights and the same applied pressure. The velocity is zero at the walls and maximum in the centre of the channel. This steady-state flow is referred to as *Poiseuille flow*.

5.2.2 Entry length

Poiseuille fluid flow occurs at some distance from the entry point of the fluid into a channel. At the entry point the fluid flow can be significantly different and there is a characteristic distance before the parabolic flow profile is established. The distance between the point of entry and the establishment of Poiseuille flow is called the *entry length*. Further detailed information on this can be found in textbooks on fluid dynamics (*e.g.* Tritton 1988).

For low Reynold's numbers, as in microsystems, the entry length is so short that it can be ignored but assuming an entry length of the order of the width/height of the channel would be suitable. For high Reynold's numbers (not relevant to microsystems), the length after which the flow profile is within 5% of the Poiseuille flow in a cylindrical pipe of radius r is

$$l_{entry} = \frac{2}{15} \frac{\rho_m u r^2}{\eta}$$ (5.15)

where u is the constant entry velocity. Taking typical values for microsystems with $r = 100$ μm, $u = 100$ μm s^{-1}, $\rho_m = 10^3$ kg m^{-3} and $\eta = 10^{-3}$ kg m^{-1} s^{-1}, the entry length is $\sim 10^{-7}$ m. As a result, the fluid can be assumed to move in the

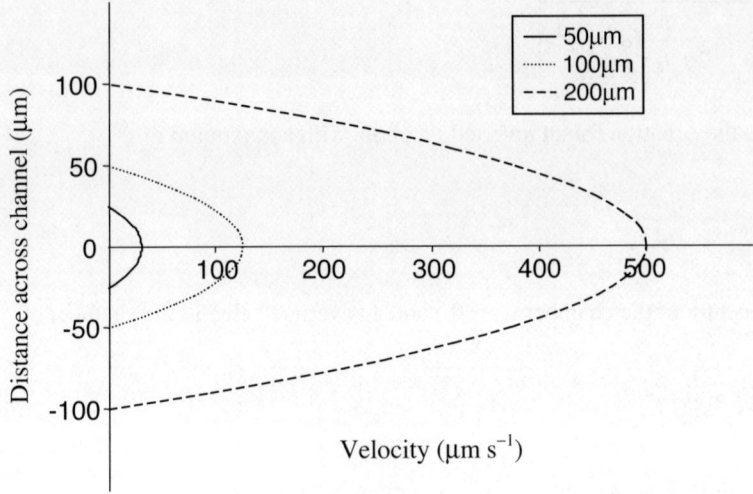

Fig. 5.2 Plot of the variation in laminar fluid velocity as a function of distance across the channel for the same applied pressure difference, for three different channel heights as indicated.

parabolic profile given by equation (5.11) almost immediately after entry into a channel.

5.2.3   Numerical simulation of entry dynamics

The description of entry length and the movement of fluid in a microsystem are more complicated than this simple picture. As an illustration of fluid flow in microsystems with multiple channels and junctions the fluid flow profile has been solved for a simple change in channel geometry, as shown in figure 5.3. Equation (5.7) was solved numerically using a commercial solver, FlexPDE (see appendix A). In the geometry shown in figure 5.3, a constant pressure difference is applied along the system. The fluid streamlines are drawn in figure 5.3, whilst figure 5.4 shows the magnitude of the fluid flow across the junction, indicating how the fluid velocity increases as the channel narrows. The flow in the narrower region of the channel reaches the steady-state parabolic profile very rapidly, as expected from the order of magnitude calculation described above. However, examination of the fluid profile in the wider channel shows how the velocity profile in this region is affected by the change in channel width some distance before the entry point. This example shows that an accurate picture of how the fluid moves is an important aspect for determining particle dynamics in microsystems.

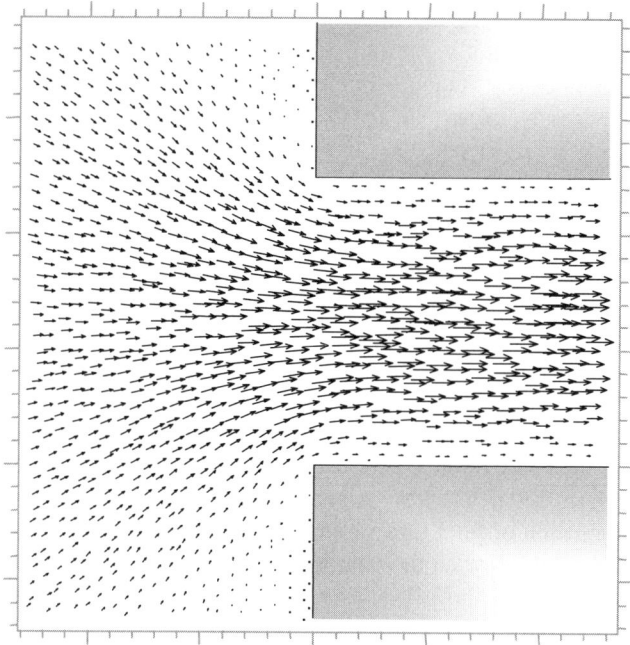

Fig. 5.3 Numerical simulation of fluid flowing into a narrow channel
calculated using equation (5.3) in the low Reynold's number
limit.

Fig. 5.4 Plot of the magnitude of
the fluid flow shown by
the streamlines of
figure 5.3. The velocity
reaches the expected
parabolic profile at the
start of the channel,
with the region where
the velocity profile
changes extending back
into the larger channel.

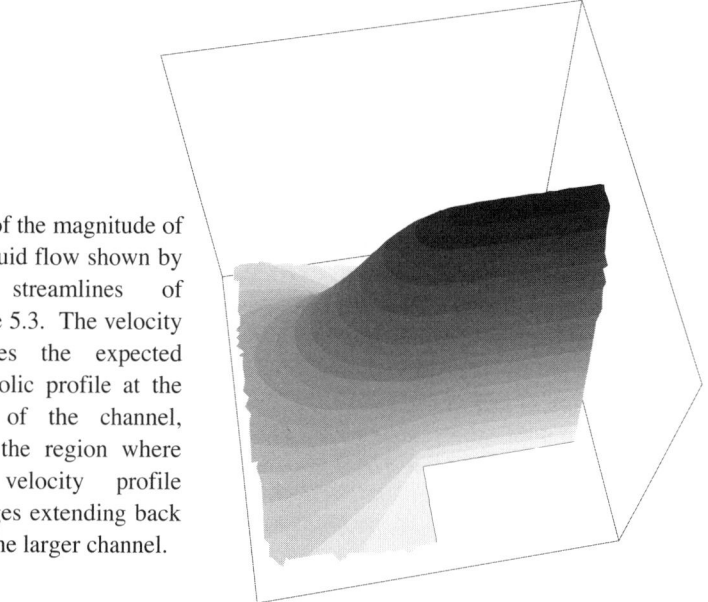

5.3   External forces on the fluid

Understanding and controlling fluid motion is central to the design and operation of electrokinetic devices. Often, particles are transported through measurement or separation devices by means of a pressure gradient generated from an external pump, but there are other external forces that can produce fluid motion inside the microsystem. Gravity produces a body force when there are gradients in the density of the fluid, which if the fluid is incompressible, can be produced by non-uniform heating. This is convection, which is not a significant effect in microdevices (Ramos *et al.* 1998).

There are also two main forces related to the action of the electric field on the fluid: (a) Electrothermal, where the electric field acts on gradients in permittivity and/or conductivity produced by non-uniform heating, and (b) Electroosmosis, where the electric field acts on free charge in the electrical double layer. In this section we will discuss electrothermal forces and their relative importance in the generation of fluid motion in microsystems. Electroosmosis and its effect on the fluid will be discussed in Chapter Six and Eight.

The examination of fluid flow in microsystems is important from two aspects: the effect of the fluid flow on the motion of the particles in the system and the use of electric field driven fluid flow as a micropump, capable of integration in a microsystem.

5.3.1   Gravity, buoyancy and natural convection

Natural convection is the result of changes in fluid density produced by gradients in temperature. When a fluid is subjected to non-uniform heating, a buoyancy force causes the lighter (hotter) parts of the fluid to rise and the denser (colder) parts to fall. The buoyancy body force f_g is given by

$$f_g = \Delta\rho_m \mathbf{g} = \frac{\partial\rho_m}{\partial T}\Delta T\mathbf{g} \qquad (5.16)$$

where T is the temperature and $\mathbf{g}$ is the acceleration due to gravity. Dimensional analysis of the relative sizes of the forces in microsystems (Ramos *et al.* 1998) shows that the buoyancy force is unimportant in comparison with the electrical forces, when the latter are present. However, since the electrical forces are frequency dependent, there are frequencies at which these forces could be less important than the buoyancy force.

5.3.2   Electrical forces

In general, the application of electric fields to a fluid produces body forces and therefore fluid flow. However, in sub-micrometre AC electrokinetics the electric fields are strong; the fluid flows are more noticeable and fluid velocities higher. The electrical body force is given by (Stratton 1941, Melcher 1981)

$$f_e = \rho \mathbf{E} - \frac{1}{2}|\mathbf{E}|^2 \nabla \varepsilon + \frac{1}{2} \nabla \left(\rho_m \left(\frac{\partial \varepsilon}{\partial \rho_m} \right)_T |\mathbf{E}|^2 \right) \qquad (5.17)$$

where ρ and ρ_m are the charge and mass densities respectively. On the right hand side of this expression, the first term is the Coulomb force, the second is the dielectric force and the last term is the electrostriction pressure. For an incompressible fluid, the electrostriction pressure can be ignored since it is the gradient of a scalar and has no effect on the dynamics of the system.

Equation (5.17) defines the electrical body force in terms of local variations in charge density. Localized heating of the fluid gives rise to gradients in the permittivity and conductivity, which in turn produce a body force on the fluid. To determine the frequency dependence and magnitude of the body force, equation (5.17) can be rewritten in terms of gradients in permittivity and conductivity as follows.

For water, the relative changes in permittivity and conductivity are small (Lide 1994). This means that the changes in the charge density and the electric field are small and we can perform a perturbative expansion of the electric field. The electric field is the sum of the applied field $\mathbf{E}_0$ and the perturbation field $\mathbf{E}_1$ so that $\mathbf{E} = \mathbf{E}_0 + \mathbf{E}_1$ with $|\mathbf{E}_1| \ll |\mathbf{E}_0|$.

We can now expand Gauss's Law, written for inhomogeneous media as $\rho = \nabla \cdot (\varepsilon \mathbf{E})$, using the above expression for $\mathbf{E}$ and $\nabla \cdot \mathbf{E}_0 = 0$

$$\rho = \nabla \varepsilon \cdot \mathbf{E}_0 + \varepsilon \nabla \cdot \mathbf{E}_1 \qquad (5.18)$$

Using this expression for the charge density, the electrical body force becomes

$$f_E = (\nabla \varepsilon \cdot \mathbf{E}_0 + \varepsilon \nabla \cdot \mathbf{E}_1) \mathbf{E}_0 - \frac{1}{2}|\mathbf{E}_0|^2 \nabla \varepsilon \qquad (5.19)$$

The next step is to obtain an expression for the divergence of the perturbation field, $\nabla \cdot \mathbf{E}_1$, in terms of the conductivity, permittivity and the applied field by using the charge conservation equation

$$\frac{\partial \rho}{\partial t} + \nabla \cdot (\rho \mathbf{u}) + \nabla \cdot (\sigma \mathbf{E}) = 0 \qquad (5.20)$$

Here $\rho \mathbf{u}$ is the convection current, the current arising from the movement of charge due to the motion of the fluid, and $\sigma \mathbf{E}$ is the conduction current, defined in Chapter Two. The ratio of the magnitude of the convection current to the magnitude of the conduction current, is the *electrical Reynold's* number (Castellanos 1998)

$$\frac{u_o \varepsilon}{l_o \sigma} \qquad (5.21)$$

where the scales are defined as before. For typical values in microsystems, this is 10^{-7}, *i.e.* the velocity of the charge in response to the electric field is very much greater than the velocity of the fluid. Therefore, we can neglect the convection current in equation (5.20) which, substituting for **E**, becomes

$$\nabla \sigma \cdot \mathbf{E}_0 + \sigma \nabla \cdot \mathbf{E}_1 + \frac{\partial}{\partial t}(\nabla \varepsilon \cdot \mathbf{E}_0 + \nabla \varepsilon \cdot \mathbf{E}_1) = 0 \qquad (5.22)$$

Assuming we have a harmonic time varying field of single frequency ω, $\partial / \partial t = i\omega$ and

$$\nabla \sigma \cdot \mathbf{E}_0 + i\omega \nabla \varepsilon \cdot \mathbf{E}_0 + \sigma \nabla \cdot \mathbf{E}_1 + i\omega \varepsilon \nabla \cdot \mathbf{E}_1 = 0 \qquad (5.23)$$

This equation can be re-arranged to give the divergence of the perturbation field

$$\nabla \cdot \mathbf{E}_1 = \frac{-(\nabla \sigma + i\omega \nabla \varepsilon) \cdot \mathbf{E}_0}{\sigma + i\omega \varepsilon} \qquad (5.24)$$

From equation (5.19), the time average force per unit volume is

$$\langle f_e \rangle = \frac{1}{2} \mathrm{Re}[(\nabla \varepsilon \cdot \mathbf{E}_0 + \varepsilon \nabla \cdot \mathbf{E}_1)\mathbf{E}_0^* - \frac{1}{2}|\mathbf{E}_0|^2 \nabla \varepsilon] \qquad (5.25)$$

Substituting equation (5.24) into equation (5.25) gives the body force in terms of the applied field

$$\boxed{\langle f_e \rangle = \frac{1}{2} \mathrm{Re}\left[\left(\frac{(\sigma \nabla \varepsilon - \varepsilon \nabla \sigma) \cdot \mathbf{E}_0}{\sigma + i\omega \varepsilon}\right)\mathbf{E}_0^* - \frac{1}{2}|\mathbf{E}_0|^2 \nabla \varepsilon\right]} \qquad (5.26)$$

This defines the body force in terms of the permittivity and conductivity gradients. These are related to the temperature gradient according to $\nabla \varepsilon = (\partial \varepsilon / \partial T)\nabla T$ and $\nabla \sigma = (\partial \sigma / \partial T)\nabla T$. Equation (5.26) can be written in terms of the temperature gradient (Castellanos 1998)

$$\boxed{\langle f_e \rangle = \frac{1}{2} \mathrm{Re}\left[\frac{\sigma \varepsilon (\alpha - \beta)}{\sigma + i\omega \varepsilon}(\nabla T \cdot \mathbf{E}_0)\mathbf{E}_0^* - \frac{1}{2}\varepsilon \alpha |\mathbf{E}_0|^2 \nabla T\right]} \qquad (5.27)$$

where $\alpha = (1/\varepsilon)(\partial \varepsilon / \partial T)$ and $\beta = (1/\sigma)(\partial \sigma / \partial T)$. For an aqueous electrolyte solution, typically $\alpha \approx -0.4\% \, \mathrm{K}^{-1}$ and $\beta \approx +2\% \, \mathrm{K}^{-1}$ (Lide 1994).

For an AC signal of frequency ω, the electric field and the charge density oscillate with the same frequency. The time dependent force (equation (5.19)) is the sum of the product of these terms and therefore contains both a steady term (given by equation (5.27)) and a term that oscillates at 2ω. The resulting fluid velocity will also have both steady and oscillating terms but at sufficiently high frequencies, the oscillatory terms are much smaller than the steady.

As discussed by Ramos et al. (1998), the body force is frequency dependent and has two distinct limits: at low frequencies the Coulomb force (the first term on the right hand side of equations (5.26) and (5.27)) dominates, and at high frequencies, the dielectric force dominates. Typically, these two forces act in different directions and over a certain range of frequencies they compete, resulting in a changing flow pattern (Ramos et al. 1998).

5.3.3 Sources of heat
In general, for electrothermal force calculations in electrohydrodynamics, the electric, temperature and velocity fields are coupled. The temperature balance equation describes the relationship between the generation and dissipation of heat

$$\rho_m c_p \frac{\partial T}{\partial t} + \rho_m c_p \mathbf{u} \cdot \nabla T = k\nabla^2 T + \sigma |\mathbf{E}|^2 \tag{5.28}$$

In this expression, c_p is the specific heat at constant pressure, k the thermal conductivity and $\sigma |\mathbf{E}|^2$ is the Joule heating term. The viscous dissipation term can be neglected, since it is of the order of 10^{-10} times smaller than the Joule heating term (Castellanos 1998). For electric fields of sufficiently high frequencies, we can simplify this equation to the steady-state and neglect the first term on the lefthand side.

More importantly, for microsystems the problem can be further simplified, leading to a decoupling of the thermal problem from the velocity problem and therefore from the electrical one. Following the same dimensional arguments as previously, the ratio of convection of heat ($\rho_m c_p \mathbf{u} \cdot \nabla T$) to heat diffusion ($k\nabla^2 T$) is given by $\rho_m c_p u_o l_o / k$. For scales typical of microsystems this ratio is $\sim 3\times10^{-3}$, so that the convection term can be neglected in the calculation of the temperature field. In other words, this means that the motion of the fluid does not substantially affect the distribution of the temperature, implying that the temperature and velocity problems are decoupled.

As stated previously, the electric field is also independent of the fluid velocity, so that the electric field is calculated independently. The temperature is then calculated by solving the diffusion equation with a source term given by the time average of the Joule heating from the electric field solution. This gives a simple expression relating the Joule heating to the temperature gradient.

$$k\nabla^2 T + \left\langle \sigma |\mathbf{E}|^2 \right\rangle = 0 \tag{5.29}$$

Finally, the electrothermal force is calculated from the solution of the electric and thermal fields, to give the body force in the Navier-Stokes equation, from which the velocity of the fluid can be calculated.

Examples of fluid flow, calculated both analytically and numerically, are described in Chapter Eight where electric field-driven fluid flow in microsystems is discussed further.

5.4   Effect of fluid on a moving particle

In experimental systems, one of the more important aspects of analysis is consideration of the effect that the fluid has on the particle. The fluid exerts a drag force on the particle that affects the velocity of the particle. If the fluid is in motion, then the drag force pulls the particle along and can influence the measurement of other effects.

In general, when a particle is moving relative to the fluid, it experiences a viscous drag force due to the action of the fluid on the particle. The force is proportional to, and acts against the relative velocity of the particle

$$\mathbf{F}_\eta = -f\mathbf{v} \tag{5.30}$$

The constant f is referred to as the friction factor and depends on a range of particle parameters such as size, shape and surface characteristics. Table 5.1 summarises some friction factors for different particle geometries.

When a particle is suspended in a fluid and is accelerated by a deterministic force, it experiences an increasing drag force. For a constant applied force, the particle reaches a terminal velocity beyond which it does not accelerate. If the fluid is in motion itself, this terminal velocity also depends on the velocity of the fluid. The dynamics of this system can be examined in a simple manner as follows. The equation of motion for the particle is Newton's second law

$$m\frac{d\mathbf{v}}{dt} = \mathbf{F} \tag{5.31}$$

where m is the mass of the particle, $\mathbf{v}$ is the velocity of the particle and $\mathbf{F}$ is the total force acting on the particle. For the general case of an arbitrary force $\mathbf{F}_{Arb}$ applied to a particle, where the fluid is moving with velocity $\mathbf{u}$, the drag force experienced by the particle is proportional to its velocity relative to the fluid *i.e.* $\mathbf{F}_\eta = f(\mathbf{u} - \mathbf{v})$. The equation of motion for the particle becomes

$$m\frac{d\mathbf{v}}{dt} = \mathbf{F}_{Arb} + f(\mathbf{u} - \mathbf{v}) \tag{5.32}$$

which can be re-written as $\dfrac{d\mathbf{v}}{dt}+\dfrac{f}{m}\mathbf{v}=\dfrac{\mathbf{F}_{Arb}}{m}+\dfrac{f\mathbf{u}}{m}$. Assuming that the particle was initially at rest, the solution to this equation is

$$\mathbf{v}=\left(\dfrac{\mathbf{F}_{Arb}}{f}+\mathbf{u}\right)(1-e^{-(f/m)t}) \tag{5.33}$$

This expression gives the velocity for the particle at times greater than zero. The exponential term describes the acceleration phase of the particle's motion and has a characteristic time constant $\tau_a=m/f$. For times much greater than τ_a, the particle moves at the terminal velocity

$$\mathbf{v}_T=\dfrac{\mathbf{F}_{Arb}}{f}+\mathbf{u} \tag{5.34}$$

Table 5.1 Translational friction factors for some simple smooth particle shapes based on an ellipsoid with semi axis lengths a_1, a_2, a_3 (from Berg 1993).

Particle	Friction factor
Sphere: $a_1=a_2=a_3$	$f=6\pi\eta a_1$
Disk: $a_1=a_2\gg a_3$ moving face on (parallel to axis a_3)	$f=16\eta a_1$
Disk: $a_1=a_2\gg a_3$ moving edge on (perpendicular to axis a_3)	$f=\dfrac{32}{3}\eta a_1$
Disk: $a_1=a_2\gg a_3$ moving at random	average $\langle f\rangle=12\eta a_1$
Prolate ellipsoid: $a_1\gg a_2=a_3$ moving lengthways (parallel to a_1)	$f=\dfrac{8\pi\eta a_1}{2\ln(2a_1/a_2)-1}$
Prolate ellipsoid: $a_1\gg a_2=a_3$ moving sideways (perpendicular to a_1)	$f=\dfrac{16\pi\eta a_1}{2\ln(2a_1/a_2)-1}$
Prolate ellipsoid: $a_1\gg a_2=a_3$ moving at random	average $\langle f\rangle=\dfrac{6\pi\eta a_1}{\ln(2a_1/a_2)}$

If the fluid is at rest, the terminal velocity is simply proportional to the applied force. A useful concept for experimentation is the time of observation τ_o, which characterises the minimum period of time that an experimental system (particle moving in fluid) can be observed. For example, for observation by eye $\tau_o \sim 1/30$ s. If $\tau_a \ll \tau_0$ then the acceleration phase is not observed and the particle appears to always move at the terminal velocity. For particles of the order of size of a cell or sub-micrometre particles, $\tau_a < 10^{-6}$ sec. This means that the velocity of sub-micrometre particles moving in microsystems is everywhere proportional to the applied force if the fluid is at rest; or given by equation (5.34), if the fluid is in motion.

5.4.1   Steady-state Dielectrophoresis

If a particle moves under the influence of a DEP force, then ignoring Brownian motion and the buoyancy force, the equation of motion is simply

$$m\frac{d\mathbf{v}}{dt} = \mathbf{F}_{DEP} - \mathbf{F}_\eta \qquad (5.35)$$

As described above, we can assume that the instantaneous velocity is proportional to the instantaneous dielectrophoretic force so that $\mathbf{v}_{DEP} = \mathbf{F}_{DEP}/f$ and, substituting for the DEP force (from Chapter Four), we obtain the steady-state particle velocity as

$$\mathbf{v}_{DEP} = \frac{\upsilon \, \mathrm{Re}[\tilde{\alpha}]\nabla|\mathbf{E}|^2}{f} \qquad (5.36)$$

which for a spherical particle is

$$\mathbf{v}_{DEP} = \frac{\pi a^3 \varepsilon_m \, \mathrm{Re}[\tilde{f}_{CM}]\nabla|\mathbf{E}|^2}{6\pi\eta a} \qquad (5.37)$$

We can define at this point the dielectrophoretic mobility as $\mathbf{v}_{DEP} = \mu_{DEP}\nabla|\mathbf{E}|^2$, which for a spherical particle is

$$\boxed{\mu_{DEP} = \frac{a^2 \varepsilon_m \, \mathrm{Re}[\tilde{f}_{CM}]}{6\eta}} \qquad (5.38)$$

It can be seen that for a spherical particle the dielectrophoretic mobility depends on the radius of the particle squared and the real part of the Clausius-Mossotti factor, together with the permittivity and viscosity of the fluid.

5.4.2   Steady-state Electrorotation

When a particle rotates, it experiences a drag torque, characterised by a friction coefficient. This depends on the viscosity of the medium and the geometry and surface properties of the particle. For slow rotation rates (a few rad s^{-1}) the rotational frictional factor for a spherical stationary body rotating about its centre is

$$f_\theta = 8\pi\eta a^3 \qquad (5.39)$$

In the case of electrorotation, the angular velocity (rate of rotation) of the particle is given by the electrorotational torque, divided by this factor

$$\mathbf{R}(\omega) = \frac{\Gamma_{ROT}}{f_\theta} \qquad (5.40)$$

Substituting for the induced torque (equation (4.16)), we obtain the steady-state electrorotational angular velocity for a particle

$$\mathbf{R}(\omega) = -\frac{\varepsilon_m |\mathbf{E}|^2}{2\eta} \operatorname{Im}\left[\frac{\tilde{\varepsilon}_p - \tilde{\varepsilon}_m}{\tilde{\varepsilon}_p + 2\tilde{\varepsilon}_m}\right] \qquad (5.41)$$

This is the constant rotational velocity of a particle in a rotating electric field of magnitude **E**. Again, this is a function of the permittivity of the suspending medium, and the imaginary part of the Clausius-Mossotti factor which governs the frequency dependence of the rotation rate, as described in Chapter Four. Note that unlike DEP, the rotation rate is independent of particle size.

5.5   Summary

This chapter has described elementary fluid mechanics pertinent to microsystems. We have given the equations that dictate the way in which different forces act on the fluid. We have also described how a moving fluid exerts a force on a particle and derived expressions for the steady-state velocity and rotation rate of particles experiencing DEP or electrorotation respectively. In the next two chapters we turn our attention to the way charges behave in response to electric fields and the manner in which this controls the electrokinetic behaviour of the particle.

5.6   References

Acheson D.J. *Elementary Fluid Dynamics* Oxford Uni. Press, Oxford (1990).
Berg H.C. *Random walks in biology* Princeton Uni. Press, Princeton (1993).
Castellanos A. (Ed.) *Electrohydrodynamics* Springer, Wien, New York (1998).
Koch M., Evans A. and Brunnschweiler A. *Microfluidic Technology and Applications* Research Studies Press, Herts., England (2000).
Landau L.D. and Lifshitz E.M. *Fluid Mechanics* Pergamon Press, Oxford (1959).

Lide D.R. (Ed.) *CRC Handbook of Chemistry and Physics, 74th edition*   CRC Press, London (1994).

Melcher J.R. *Continuum Electromechanics* MIT Press, Cambridge, USA (1981).

Ramos A., Morgan H., Green N.G. and Castellanos A. *AC Electrokinetics: A review of forces in microelectrode structures* J. Phys. D: Appl. Phys. **31** 2338-2353 (1998).

Stratton J.A. *Electromagnetic Theory* McGraw-Hill, New York (1941).

Tritton D.J. *Physical Fluid Dynamics* Oxford University Press, Oxford (1988).

Chapter Six

Electrokinetics in aqueous solution: Ions and the Double Layer

In this chapter, we will discuss the physical mechanisms governing charge motion in aqueous electrolytes and the behaviour of ions in the bulk electrolyte and at interfaces. We begin by exploring how the conductivity and permittivity of a suspending medium is influenced by its composition. We then explain the physical chemistry of the interface between a solid surface and an electrolyte, and the formation of the *Electrical Double Layer*. The double layer structure is examined in detail, looking at classical theories and current molecular models. We follow this with a discussion of the effect of the double layer on experimental systems, particularly electrode polarisation and electroosmosis in AC and DC fields. We conclude by considering the range of electrostatic forces that control particle behaviour, interaction and colloid stability (for further reading on this issue see for example Hunter 1989; Russell *et al.* 1989; Hiemenz 1997; Israelachvili 1992). In Chapter Seven we take these ideas further, and discuss how the double layer influences the electrokinetic behaviour of a particle, particularly with respect to the polarisation of the particle-electrolyte interface by an electric field, and the ensuing changes in the dielectric and electrokinetic properties of the particle.

For many applications of AC electrokinetics, particularly those involving the manipulation of biological particles, the suspending medium consists of an aqueous solution of ions, referred to as an electrolytic solution or electrolyte. Often, when manipulating particles such as cells, the suspending medium must also be iso-osmotic, so that the cells do not burst or shrink. Such a solution will typically contain sugars, added to balance the osmotic pressure between the inside and outside of the cell.

The amount and type of ion in the solution governs the conductivity and can affect the permittivity, whilst the molecular composition of the suspending medium governs the permittivity. The mechanics of conduction in an electrolyte are more complicated than for an ideal dielectric. The ions have a finite size, and their mobility in solution is affected by the high permittivity of the water. At any interface between a charged surface (particle or electrode) and an electrolyte, the ions collect in a thin layer in order to maintain electroneutrality. This layer is referred to as the electrical double layer. It has a complicated structure and affects the electrical forces acting on the particle, changing the effective charge on the surface of the particle. The behaviour of the double layer is therefore fundamental in determining the polarisation of a particle in an electric field. For sub-micrometre particles in particular, where the double layer size can be large compared to the particle radius, it plays a dominant role in controlling the AC electrokinetic properties of the particle. The presence of the double layer on electrodes can also affect the electric field in the bulk electrolyte and can also lead to fluid motion (electroosmosis). We begin by examining the principles of conduction and polarisation of an electrolyte.

6.1   Ionic conduction in solution

The relationship between the current density $\mathbf{J}$ and the electrical conductivity σ of a solution of ions, is described by Ohm's law (given by equation (2.7)). In describing the conductivity of an ionic solution, the situation is more complicated since the current can be carried by different types of ion, each with a different mobility. The current density for each ion of type j, with valence z_j is

$$\mathbf{J}_j = -\sigma_j \nabla \phi = \sigma_j \mathbf{E} \tag{6.1}$$

with $\sigma_j = z_j F c_j \mu_j \equiv z_j n_j q \mu_j$.

In this expression, F is the Faraday constant ($= 9.6487 \times 10^4$ C mol^{-1}), c_j the molar concentration, n_j the number density (m^{-3}) and μ_j the mobility of the ion (m^2 V^{-1} s^{-1}). A possible source of confusion often arises in defining the relationship between concentration and number density, mainly because the molar concentration is in units of moles per *litre*. In S.I. units the concentration is per m^3 and factors of 1000 are required to convert between the two. Comparison of the two expressions for σ_j in equation (6.1) demonstrates that the Faraday constant is $N_A q$, where N_A is Avogadro's number ($= 6.02252 \times 10^{23}$ mol^{-1}).

The motion of the ion in the fluid is described by its mobility, which from the Einstein equation is related to its diffusion constant D_j by

$$\mu_j = D_j \frac{q}{k_B T} = D_j \frac{F}{RT} \tag{6.2}$$

where $R = N_A k_B$ is the (molar) gas constant ($= 8.3143$ J K^{-1} mol^{-1}) and k_B is the Boltzmann constant . The diffusion constant of each ion is given by

$$D_j = \frac{k_B T}{6\pi \eta a_j} = \frac{qRT}{6\pi F \eta a_j} \quad (\text{m}^2 \text{s}^{-1}) \tag{6.3}$$

where a_j is the radius of the ion (see section 6.1.1).

For an ionic species, j, the ionic conductivity λ_j is defined as

$$\lambda_j = z_j F \mu_j \qquad (\text{S m}^2 \text{mol}^{-1}) \tag{6.4}$$

so that for an ideal system the conductivity of the solution is given by the sum of the contributions of each ion

$$\sigma = \sum_j \lambda_j c_j \qquad (6.5)$$

For a symmetrical electrolyte (such as KCl) with ionic conductivities λ_+ and λ_-, the conductivity is given by

$$\sigma = (\lambda_+ + \lambda_-)c = \Lambda c \qquad (6.6)$$

where Λ is the *molar conductivity* (S m^2 mol^{-1}). At low concentrations ($c \to 0$), the ions can move freely without interaction and the molar conductivity tends to a constant value, referred to as the *limiting conductivity* $\Lambda_{c=0}$

$$\Lambda_{c=0} = F(\mu_+ + \mu_-) \qquad (6.7)$$

Λ is constant for molarities up to approximately 10^{-3} M.   For higher concentrations, the ions interact with each other and the molar conductivity is influenced by higher order effects.   This means that at low concentrations, the conductivity is directly proportional to the concentration.   A graph showing the variation in conductivity with concentration for KCl is shown in figure 6.1.   This data was measured at 20° C, and shows that within the precision of the experiment it can be seen that σ is proportional to c.

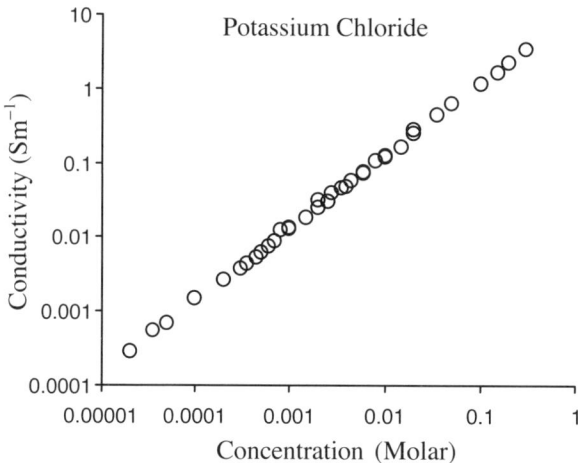

Fig. 6.1 Experimental data showing the variation of the conductivity of an aqueous solution of potassium chloride (KCl) with concentration, at 20° C.

More accurate experiments show that the situation is not that simple. Atkins (1978) quotes an equation, which was derived by Kohlrausch following a series of exhaustive measurements on the conductivities of strong electrolytes. It was found that the molar conductivity is given by the following function

$$\Lambda = \Lambda_{c=0} - \Im c^{1/2} \tag{6.8}$$

where $\Im$ is a constant that depends more on the structural makeup of the salt rather than the specific atoms comprising it.

For illustration of equations (6.1) to (6.7), we can examine the case of a simple electrolyte such as potassium chloride (KCl). This is referred to as a *symmetrical* electrolyte since the mobilities of the potassium K^+ and the chloride Cl^- ions are almost the same. The mobilities of ions at infinite dilution are quoted in numerous textbooks, a good example of which is the CRC handbook (Lide 1994) and the mobilities of some common ions are listed in Table 6.1. From the CRC handbook, $\mu_{K^+} = 7.62 \times 10^{-8} \, m^2 \, V^{-1} \, s^{-1}$, and $\mu_{Cl^-} = 7.91 \times 10^{-8} \, m^2 \, V^{-1} \, s^{-1}$, so that for KCl, $\Lambda_{c=0} = F(\mu_+ + \mu_-) = 9.6485 \times 10^4 \times (7.62 \times 10^{-8} + 7.91 \times 10^{-8}) = 0.01498 \, (S \, m^2 \, mol^{-1})$. Therefore 1 mM KCl has a conductivity of 14.98 mS m^{-1} at 298K, confirmed by figure 6.1. (For comparison, $\Lambda_{c=0}$ for NaCl $= 12.64 \times 10^{-3} \, S \, m^2 \, mol^{-1}$, at 298K.)

Table 6.1   Mobilities (at 25°C) and bare ion radii for some common ions (from Hille 1984; Pauling 1960; Atkins 1978).

Ion	Ion mobility ($\times 10^{-8} \, m^2 \, V^{-1} \, s^{-1}$)	Bare ion radius (nm)
Na^+	5.19	0.095
K^+	7.62	0.133
Ca^{2+}	6.17	0.099
OH^-	20.48	0.176
Cl^-	7.91	0.181
H^+	36.3	

6.1.1   Hydration radius

When an ion is placed in water, ion-water interactions produce a change in the properties of the medium close to the ions. This interaction is responsible for the ion remaining in solution, and when the suspending medium is water, the process is known as solvation or hydration (Hasted 1973).

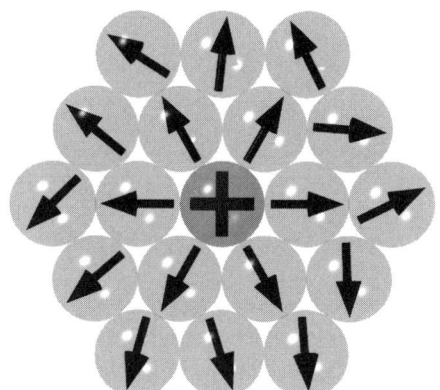

Fig. 6.2 Schematic diagram showing the polarisation of the water molecules around a hydrated positive ion. The arrows indicate the direction of the dipole moments of the water molecules.

For a simple view of this mechanism, consider the model of a water molecule, as shown in Chapter Two, section 2.2.2. Water molecules have a dipole moment and are highly polarisable; this is responsible for the high dielectric constant of water. It is because of this high polarisability that water is such a good solvent for ions. When an ion is placed in water, the charge produces a local field (as described in section 2.1.2) which polarises the water molecules around it. This region of polarised water is referred to as the *ionic atmosphere*: a region that screens the ion from the other ions in the medium, ensuring electroneutrality on the global scale. Locally, the electrical potential resulting from the ion falls off exponentially with distance, and the length scale is given by the *Debye* length (the distance at which the potential falls to $1/e$ of its maximum value), written as λ_D or κ^{-1} where, for a monovalent ion

$$\kappa = \sqrt{\frac{q^2 n_o}{\varepsilon k_B T}} \equiv \sqrt{\frac{1}{D} \frac{\sigma}{\varepsilon}} \qquad (6.9)$$

In this expression $\varepsilon = \varepsilon_o \varepsilon_r$, and n_o is the number density of ions in the bulk (m^{-3}).

Figure 6.2 shows a schematic diagram of how the water molecules orient themselves around an ion. It can be seen from this figure that the dipoles point away from the positive charge. The nearest neighbour water molecules can be considered to be irrotationally bound to the ion and in this region, the dielectric constant falls rapidly from the bulk value with decreasing radial distance.

The creation of the polarised ionic atmosphere produces a region around the ion that has very different properties from the bulk fluid. Since the polar molecules are prevented from responding to an applied field (because of the local potential), the local permittivity is reduced. The water molecules are also structured and, as the ion moves, its atmosphere moves with it, so that the moving object is bigger than the naked ion. The radius of the solvated or hydrated ion is called the hydration radius and this value can be calculated from the equations (6.2) and (6.3) using the mobility data shown in Table 6.1. (Also shown in this table are the bare ion radii determined from crystal structures.) The hydration radius is therefore a representation of the actual size of a moving entity and is the parameter used when considering the viscous drag exerted on the ion by the fluid when it moves, *i.e.* the radius in equation (6.3). This simple picture of the solvated ion is accurate as long as any externally applied field is weak enough so that the molecules around the ion do not respond to it. If the applied field is strong in comparison to the local field, other effects occur.

6.1.2   The charge relaxation time

An important characteristic of an electrolyte is the time required for an ion to move a distance of the order of the Debye length, κ^{-1}, by diffusion (Bockris and Reddy 1973). This is called the charge relaxation time and is given by

$$\tau_q = \frac{1}{D\kappa^2} \tag{6.10}$$

Substituting for the Debye length (equation 6.9) and the conductivity (equation 6.1) in this expression, the charge relaxation time is also

$$\boxed{\tau_q = \frac{\varepsilon}{\sigma}} \tag{6.11}$$

The angular frequency associated with this time $\omega_q = 1/\tau_q$, is referred to as the charge relaxation frequency.

6.1.3   The permittivity of a solution

The molecular composition of a solution has a strong influence on its permittivity. At high ion concentrations, the static dielectric constant of an electrolyte is lower than that of pure water. For example, the dielectric constant of 1M NaCl is 67 compared with water, which is ~80. Using microwave spectroscopy, Hasted (1973) studied this effect in detail and found a linear relationship between the dielectric constant of the solution ε_{Soltn} and the ion concentration

$$\varepsilon_{Soltn} = \varepsilon_{StaticW} + \delta c$$

where $\varepsilon_{StaticW}$ is the static dielectric constant of water and c is the molar concentration of ions. The origin of the reduction in the dielectric constant is the orientation of the water dipole by the ions, thereby reducing their ability to orient in an electric field. This reduces the dielectric constant of the bulk. The dielectric *decrements* of some common ions are presented in Table 6.2.

Table 6.2   The dielectric decrements for various ions (Hasted 1973). Note that the total decrement is the average of the separate decrements, so that for example $\bar{\delta}$ (KCl) = 5.5.

Ion	H^+	Na^+	K^+	Mg^{2+}	Cl^-	OH^-	SO_4^{2-}
δ_+, δ_- (mol l^{-1})	−17	−8	−8	−24	−3	−13	−7

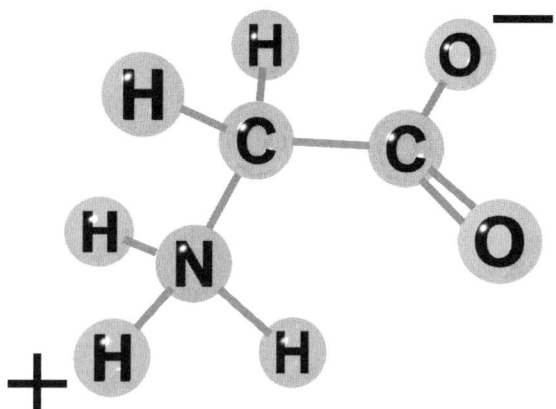

Fig. 6.3 Schematic diagram of the structure of the Glycine molecule showing the approximate arrangement of the atoms and the fixed dipole moment.

The dielectric constant of a suspending medium can also be *increased* by adding ionic molecules of high polarisability to the water. An example of such a molecule is Glycine whose structure is shown in figure 6.3. This is a common polar amino acid, and has a high dipole moment. The separation between the oxygen atom and the carbonyl group in this molecule is approximately 0.33 nm, so that the dipole moment of the molecule can be very approximately calculated as $|\mathbf{p}| = 1.6\times10^{-19}\times3.3\times10^{-10} = 5.3\times10^{-29} = 15.9\,\text{Debye}$, which compares favourably with experimental estimates. Other polar molecules can also increase the dielectric constant of water and the dielectric *increment* for water containing polar molecules is given by the following empirical relationship (Pethig 1979)

$$\varepsilon_{Soltn} = \varepsilon_{StaticW} + \delta\, c$$

where c is the concentration of the solute (*e.g.* the amino acid). The dielectric increments for some common polar solutes are given in Table 6.3. Note that the

Table 6.3 Dielectric increments for some common solutes (from Pethig 1979; Arnold and Zimmerman 1993).

Solute	Glycine (Gly)	Gly-Gly	Gly-Gly-Gly	l-Glutamine	α-Alanine	MOPS	HEPES
δ (mol l^{-1})	22-26	72	126	21	16-17	40	90

medium will then exhibit a dielectric dispersion which in some circumstances must be taken into account when calculating electrokinetic forces (Arnold 2001).

For electrokinetic manipulation of particles, the solution viscosity is also important. Arnold and Zimmerman (1993) proposed a figure of merit for suspending solutions, derived from the ratio of the liquid permittivity to its viscosity. This value reaches a maximum at some concentration then declines. The highest values were found for the glycine peptides.

Sugars are often used in cell suspending media. These are non-polar and dispersion free. The dielectric decrements of some common electromanipulation media have been studied in detail by Arnold $et\ al.$ (1993), and the data was fitted to a second order polynomial of the form $\varepsilon_{Soln} = \varepsilon_{StaticW} + \delta_1 c + \delta_2 c^2$. The decrements for some media are given in Table 6.4.

6.1.4   High electric field effects
As mentioned in the previous section, high electric field strengths can produce changes in the conductivity of the solution. While most dielectrophoretic experiments are conducted in the linear regime, it is useful to examine potential non-linear effects.

Table 6.4 Dielectric decrements for some common sugars (data from Arnold $et\ al.$ 1993).

Sugar	Molecular weight	δ_1 (mol l^{-1})	δ_2 (mol l^{-1})
Glycerol	92.1	-1.73	-0.05
Mannitol	182.2	-2.63	$+0.13$
Sorbitol	182.2	-2.51	-0.14
Sucrose	342.3	-7.69	-0.19
Raffinose (5H$_2$O)	504.4 + 90.1	-8.62	-10.1
Nycodenz	821	-9.11	-2.25

(i) *The Wien effect*
At very high values of electric field, the ions in the electrolyte can move so fast that there is no time for the ionic atmosphere to re-establish itself around the ion. In other words, the water molecules around the ion do not have sufficient time to polarise before the ion has moved on. The net result is that the conductivity of the electrolyte increases, and this is called the Wien effect (Wien 1929, Hasted 1973). The condition for the onset of the Wien effect is when the velocity of the ion, μE, is greater than $1/\kappa\tau_q$, where τ_q is the relaxation time for the ion in solution. Figure 6.4 shows how the threshold field for the onset of the Wien effect varies with electrolyte concentration (for KCl). For low electrolyte concentrations (less than 1mM) the threshold field is 10^6 Vm^{-1}, rising to 10^7 Vm^{-1} for 10mM concentration, so that for most cases the Wien effect is insignificant. At the onset of the Wien effect, the electrolyte molar conductivity reaches a maximum value equal to that at infinite dilution, $\Lambda_{c=0}$. To all intents and purposes, this effect can be ignored in DEP experiments since, even if the condition for onset is met, the change in conductivity will be very small.

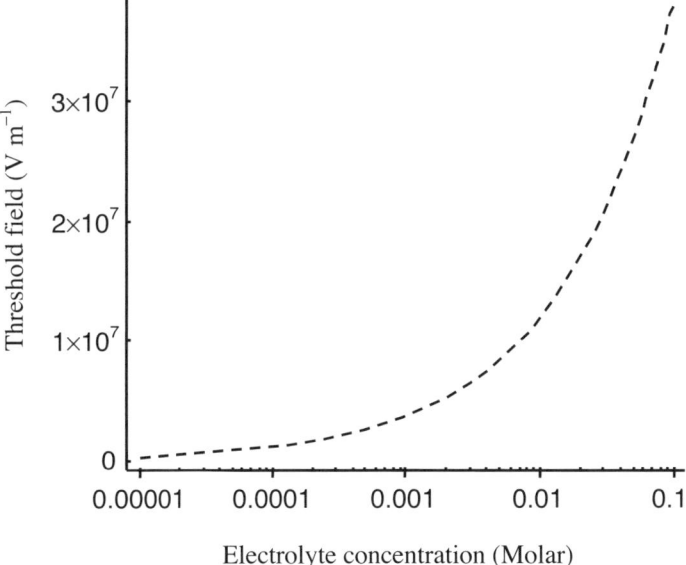

Fig. 6.4 Threshold field strength for the onset of the Wien effect, as a function of the electrolyte concentration in moles.

(ii) *The Debye-Falkenhagen condition*
The Debye-Falkenhagen condition describes the situation which occurs when the frequency of the applied field is high (Debye and Falkenhagen 1928). In this case the ion moves so fast that the ionic cloud surrounding it cannot polarise in response

to the ion.   The condition for the onset of this effect is that the frequency must be greater than the charge relaxation frequency of the electrolyte, *i.e.* $\omega\tau_q > 1$; in most practical cases for frequencies above 10MHz.

6.2   The electrical double layer

A detailed treatment of the electrical double layer can be found in a number of excellent texts (Bockris and Reddy 1973; Dukhin and Shilov 1974; Hiemenz 1977; Hunter 1989; Russel *et al.* 1989; Israelachvili 1992; Lyklema 1995). The double layer plays a fundamental role in defining the electrokinetic behaviour of colloidal or sub-micrometre particles.   In particular, the perturbation of the double layer charge by the applied electric field affects the polarisation of the particle.   In the following sections we will describe the physical characteristics of the double layer and summarise some of the pertinent equations required to model its behaviour.

In general, a surface carries a net charge which comes about either through dissociation of the chemical groups on the surface or by adsorption of ions or molecules from the solution onto the surface.   This charge creates an electrostatic surface potential ϕ_o local to the interface.   When the surface is immersed in an electrolyte, this electrostatic potential attracts ions of opposite charge (counterions) from the solution and repels ions with like charge (co-ions), as shown in figure 6.5. The figure shows the distribution of free charges in an electrolyte, close to a charged surface.   The region of liquid near to the interface has a higher density of counterions and a lower density of co-ions than the bulk.   This region is referred to as the *diffuse* region of the electrical double layer.   The resulting change in the distribution of ions near the surface is governed by the spatial distribution of the surface electrostatic potential.   For example, when a particle or an electrode is immersed in an electrolyte, the surface charge is balanced by an equal (and opposite) amount of excess charge in the double layer.   The net result is that the countercharge from the solution effectively screens the surface charge so that on the global scale (of the bulk medium), the overall charge is zero.

In a thin region between the surface and the diffuse layer, there is a layer of bound or tightly associated counterions, generally referred to as the Stern layer (Stern 1924), as shown in figure 6.6. This region is of the order of one or two solvated ions thick and is also referred to as the bound part of the double layer.   In this region it is assumed that the potential falls linearly from the surface value ϕ_o to ϕ_d, the value of the potential at the *interface* between the diffuse layer and the Stern layer. The potential decays across the diffuse layer exponentially with a characteristic distance given by the Debye length, κ^{-1}, as shown in figure 6.5.

In a further refinement to this model, the Stern layer can be subdivided into two regions, as shown in figure 6.7 (Bockris and Reddy 1973).   The inner layer, the outer surface of which is referred to as the inner Helmholtz plane, consists of ions which are not hydrated and have specifically adsorbed to the surface.   The outer layer consists of a layer of bound hydrated ions as for the simple Stern layer; the interface between this and the diffuse layer is referred to as the outer Helmholtz plane.   The inner layer can contain both co- and counter ions, whilst the outer layer contains counterions. Again, the potential obeys a linear rule across the inner and outer layers, decreasing across the outer region to ϕ_d. When the Stern layer is included, the potential in the diffuse layer again falls exponentially with a characteristic distance κ^{-1}, but in this case the decrease begins from the potential at the outer edge of the Stern layer, ϕ_d.

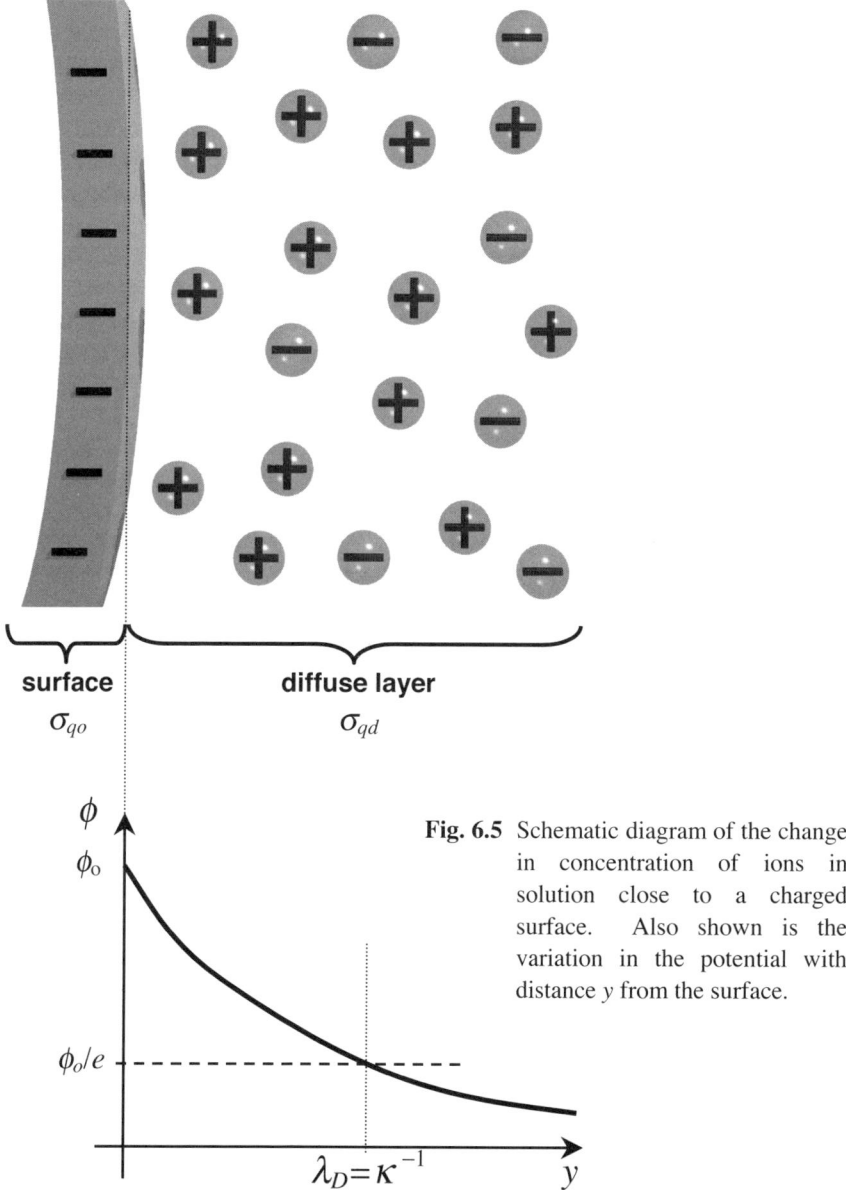

Fig. 6.5 Schematic diagram of the change in concentration of ions in solution close to a charged surface. Also shown is the variation in the potential with distance y from the surface.

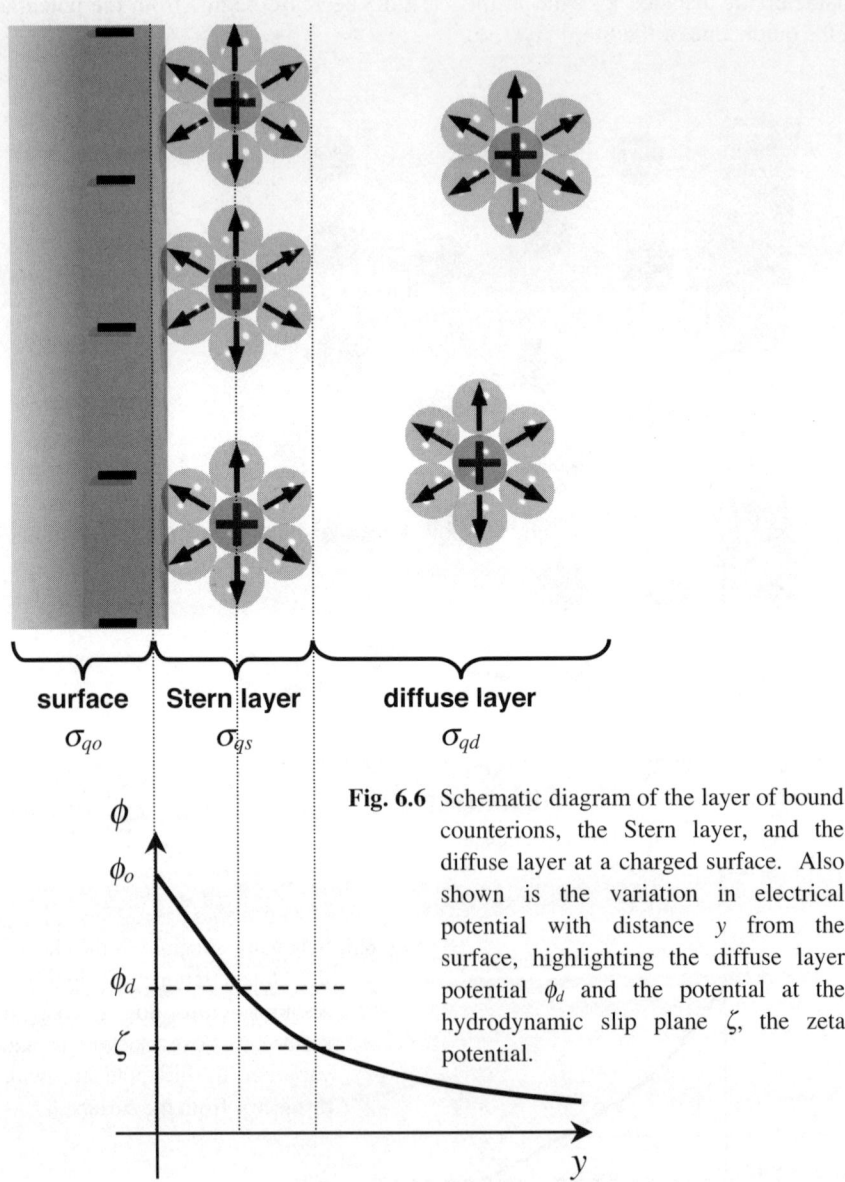

Fig. 6.6 Schematic diagram of the layer of bound counterions, the Stern layer, and the diffuse layer at a charged surface. Also shown is the variation in electrical potential with distance y from the surface, highlighting the diffuse layer potential ϕ_d and the potential at the hydrodynamic slip plane ζ, the zeta potential.

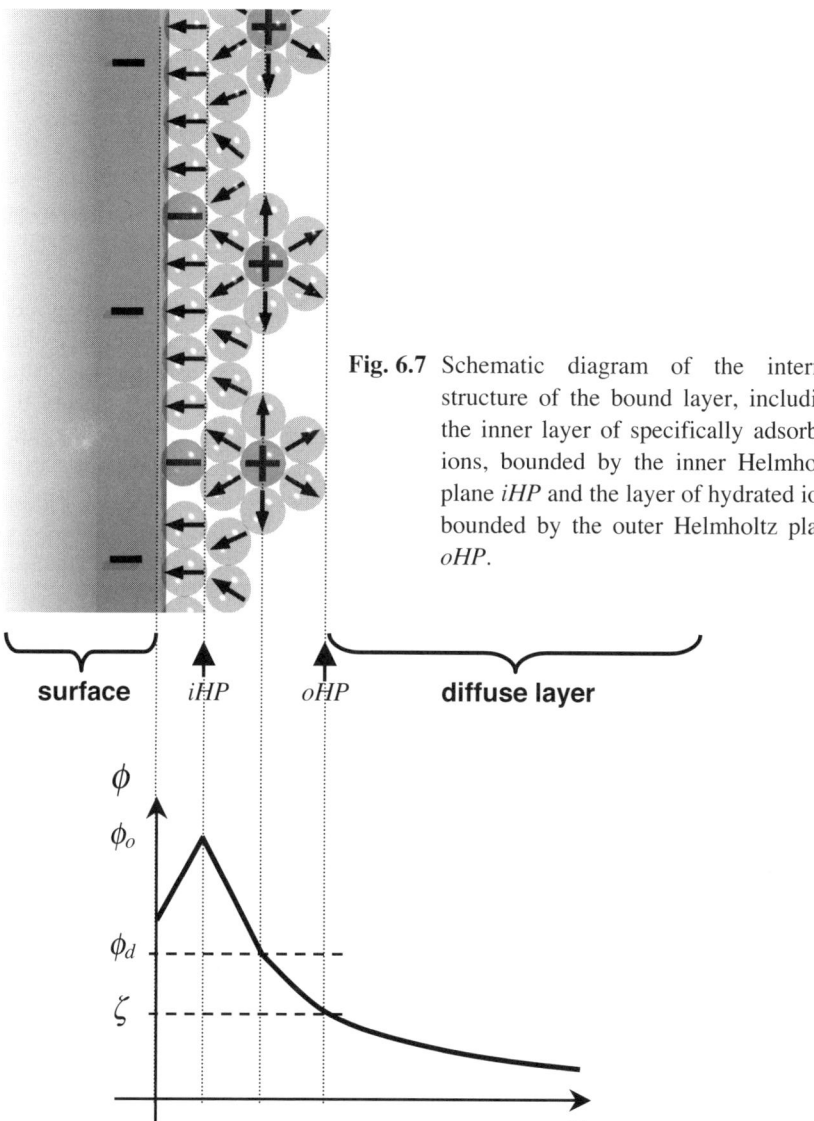

Fig. 6.7 Schematic diagram of the internal structure of the bound layer, including the inner layer of specifically adsorbed ions, bounded by the inner Helmholtz plane *iHP* and the layer of hydrated ions bounded by the outer Helmholtz plane *oHP*.

A simplified analysis of the double layer begins by defining a surface charge density as σ_{qo} (C m^{-2})[†], (note that in many publications, surface charge densities are often quoted in μC cm^{-2}). When a surface is immersed in an electrolyte, the excess charge density in the diffuse part of the double becomes equal and opposite to the charge density on the surface, *i.e.* the system is globally electroneutral. Clearly, the double layer charge is a volume charge ρ_d and does not exist as a single sheet of charge on the surface. Considering only the diffuse layer, the equivalent surface charge density can be found by integrating the charge from the surface to infinity

$$\sigma_{qd} = \int_0^\infty \rho_d(y)dy \tag{6.12}$$

As shown in figure 6.6 and 6.7, in addition to the diffuse layer there is often a Stern layer next to the surface. This consists of bound ions and water molecules, which are oriented not only by the hydrated ions but also by the charged surface. Since the charge is strongly associated with the surface and is contained in a very thin layer, this can be considered equal to a surface charge density σ_{qo}. Again, when the Stern layer is included the system is electroneutral so that $\sigma_{qo} + \sigma_{qd} + \sigma_{qs} = 0$.

6.2.1   The diffuse layer

Assuming that $y = 0$ defines the surface, we start from the Poisson equation, (equation (2.3))

$$\nabla^2\phi = -\frac{\rho_d}{\varepsilon} \tag{6.13}$$

where ρ_d is the non-uniform charge density in the diffuse layer and ϕ a potential. The charge density is obtained from the Boltzmann equation, which relates the number density of ions $n(y)$ at y to the bulk concentration n_o. For a single ion type of valence z_j, this is

$$n_j(y) = n_{o,j}e^{-\phi(y)\frac{qz_j}{k_BT}} \tag{6.14}$$

For several ion types j, the total charge density is given by

$$\rho_d = q\sum_j z_j n_j \tag{6.15}$$

[†] Here we have used the subscript q to avoid confusion with conductivity σ, and to make it explicit that we are dealing with charge densities.

from equations (6.13) and (6.14) we can write the volume charge density as

$$\rho_d(y) = q \sum_j n_{o,j} z_j e^{-\phi(y)\frac{qz_j}{k_BT}} \tag{6.16}$$

Inserting this expression for the charge density into equation (6.12) gives the Poisson-Boltzmann equation

$$\nabla^2\phi(y) = -\frac{q}{\varepsilon} \sum_j n_{o,j} z_j e^{-\phi(y)\frac{qz_j}{k_BT}} \tag{6.17}$$

The solution to this equation is the potential as a function of distance from the surface.

(i) *The Debye-Hückel theory*
For spherical geometry (*e.g.* a small particle or an ion), there is no analytical solution to this equation. However, the Debye-Hückel theory (Debye and Hückel 1923) was developed to give an approximate analytical solution for low surface potentials. This is equivalent to the thick double layer limit, *i.e.* $\kappa a \ll 1$. In this case the exponential term can be replaced by the first two terms in the series expansion: $1 - (qz/k_BT)\phi(r)$, where r is the radial distance from the particle (in place of y). With this approximation a linear differential equation in ϕ is obtained

$$\nabla^2\phi(r) = \kappa^2\phi(r) \tag{6.18}$$

where

$$\kappa^2 = q^2 \sum_j \frac{z_j^2 n_o}{\varepsilon k_BT} \tag{6.19}$$

The general solution to equation (6.17) is $\phi = \phi_1 e^{\kappa r} + \phi_2 e^{-\kappa r}$. There are two boundary conditions: (a) the potential goes to zero as $r \to \infty$ implies that $\phi_1 = 0$ and (b) the potential at $r = 0$ is ϕ_o so that $\phi_2 = \phi_o$. The potential function is therefore given by

$$\boxed{\phi = \phi_o e^{-\kappa r}} \tag{6.20}$$

This is an exponentially decaying potential with characteristic decay length κ^{-1}, which we have already referred to as the Debye length. For a symmetrical electrolyte with $z = z^+ = z^-$, the Debye length, as given by equation (6.19) becomes

$$\kappa^{-1} = \sqrt{\frac{\varepsilon k_B T}{2z^2 q^2 n_o}} \qquad (6.21)$$

Inserting numerical values for the constants for KCl in water, the Debye length can be written in terms of molar concentration c as

$$\kappa^{-1} = 1.764 \times 10^{-11} \sqrt{\frac{T}{c}} \qquad (6.22)$$

The Debye length is a measure of the thickness of the ionic atmosphere around the particle that screens the charge, and is typically a few nm thick.

(ii) *The Gouy-Chapman theory*
For planar surfaces, which is the thin double layer limit for a particle, Gouy (1910) and Chapman (1913) solved equation (6.17) analytically. Again equation (6.13) is solved for the diffuse part of the double layer only, or more strictly from the outer Helmholtz plane when the Stern layer is included. Defining a dimensionless potential function $\psi = (q/k_B T)\phi$, the Boltzmann equation becomes $n = n_o e^{-\psi z}$. Combining the Poisson equation, the equation for the space charge density and the Boltzmann equation gives

$$\frac{d^2 \psi}{dy^2} = -\frac{q^2}{k_B T} \frac{z n_o}{\varepsilon} (e^{-z\psi} - e^{+z\psi}) \qquad (6.23)$$

where it is assumed that the ion concentrations for the two species are the same, i.e. $n_o = n_o^+ = n_o^-$ and the electrolyte is symmetrical $z = z^+ = z^-$. This equation can be re-written as

$$\frac{d^2 z\psi}{dy^2} = \kappa^2 \sinh(z\psi) \qquad (6.24)$$

Again, with appropriate boundary conditions (at $y = \infty$, $d\psi/dy = 0$ and $\psi = 0$), the equation can be solved to give the solution for the electric field

$$E = -\frac{d\phi}{dy} = \sqrt{\frac{8k_B T n_o}{\varepsilon}} \sinh\left(\frac{zq}{2k_B T}\phi\right) \qquad (6.25)$$

Integrating this equation gives the expression for the potential in terms of the ion concentration and distance y

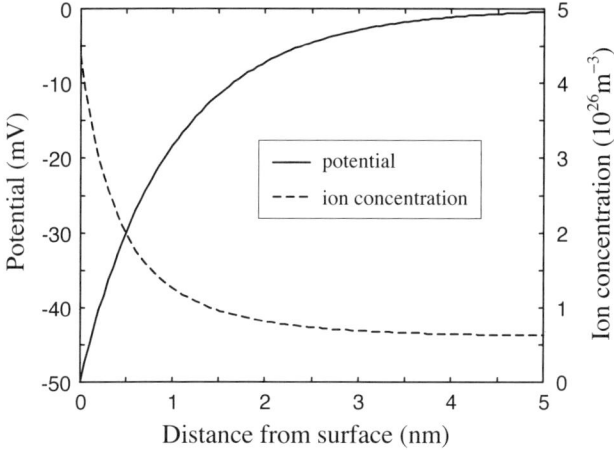

Fig. 6.8 Plot of the drop in electrical potential and ion concentration as a function of distance from the surface. The plots were calculated from equations (6.14) and (6.26) for 100mM KCl and a surface of potential −50mV.

$$\phi = \frac{2k_BT}{q}\ln\left(\frac{1+\gamma e^{-\kappa y}}{1-\gamma e^{-\kappa y}}\right) \approx \frac{4k_BT}{q}\gamma e^{-\kappa y} \qquad (6.26)$$

where $\gamma = \tanh(q\phi_o / 4k_BT)$. The variation in ion concentration with distance can be derived by substituting equation (6.26) into the Boltzmann equation (6.14). As an example, a plot of ion concentration and potential with distance is shown in figure 6.8. This was calculated for a surface potential $\phi_o = -50$ mV and for an ion concentration of 100mM. Note: The low potential limit of expression (6.26) ($\phi_o < 50$ mV) gives the Debye-Hückel approximation (equation (6.20)).

These two examples, Debye-Hückel for a point charge and Gouy-Chapman for a planar surface, represent the two limiting cases for thick and thin double layers around a colloidal particle. As such, they form the basis of many of the theories which describe the polarisation of colloidal particles in electrolytic solutions.

6.2.2 The Grahame equation

The Grahame equation (Grahame 1947) relates the potential at the surface to the surface charge density. The charge in the diffuse layer σ_{qd}, is related to the field at the surface, according to Gauss's law

$$\left(\frac{d\phi}{dy}\right)_{y=0} = \frac{\sigma_{qd}}{\varepsilon} = -\frac{\sigma_{qo}}{\varepsilon} \qquad (6.27)$$

For a flat surface, and using equation (6.25), we obtain the Grahame equation

$$\sigma_{qd} = -\sqrt{8\varepsilon n_o k_B T}\,\sinh\left(\frac{z\psi_d}{2}\right) = -\varepsilon\kappa\phi_d\,\frac{\sinh(z\psi_d/2)}{(z\psi_d/2)} \tag{6.28}$$

where ψ_d is the dimensionless diffuse layer potential. At 25°C, this can be written as: $\sigma_{qd} = -0.1174\sqrt{c}\,\sinh(19.5z\phi_d)$ with c in mol l^{-1}, surface potential in Volts and the diffuse layer charge density in C m^{-2}. For small double layer potentials, equation (6.28) simplifies to $\sigma_{qd} = -\varepsilon\kappa\phi_d$.

For most practical purposes the diffuse layer potential is equivalent to the potential at the slip plane, known as the zeta (ζ) potential, see figure 6.6. The slip plane is used to define the interface between a moving fluid and a stationary or immobile region. This interface is generally equated to the Stern layer-diffuse layer interface. The potential at the slip plane can be measured by experiment, and for low surface charge densities or potentials this is often equated to the diffuse layer potential. Therefore, measurements of the zeta potential are used to quantify the surface charge density of a range of colloidal particles.

In a simplified situation, the Stern layer can be ignored so that the diffuse layer effectively exists right up to the surface. Therefore, given that $\sigma_{qo} = -\sigma_{qd}$, the surface charge density and surface potential are related according to

$$\sigma_{qo} = \sqrt{8\varepsilon n_o k_B T}\,\sinh\left(\frac{z\psi_o}{2}\right) \tag{6.29}$$

where the absolute surface potential is $\phi_o = (q/k_B T)\psi_o$. In practice, this equation only holds for regions of small surface charge densities, since at high surface charge densities the effect of the Stern layer can be pronounced.

6.2.3   The Stern layer: current thinking

Helmholtz and Stern developed improved models of the double layer that, in particular, accounted for the specific adsorption of ions onto the surface. Because the Stern layer consists of immobile molecules and ions on the surface, it is variously referred to as the bound layer or the stagnant layer. The latter term is perhaps quite revealing since it implies a hydrodynamically immobile region with a very high viscosity. At the outer surface of this layer lies the *slip plane i.e.* the point at which the fluid *can* move relative to the surface. Current thinking is that this slip region is only one or two molecules thick (Lyklema *et al.* 1999). The slip plane is also assumed to coincide with the outer Helmholtz plane so that the zeta potential (potential at the slip plane) can be directly related to the diffuse layer potential *i.e.* $\zeta = \phi_d$. Stern layers are present on both charged and uncharged surfaces and the formation does not depend on hydrogen bonding since they are also present on hydrophobic and hydrophilic surfaces (Lyklema *et al.* 1999).

This picture of a *stagnant* layer is contradicted by the observation that a considerable component of charge movement in the double layer can be attributed to the charge in the Stern layer (see Chapter Seven). Experiments have shown that this charge moves with a mobility almost equal to that of the bulk (Lyklema 1995), but how can a layer with infinite viscosity also support the movement of ions tangential to the surface with such a high mobility? As illustrated by Lyklema *et al.* (1999), the Stern layer can be thought of as a two-dimensional gel, where the solvent is immobile but the ions and low molecular weight solute molecules diffuse with near bulk diffusion constants. Sophisticated molecular dynamic techniques have been used to model the Stern layer (Lyklema *et al.* 1998) and simulate the behaviour of both solvent and ions close to the surface. In these calculations, the atoms were modelled using Lennard-Jones interactions, and individual ion and molecule trajectories were determined as well as local variations in viscosity. The number density and distribution of molecules and ions obtained from such calculations is shown in figure 6.9. It can be seen from this simulation that the solvent molecules mainly occupy a region close to the surface of the order of one or two molecular diameters. The ion distribution has a peak close to the peak for

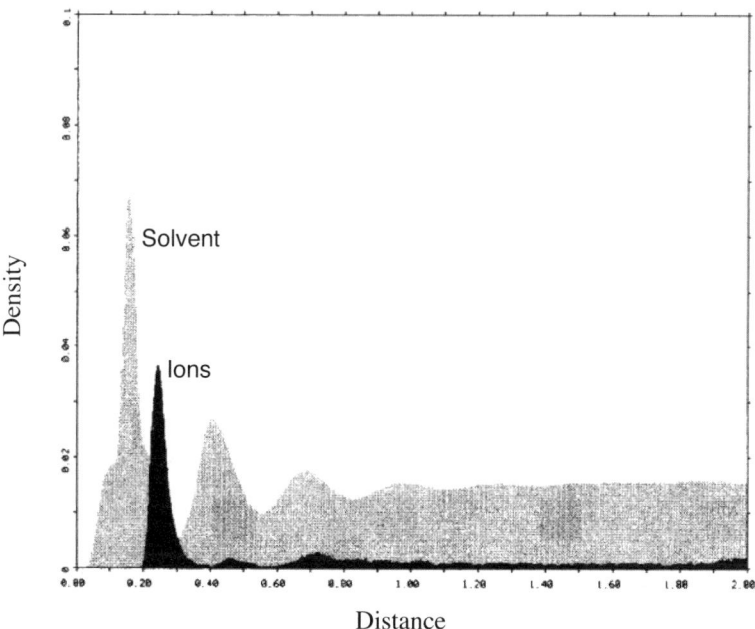

Fig. 6.9 Plot of the density profile for the solvent molecules and the ions as a function of height above the surface. This plot was obtained by using a Molecular Dynamics procedure to simulate the behaviour of the individual solvent molecules and the ions. (Reprinted with permission from Lyklema *et al.* (1998). Copyright (1998) American Chemical Society.)

the solvent molecules (the ions are larger), leading to the reasonable conclusion that this region can be identified with the outer layer of the Stern layer (bounded by the outer Helmholtz plane).

The result of adding a constant tangential force on the system was also simulated. The trajectories of the ions over time frames of the order of 0.1ns were calculated. Both the ions and the solvent molecules moved tangential to the surface, hopping out of the Stern layer, moving parallel to the surface and either returning to the Stern layer or moving away into the bulk and being replaced by other molecules or ions. No accumulation of these ions in the diffuse region was observed. The conclusion of this simulation is that those ions which move from the Stern layer to the bulk give rise to the observed Stern layer charge transport. From the simulation, it was also concluded that the ratio of Stern layer to bulk mobility was almost unity (0.96) and an anisotropic viscosity was demonstrated. Close to the surface, the viscosity parallel to the surface is almost the same as for the bulk (or slightly higher, by a factor of two), with ions in this region and direction slowed but not immobilised. The perpendicular viscosity at the surface is higher by a factor of four to five, but is by no means infinite. Lyklema points out that the phenomenological interpretation of a "stagnant" layer is related to the low probability of solvent molecules leaving and re-entering the region within nanometres of the surface.

6.3   Electrode polarisation

When charge accumulates at the interface between an electrode and an electrolyte, the system behaves like a capacitor with a non-uniform charge density. The potential across the charged layer decays exponentially from a maximum at the electrode to nearly zero in the bulk. Application of a constant potential to the electrode (*e.g.* from a battery) changes the charge distribution, analogous to the charging of a capacitor; this is referred to as *electrode polarisation* (Schwan 1968; 1992; Bard and Faulkener 1980). The consequence is that most of the applied potential is dropped across this capacitor so that the potential in the bulk electrolyte may only be a fraction of that applied to the electrode (in the absence of electrode reactions).

It is well known that dielectric measurements of aqueous suspensions of particles show large capacitances at low frequencies, particularly at high suspending medium conductivities. This capacitance is due to electrode polarisation. Similarly, in dielectrophoretic measurements, as the frequency of the applied field is decreased the particles move more slowly since the potential in the suspending medium is reduced and the electric field is weaker.

6.3.1   Double layer capacitance

The double layer capacitance can be thought of as a reactance of value $1/i\omega C_{DL}$ in series with the suspending medium impedance. As $\omega \to 0$, the value of this reactance tends to infinity and the voltage across the suspending medium goes to zero. However, this simple picture is not an accurate representation of the system;

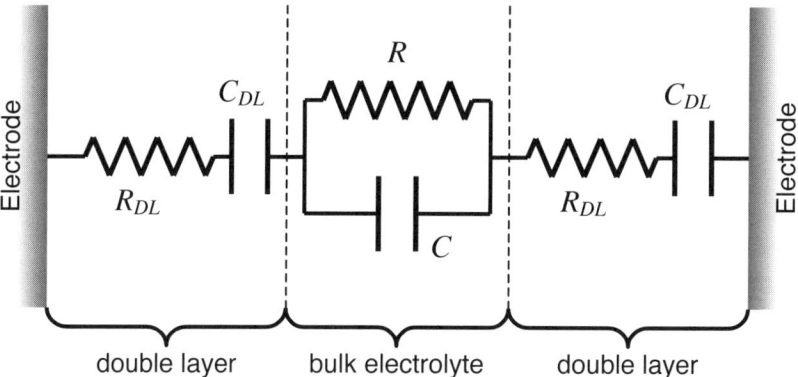

Fig. 6.10 Approximate circuit diagram for the system comprising an electrolyte between two metal electrodes, each with an associated double layer. The bulk electrolyte can be considered to consist of a capacitor C and a resistor R in parallel. This is in series with the double layer impedance, represented here by resistor R_{DL} and capacitor C_{DL}. This simplified circuit model assumes that electrochemical reactions at the electrode surface can be ignored.

the double layer is a complex structure and also electrode reactions can (and often) occur at low frequencies.

The behaviour of an electrode-electrolyte system can be understood by analogy with capacitors and resistors in an electrical circuit, as shown in figure 6.10. At high frequencies, the total capacitance tends to a constant value (proportional to the permittivity of the suspending medium) and the total resistance is constant, approximately inversely proportional to the conductivity of the solution. For a given electrolyte, these values correspond to the impedance of the bulk electrolyte outside the double layer.

At low frequencies both the resistance and the capacitance are affected by the double layer. Assuming that the impedance of the double layer and the bulk electrolyte are in series, as shown in the figure, the constant value of the bulk impedance can be subtracted from the total measured impedance, to give the frequency dependent double layer impedance Z_{DL}. Generally, the double layer polarisation impedance is of the form

$$Z_{DL} = \frac{A}{(i\omega)^{\beta}} = \frac{A}{\omega^{\beta}}\left[\cos\left(\frac{\pi}{2}\beta\right) - i\sin\left(\frac{\pi}{2}\beta\right)\right] \qquad (6.30)$$

where A and β are constants. This type of polarisation impedance is referred to as *constant phase angle (CPA) impedance*. This expression satisfies the Kramers-Krönig relations for the impedance of a linear system. For a perfect capacitor,

$\beta = 1$ and for a perfect resistor, $\beta = 0$. For double layer impedance measurements on solid electrodes, β is typically found to be between 0.7 and 0.9 (Bard and Faulkener 1980; Liu 1985; Bates *et al.* 1998; Kerner and Pajkossy 1998).

Defining the capacitance as the ratio of charge to potential, the specific capacitance (capacitance per unit area) of a parallel plate capacitor is $C' = \sigma_{qo} / \Delta\phi$, where σ_{qo} is the surface charge density on one plate and $\Delta\phi$ is the potential difference between the plates. In the absence of volume charge the electric field is $\Delta\phi / d = -\sigma_{qo} / \varepsilon$, so that the specific capacitance is ε / d, where d is the distance between the plates.

Diffuse layer capacitance
Considering only the diffuse layer part of the double layer, the specific capacitance is $\sigma_{qo} / \Delta\phi = \sigma_{qd} / \phi_d$, since $\phi_{y\to\infty} = 0$. However, owing to the fact that charges are distributed over the volume of the layer, this is not equal to ε / d. It is common practice to define a differential capacitance C_d' for the diffuse part of the double layer, which is found by differentiating equation (6.28)

$$C_d' = \frac{d\sigma_{qd}}{d\phi_d} = \varepsilon\kappa \cosh\left(\frac{z\psi_d}{2}\right) \qquad (6.31)$$

For low surface potentials, the surface charge and potential are related through $\sigma_d = -\varepsilon\kappa\phi_d$, equation (6.28). Therefore, the specific capacitance of this layer becomes $C_d' = \varepsilon\kappa$, *i.e.* that of a parallel plate capacitor of thickness κ^{-1} with a dielectric of permittivity $\varepsilon = \varepsilon_o\varepsilon_r$. This is also called the integral capacitance.

As an example, let us calculate the double layer capacitance at the surface of an electrode immersed in 1mM KCl. The Debye length, given by equation (6.21), is $\kappa^{-1} = 10$ nm in this case, and taking a value for the permittivity of the diffuse part to be the same as the bulk, the capacitance can be calculated as 70 mF m^{-2}. It is clear that in order to maximise the voltage in the suspending medium, the value of this capacitance must be as big as possible. Now, since the Debye length is proportional to the inverse of the square root of the ion concentration, the capacitance is clearly maximum at low ion concentrations, so that for a given frequency the voltage in the medium will be highest at lower conductivities.

The actual (rather than specific) capacitance of the double layer on an electrode is $C = \varepsilon\kappa A$, *i.e.* it is proportional to A, the electrode area. In order to reduce the voltage drop across the interface, the capacitance must be maximised and one way of doing this is to make the area of the electrode as large as possible. This is often done by coating the electrode with a porous, conducting material such as platinum black (Schwan 1963, 1968); the traditional way of reducing electrode polarisation for dielectric spectroscopy. Such methods have also been used in AC electrokinetic experiments to reduce electrode polarisation and maximise the voltage in the suspending medium at low frequencies (*e.g.* Huang *et al.* 1997).

Stern layer capacitance

The effective capacitance of the Stern layer is easier to model, since the potential drops linearly in this region. This region is thus exactly equivalent to a capacitor with capacitance $C'_s = \varepsilon_s / d_s$, where d_s is the width of the layer (typically 0.5nm) and ε_s the relative permittivity of the layer (somewhere in the range 6-20). As we have seen, the Stern layer can be further divided into inner and outer Helmholtz layers, which we will refer to as 1 and 2. These can then be treated as separate capacitors so that the total capacitance of the bound layer is $1/C'_s = 1/C'_{s1} + 1/C'_{s2}$.

The *total* capacitance of the double layer is the series sum of the Stern layer and diffuse layer capacitances, $1/C'_{DL} = 1/C'_s + 1/C'_d$. Since the diffuse layer capacitance depends on the Debye length, the capacitance of the Stern layer is most influential at high ion concentrations, where the diffuse layer collapses and its thickness approaches that of the Stern layer.

6.3.2   Frequency dependence of the double layer potential

When a potential is applied to an electrode immersed in an electrolyte, the double layer takes a finite time to form, since the ions have a finite mobility and the effects of diffusion are important. If the applied potential is alternating, then there is a certain frequency above which there is insufficient time for the double layer to form before the potential switches sign. This is the physical origin of the frequency dependence of the double layer impedance. There are two limiting cases to consider. At low frequencies, in each half of the cycle of the applied potential the double layer has sufficient time to completely form, with the consequence that the potential is dropped entirely across the double layer. At high frequencies, the double layer has no time to form, so that the potential across the medium is the same as the potential applied to the electrodes. At intermediate frequencies, there is a transition between these two limiting cases with the potential across the medium decreasing as the frequency of the applied potential is reduced. In actual fact, the system is not this simple and empirically the double layer capacitance is also a function of the applied frequency. For further details see, for example, Bard and Faulkener (1980) and Sluyters-Rehbach and Sluyters (1970).

In microsystems, the electrodes that are used to generate the non-uniform electric fields cannot be modelled as simple parallel plate capacitors because of the non-uniform geometry. As an example, consider a simple electrode geometry consisting of two thin parallel plates fabricated on an insulating substrate and separated by a gap (small compared with their width and length), as shown in figure 6.11. The dimensions of the system are such that it can be considered two-dimensional. In this case, a simple model of the double layer impedance can be constructed, as shown in the figure. The electrodes are covered in an aqueous medium which can be thought of as a series of concentric, semicircular resistors with a value $\Delta R = \pi x / \sigma d \Delta x$. These are terminated at either end on the double layer, which is modelled as a series of distributed capacitors with values $\Delta C = \varepsilon \kappa d \Delta x$ (Ramos *et al.* 1999). Here, x is the horizontal distance (across the electrode) with the origin at the centre of the gap between the electrodes. These

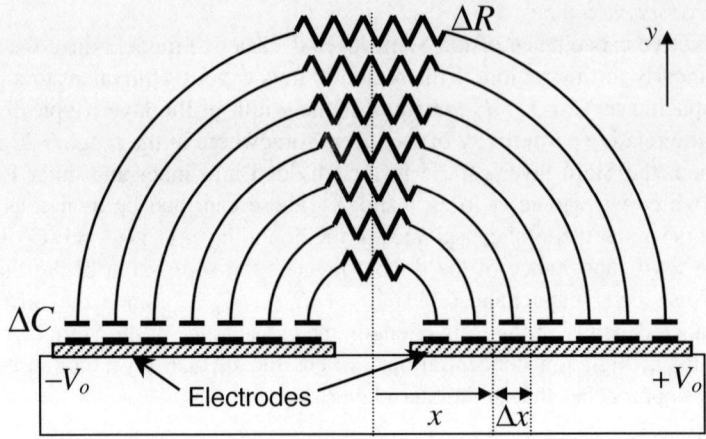

Fig. 6.11 Approximate circuit diagram for two long coplanar plate electrodes separated by a narrow gap. The semi-circular resistive paths are terminated at the electrode at either end by a capacitive element representing the double layer.

expressions for the resistance and capacitance include the length of the electrodes into the page in the figure, given as d. (This parameter cancels in the derivation of the final expression for the potential.) Each series ΔC-ΔR-ΔC circuit has a different time constant, since the resistors have different values because of the varying path lengths (the capacitors are all the same value). This gives rise to a spread of relaxation times for the polarisation of the double layer capacitance. The potential dropped across the double layer can be derived as (Ramos *et al.* 1999):

$$\Delta\phi_{DL} = \frac{V_o}{2 + i\omega\pi(\varepsilon/\sigma)\kappa x} \tag{6.32}$$

In deriving this expression it is assumed that the frequency of the applied potential is much smaller than the charge relaxation frequency *i.e.* $\omega \ll \sigma/\varepsilon$.

This is, of course, a simplified expression for the system, and complete expressions have been defined analytically (Gonzalez *et al.* 2000), but it can be used to illustrate some important points. First, the potential dropped across the double layer $\Delta\phi_{DL}$ and the potential across the medium $\Delta\phi_m = 1 - 2\Delta\phi_{DL}$ (there is a double layer on both electrodes) are frequency dependent. The second is that the potential across the medium is complex and, as mentioned in Chapter Four, real fields produce dielectrophoresis but complex fields can also produce travelling wave effects. Finally, the potential across the electrolyte (or rather the potential at the outer edge of the double layer), also depends on the position on the electrode surface. The variation of the potential across the double layer and the medium as a function of frequency is plotted in figure 6.12.   These figures demonstrate that in order to achieve maximum field in the bulk, the frequency must be higher than the inverse of the time constant given by the product RC.

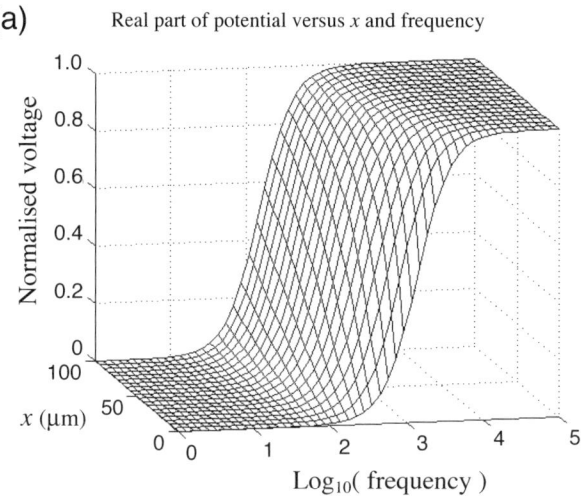

Fig. 6.12   The real (a) and imaginary (b) parts of the normalised potential across the electrolyte (outside the double layer) as a function of frequency (in Hz) and distance from the edge of the electrode x (in micrometres). These plots were calculated for KCl of conductivity 1 mS m^{-1} and assuming the capacitor-resistor model shown in figure 6.11.

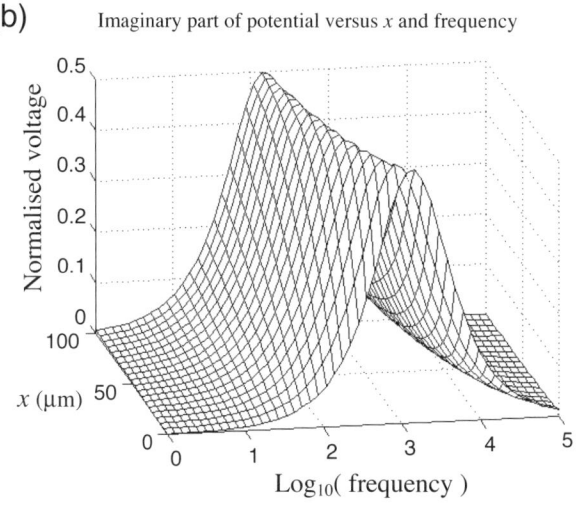

6.4   Electroosmosis

If an electric field is applied tangential to a surface bathed in electrolyte, the charges in the double layer between the surface and the electrolyte experience a force. The consequence of this is that these double layer charges move, pulling the fluid along and generating a flow. This flow is zero immediately at the surface and rises to a maximum (and constant thereafter) at the slip plane. This flow is called electroosmosis.

6.4.1   Electroosmosis in a DC field

We can calculate the two-dimensional electroosmotic velocity of the fluid in an infinitely wide planar channel, as shown in figure 6.13, by considering the balance of forces acting on an element of fluid with volume Ady (where the y-direction is normal to the surface). The electrical force acting on the liquid is given by Coulomb's law. This force is opposed by frictional forces, obtained from Newton's second law. Therefore, we can write

$$F_x = E_x \rho \, A dy = \eta A \left[\left(\frac{du_x}{dy} \right)_y - \left(\frac{du_x}{dy} \right)_{y+dy} \right] \tag{6.33}$$

where $\mathbf{u} = (u_x, 0)$ is the 2D fluid velocity. This expression can be re-written as

$$E_x \rho dy = -\eta \left(\frac{d^2 u_x}{dy^2} \right) dy \tag{6.34}$$

Using Poisson's equation to substitute for the charge density, we can write

$$E_x \varepsilon \left(\frac{d^2 \phi}{dy^2} \right) dy = \eta \left(\frac{d^2 u_x}{dy^2} \right) dy \tag{6.35}$$

Integrating from $y = \infty$ to the shear plane where the potential becomes equal to the zeta potential and the fluid velocity is zero, the final result is

$$u_x = -E_x \frac{\varepsilon \zeta}{\eta} \tag{6.36}$$

or in terms of electroosmotic mobility

$$\boxed{\mu_x = \frac{u_x}{E_x} = -\frac{\varepsilon \zeta}{\eta}} \tag{6.37}$$

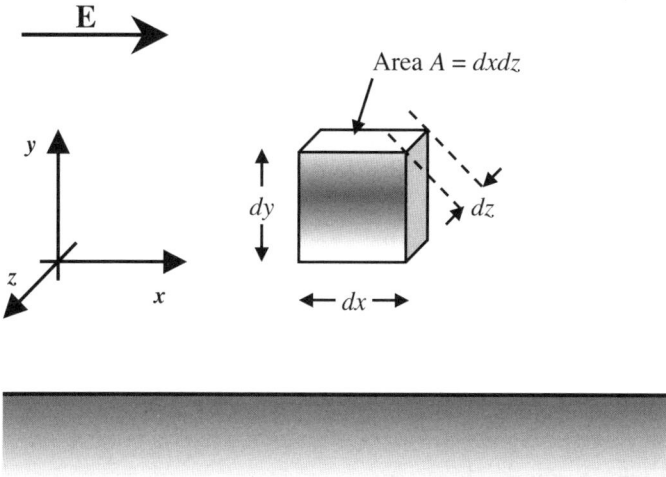

Fig. 6.13 Schematic diagram of the model system for electroosmosis.

The velocity of the fluid is therefore directly proportional to the electric field, and the charge density (the zeta potential) in the double layer.   Equation (6.36) can be re-written as

$$u_x = \frac{E_x \sigma_q}{\kappa \eta}$$ (6.38)

Of course, for an AC field and with this geometry, the fluid would move in an oscillatory manner with zero time-averaged displacement.

6.4.2   AC electroosmosis

Experiments with microelectrodes have shown that AC fields can generate local fluid motion (Müller *et al.* 1996; Trau *et al.* 1997; Yeh *et al.* 1997; Ramos *et al.* 1998; Green *et al.* 2000a; 2000b), related to the charging of the double layer. Experimental and theoretical work performed using simple parallel finger micro-electrodes demonstrated this fluid flow to be the AC analogue of electroosmosis (Ramos *et al.* 1998; Ramos *et al.* 1999; Green *et al.* 2000a; González *et al.* 2000). Since the electric field is non-uniform, there is a non-zero time-averaged flow. This does not occur in a uniform AC field, just as a non-uniform field results in dielectrophoresis but a uniform field does not.

An appreciation of the mechanism responsible for electrode polarisation is central to understanding AC electroosmosis. The polarisation process governs both the potential at the outer edge of the double layer (which gives the tangential field) and the potential dropped across the double layer (which gives the charge at the surface). The mechanism can be understood by reference to figure 6.14. As before, voltages $\pm V$ are applied to either electrode, which give rise to the electric field $\mathbf{E}$ with tangential components $\mathbf{E}_t$ outside the double layer and an induced charge on each electrode. The induced charge experiences a force $\mathbf{F}_q$ due to the action of the tangential field, resulting in fluid flow. Figure 6.14(a) shows the system for one half-cycle of an AC potential/field. In the other half-cycle the sign of potential, the direction of the tangential field and the sign of the induced charge are all opposite. As a result, the direction of the force vector remains the same giving a non-zero time-averaged force and a steady-state fluid flow occurs, as shown in figure 6.14(b).

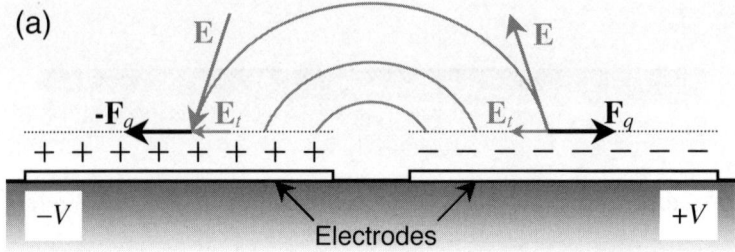

Fig. 6.14   (a) Schematic diagram outlining the mechanism of AC electroosmosis. (b) The interaction of the tangential field at the surface with the charge in the double layer gives rise to a surface fluid velocity u_x and a resulting bulk flow.

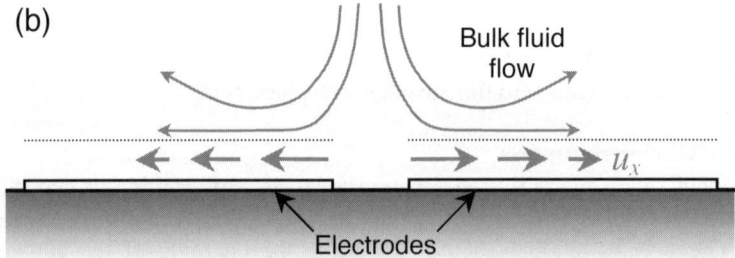

From equation (6.38), we expect the velocity of the fluid close to the surface to be proportional to the tangential field **and** the charge in the double layer. In the relatively simple electrode geometry shown in figure 6.14(a), the tangential component of the field close to the electrode surface, E_t, is proportional to the potential dropped across the medium. The charge in the diffuse double layer can be represented as the sum of a constant term and a time-dependent excess charge, $\sigma_q(t)$, which varies with the electric field. Assuming a linear approximation and ignoring the Stern layer, the time varying charge is $\sigma_q = \varepsilon \kappa \phi_d$.

The velocity of AC electroosmotic flow is frequency dependent. At low frequencies, the potential across the suspending medium, and therefore E_t , are zero. Since the tangential field must be continuous and the double layer is very thin, $E_t = 0$ in the double layer and the fluid velocity is zero. At high frequencies the potential across the double layer and the induced charge are both zero, and again there is no AC electroosmotic flow. Therefore, the velocity profile for AC electroosmosis on perfectly polarisable electrodes is zero at high and low frequencies, and maximum at an intermediate characteristic frequency. This is the same behaviour that is observed experimentally (as discussed in Chapter Eight).

Equation (6.32) can be simplified by introducing the non-dimensional frequency $\Omega = (1/2)\pi\kappa x(\varepsilon/\sigma)\omega$. The potential across the double layer is then

$$\phi_d = \frac{V_o}{2(1+i\Omega)} \tag{6.39}$$

Using the expressions for the surface charge $\sigma_{qo} = \varepsilon\kappa\phi_d$ (low potential limit) and the tangential field $E_t = -\partial\phi_d/\partial x$, the time averaged fluid velocity is

$$\langle u_x \rangle = \frac{1}{2}\text{Re}\left[\frac{\sigma_{qo}E_t^*}{\eta\kappa}\right] = \frac{1}{8}\frac{\varepsilon\phi_o^2\Omega^2}{\eta x(1+\Omega^2)^2} \tag{6.40}$$

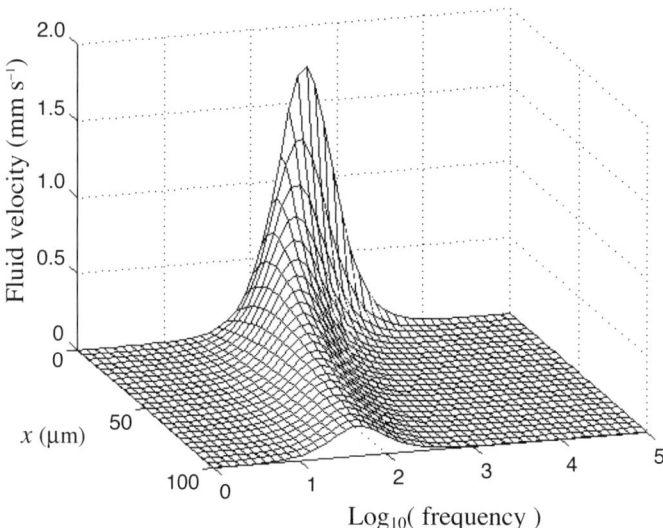

Fig. 6.15 AC electroosmotic fluid flow at the surface of an electrode calculated from the circuit model. The data is plotted as a function of distance from the electrode edge, x, and the frequency of the applied signal. The magnitude of the applied potential was 1 Volt and the separation of the two electrodes was 25 µm.

Figure 6.15 shows how the time-averaged fluid velocity varies as a function of ω and x calculated using this expression. Gonzalez *et al.* (2000) have taken this analysis further, deriving an analytical expression for the AC electroosmotic fluid flow from first principles.

6.5   Particle-particle interaction

As well as the dielectrophoretic and Stokes' forces, there are other forces which act on particles or molecules in solution. These are electrostatic in nature, since they involve the action of charges on each other but have more involved definitions. These forces can be relatively important in controlling the behaviour of colloidal particles at short distances.

6.5.1   van der Waals forces or Hamaker forces

Particles suspended in a medium experience a long-range attractive force, commonly referred to as the van der Waals force or Hamaker force. A complete discussion of van der Waals force is outside the scope of this work, but the reader can find a number of excellent text books dealing with this subject in great detail (see for example Israelachvili 1992; Landau and Lifshitz 1980). These forces have an important role to play in controlling the stability of colloidal particles and must be taken into account when considering particle-particle interactions.

The van der Waals force comprises three forces, some of which may contribute to a greater or lesser extent to the total interaction force between particles, depending on the nature of the interacting molecules or particles. The first of the three forces is called the *orientation* or Keesom interaction force. This arises from the averaged interaction between two *permanent* dipoles. Take for example a water molecule possessing a permanent dipole moment, as discussed in Chapter Two. Because of the interaction between the dipoles, each water molecule attracts other molecules. Even though the thermal energy of the system causes them to move, on average the presence of this force means that they spend more time attracted to each other than in any other orientation, thereby reducing the free energy of the system.

A polar molecule (with a permanent dipole) can also interact with a non-polar molecule through dipole induced-dipole interactions. We know from Chapter Two, that a permanent dipole has an associated electric field which decays rapidly with distance. This field can induce a dipole moment in a neighbouring non-polar molecule, which then interacts with the permanent dipole. The force of interaction is referred to as the *induction* or Debye interaction and is generally weak, insufficient to orient the molecules. These two forces (orientation and induction) are polarisation forces, *i.e.* forces either between two dipoles or between a dipole and a dipole induced in a non-polar molecule.

The third and most significant force is known as the London *dispersion* force, which is an attractive force. This force is always present and acts between *all* atoms and molecules (charged or uncharged). It arises from fluctuations in the electron distribution around an atom or molecule. Over short time intervals, random fluctuations in the electron cloud of an atom leads to the formation of an instantaneous dipole moment. This dipole rapidly changes in magnitude and sign so that the time average value is zero. However, the field generated by the dipole interacts with electrons in neighbouring atoms causing an induced dipole. The two dipoles attract and the time-averaged force between the atoms is non-zero. The magnitude of this force can be derived quantum mechanically, however, the following greatly simplified approach, described by Israelachvili (1992), can be used to understand the origin of the force.

Consider the Bohr model of the atom, where we have an electron orbiting the nucleus at the first Bohr radius a_0 (the smallest distance between the orbiting electron and the proton). This is the point where the Coulomb energy is equal to the first ionisation potential, $I = 2h\nu$, where h is Plank's constant and ν the orbiting frequency. The Bohr radius is given by

$$a_o = \frac{q^2}{2(4\pi\varepsilon_o)h\nu} \tag{6.41}$$

The atom has no permanent dipole but at an instant in time, has the instantaneous dipole

$$p_1 = a_o q \tag{6.42}$$

We know from Chapter Two that if an atom has an instantaneous dipole p_1, then it generates an electric field

$$E_1 = \frac{p_1}{4\pi\varepsilon_o r^3} \tag{6.43}$$

This polarises an identical nearby atom of polarisability $\alpha_{e,2}$, which, from equation (2.20) is

$$\alpha_{e,2} = 4\pi\varepsilon_o a_o^{\,3} \tag{6.44}$$

The electric field acting on the second atom gives rise to an induced dipole

$$p_2 = \alpha_{e,2} E_1 = \frac{p_1 \alpha_{e,2}}{4\pi\varepsilon_o r^3} \tag{6.45}$$

The average potential energy of interaction between the two atoms is

$$U = -E_1 p_2 = -\frac{p_1 p_2 \alpha_{e,2}}{(4\pi\varepsilon_o)^2 r^6} = -\frac{(a_o q)^2 \alpha_{e,2}}{(4\pi\varepsilon_o)^2 r^6} \tag{6.46}$$

or more generally

$$U = -\frac{1}{2}\frac{(p_1^2 \alpha_{e,2} + p_2^2 \alpha_{e,1})}{(4\pi\varepsilon_o)^2 r^6} \tag{6.47}$$

This is the Debye interaction energy. Substituting for a_o (equation (6.41)) and using equations (6.44) and (6.46) gives the interaction energy as

$$U = -\frac{2\alpha_e^2 h\nu}{(4\pi\varepsilon_o)^2}\frac{1}{r^6} = -C\frac{1}{r^6} \tag{6.48}$$

This is an approximate expression for the dispersion interaction between two *identical* atoms. In London's famous equation there is a numerical factor so that

$$\boxed{U = -C\frac{1}{r^6} = -\frac{3}{4}\frac{\alpha_e^2 h\nu}{(4\pi\varepsilon_o)^2}\frac{1}{r^6}} \tag{6.49}$$

Although this expression has been superseded by more exact equations calculated using quantum mechanics, the important point to note is that the interaction energy falls as the inverse sixth power of distance, and also depends on the polarisability of the atoms. In order to calculate the interaction between large particles or surfaces, the force must be summed from all the individual atoms. Table 6.5 shows the result of this sum for a number of analytical cases, assuming that the interactions are purely additive. In this table C is taken from equation (6.48) and A is the Hamaker constant (Israelachvili 1992, Hamaker 1937), given by $A = \pi^2 C n_{o1} n_{o2}$, where n_o is the atom density of the solids.

These constants are calculated for vacuum, but can easily be modified for interactions in liquids or gases. The significance of this table lies in the differences in the distance-dependence of interaction energies. For two interacting atoms the force falls off very quickly. However, for two colloidal particles the energy of interaction varies with the inverse of the distance y, leading to a *long-range*

Table 6.5   Non-retarded van der Waals interaction energies between bodies of different geometries (taken from Israelachvili 1992). Note that r denotes atomic distances and y distances between solid objects.

atom-atom	sphere-surface (radius a)	sphere-sphere (radius a)	surface-surface
$U(r) = -C\dfrac{1}{r^6}$	$U(y) = -\dfrac{Aa}{6}\dfrac{1}{y}$	$U(y) = -\dfrac{Aa}{12}\dfrac{1}{y}$	$U(y) = -\dfrac{A}{12\pi}\dfrac{1}{y^2}$

interaction. In order to calculate the *force* of interaction, the energy must be differentiated w.r.t. distance, so, for example, the attractive force between two spheres is $F = Aa/12y^2$. As an example, the force of attraction between two colloidal particles of 100nm radius, with a separation of 1 nm is only 1 nN (of the order of the DEP force!) and for two cells of 5 μm radius, this is a not insignificant 50 nN.

The van der Waals force is essentially electromagnetic in origin. When atoms are some distance apart, the time taken for the field of the first atom to influence the dipole of the second atom can become appreciable. This is called the retardation effect and leads to a reduction in the dispersion energy so that the decay approaches $1/r^7$. This effect is particularly important when considering the detailed forces between macroscopic bodies.

6.5.2 Total energy of interaction. The DLVO theory

The total interaction between two surfaces (*e.g.* two colloidal particles) consists of two forces or potentials acting in opposition, the *repulsive* double layer interaction and the *attractive* van der Waals interaction. At any distance, the sum of these forces determines whether the interaction energy is positive or negative. This determines, for example, whether or not a colloidal suspension is stable.

In order to calculate the repulsive force between two particles we must return to the theory of the double layer and examine the interaction energy of two overlapping double layers. The repulsive force between two similarly-charged surfaces or particles contains an electrostatic and an osmotic force component. The net force is calculated at the mid-plane between the two surfaces, where the electric field is zero, $d\phi/dy = 0$, the electrostatic force is zero and the only force acting on the surfaces is that due to osmosis. To understand the origin of this force, consider two large particles with thin double layers coming together. As they approach, the number of counterions between the particles must remain constant to maintain charge neutrality overall. Bringing the surfaces closer and closer effectively squashes the double layer ions into a smaller space (closer to the particles surface). This gives rise to an osmotic pressure, which is proportional to the concentration of ions. The energy of repulsion can be found from the integral of the work done in bringing the particles together (from infinity) against the opposing pressure.

An expression for the repulsion potential, as a function of distance y between two flat surfaces, can be derived by considering the variation in the potential between the two approaching surfaces and the osmotic pressure difference between the mid-plane of the surfaces and that outside. Without going into details of the calculations, the result for two flat surfaces with monovalent ions is (Israelachvili 1992)

$$U_R = \frac{64 n_o k_B T \gamma^2}{\kappa} e^{-\kappa y} \tag{6.50}$$

where $\gamma = \tanh(zq\phi_o/4k_BT)$. Although this expression is derived for small surface potentials and weak interactions, it is widely used in practice. The important point to note here is that the interaction energy decays exponentially in the same manner as the double layer potential.

The corresponding expression for two spheres of radius a is

$$U_R = \frac{64\pi n_o k_B T a\gamma^2}{\kappa^2} e^{-\kappa y} \tag{6.51}$$

At low surface potentials (< 25mV) this expression becomes

$$U_R \approx \frac{2\pi R\sigma_{qd}^2}{\kappa^2 \varepsilon} e^{-\kappa y} \tag{6.52}$$

The expression for the **force** between two spheres is

$$F_R \approx \frac{2\pi a\sigma_{qd}^2}{\kappa \varepsilon} e^{-\kappa y} \tag{6.53}$$

Referring to Table 6.5, the attraction energy for two flat surfaces can be written as $U_A(y) = -A/12\pi y^2$. Writing the total interaction potential as $U_T = U_A + U_R$, we note that the total interaction depends on the surface charge density and/or the electrolyte concentration, and that the van der Waals attractive force will dominate at small distances over the double layer repulsive energy. The variation in the total potential energy can be plotted as a function of distance, as shown in figure 6.16. The plots were calculated using the energy of attraction given by Table 6.5 and the energy of repulsion from equation (6.50). The dependence of the energy on both electrolyte concentration and distance is the key to understanding colloidal stability. The quantitative explanation of these phenomena forms the basis of the famous DLVO theory named after Derjaguin, Landau, Verwey and Overbeek (Derjaguin and Landau 1941; Verwey and Overbeek 1948).

Colloids become unstable when the collision of individual particles (from Brownian motion) leads to aggregation. This process is called coagulation and can be understood by reference to figure 6.16, which shows how the energy of interaction between two surfaces depends on the separation distance, together with the surface charge density and/or electrolyte concentration. First, note that the energy of interaction is always negative (attractive force) when the surfaces are closer than a critical distance and the van der Waals force dominates. Similarly, for large distances, the long range van der Waals forces dominate and again the force is attractive (negative potential). Between these extremes the situation depends on the relative magnitudes of U_A and U_R. If the surfaces have very low or zero surface charge, then U_R is very weak and the force is always attractive, as shown by curve (a). For colloidal particles this leads to rapid coagulation. For a high surface charge density and/or low electrolyte concentration, we have curve (b) where a significant

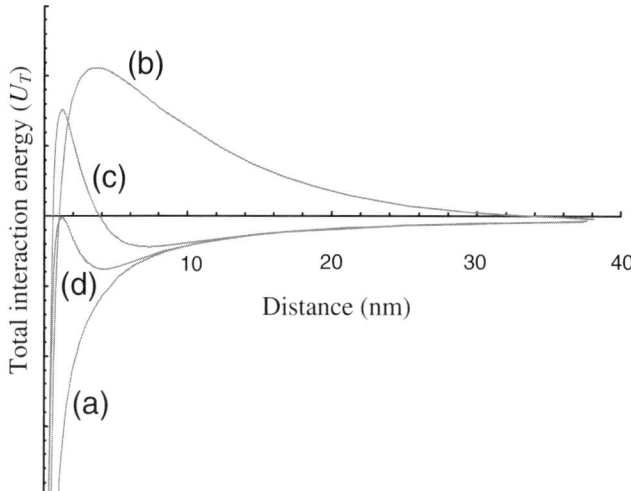

Fig. 6.16   The total interaction energy as a function of distance
for particles of different surface charged density.

repulsive force exists; the surfaces strongly repel until their approach distance is less than the order of the Debye length.   Curve (c) shows an intermediate case where the force-distance curves show a secondary minimum.   If this secondary potential is deep enough, particles can remain stable without aggregating and the colloids are kinetically stable; they can adhere, but the process is reversible.   Case (d) shows what is referred to as the "critical coagulation concentration", particles adhere and then coagulate or flocculate when the electrolyte is raised above a critical concentration.

A good illustration of the stability of colloidal particles can be found by looking at the action that salt water has on clay particles suspended in fresh water. Particles of clay are continually being washed into rivers from the surrounding land.   These are carried down the river because they are kinetically stable in water. However, once they reach the sea and the electrolyte concentration increases rapidly, the colloid becomes unstable and the particles flocculate. They sink to the bottom and give rise to the familiar mud flats or deltas seen at the mouths of rivers.

6.6   Summary
In this chapter we have laid the foundations of the behaviour of charges in liquids and at interfaces.   We have seen that the double layer plays a fundamental role in defining and controlling the physical characteristics of charged particles and surface in electrolytes.   Understanding and controlling the behaviour of the double layer is important in areas ranging from electroosmosis to colloid stability.   In the next chapter we take these ideas further and explore the role of the double layer in controlling the electrokinetic behaviour of particles.

6.7   References

Atkins P.W. *Physical Chemistry*. Oxford Uni. Press (1978).

Arnold W.M. and Zimmermann U. *Dielectric Properties of Zwitterion Solutions*. Biochem. Soc. Trans. **21** 475 (1993).

Arnold W.M., Gessner A.G and Zimmermann U. *Dielectric measurements on electro-manipulation media* Biochim. Biophys. Acta. **1157** 32-44 (1993).

Arnold W.M. *Positioning and levitation media for the separation of biological cells* IEEE Trans. Ind. Appls. **37** 1468-1475 (2001).

Bard A.J. and Faulkener L.R. *Electrochemical Methods, Fundamentals and Applications*. John Wiley and Sons (1980).

Bates J.B., Chu T.T. and Stribling W.T. *Surface Topography and Impedance of Metal-Electrolyte Interfaces*. Phys. Rev. Letters **60** 627-630 (1988).

Bockris J.O'M. and Reddy A.K.N. *Modern Electrochemistry* Plenum Press, New York (1973).

Chapman D.L. *A Contribution to the Theory of Electrocapillary* Phil. Mag. **25** 475-481 (1913).

Debye P. and Hückel E. *Zur theorie der elektrolyte II. das grenzgesetz für die elektrishe leitfähigkeit* Physik Z. **24** 305-325 (1923).

Debye P. and Falkenhagen H. *Dispersion von leitfähigkeit und dielecktrizitäts-konstante bei starken elektrolyten* Physik Z. **29** 121 401-416 (1928).

Derjaguin B.V. and Landau L. *The Theory of the Stability of Strongly Charged Lyophobic Sols and of the Adhesion of Strongly Charged Particles in Solution of Electrolytes*. Acta Phys. Chim. USSR **14** 633-662 (1941).

Dukhin S.S. and Shilov V.N. *Dielectric Phenomena and the Double Layer in Disperse Systems and Polyelectrolytes*. Wiley, New York (1974).

González A., Ramos A., Green N.G., Castellanos A. and Morgan H. *Fluid flow induced by non-uniform AC electric fields in electrolytes on micro-electrodes II: A linear double layer analysis* Phys. Rev. E. **61** 4019-4029 (2000).

Gouy G. *Sur la Constitution de la Charge Électrique à la Surface d'un Électrolyte* J. Physiol (Lond) **9** 457-468 (1910).

Grahame D.C. *The Electrical Double Layer and the theory of Electrocapillary*. Chem. Rev. **41** 441-501 (1947).

Green N.G., Ramos A., Gonzalez A., Morgan H. and Castellanos A. *Fluid flow induced by non-uniform AC electric fields in electrolytic solutions on micro-electrodes. Part I: Experimental measurements* Phys. Rev. E **61** 4011-4018 (2000a).

Green N.G., Ramos A. and Morgan H. *Ac electrokinetics: A survey of sub-micrometre particle dynamics* J. Phys. D: Appl. Phys. **33** 632-641 (2000b).

Hamaker H.C. *London-van der Waals attraction between spherical particles* Physica **4** 1058-1072 (1937).

Hasted J. *Aqueous Dielectrics* Chapman and Hall, London (1973).

Hiemenz P.C. *Principles of colloid and surface chemistry* Dekker, New York and Basel (1977).

Hille B. *Ionic Channels of Excitable Membranes* Sinauer Associates Inc., Sunderland, MA, USA (1984).

Huang Y., Wang X-B., Becker F.F. and Gascoyne P.R.C.   *Introducing dielectrophoresis as a new force field for field-flow fractionation*   Biophys. J.   **73** 1118-1129 (1997).

Hunter R.J.   *Foundations of colloid science*   Oxford Uni. Press, New York   (1989).

Israelachvili J.   *Intermolecular and Surface Forces*   Academic Press   (1992).

Kerner Z. and Pajkossy T.   *Impedance of rough capacitive electrodes electrodes: the role of surface disorder*   J. Electroanalytical Chem.   **448** 139-142   (1998).

Landau L.D. and Lifshitz E.M.   *Statistical Physics 2ⁿᵈ Edition*   Pergamon Press, Oxford   (1980).

Lide D. R. edt.   *CRC Handbook of Chemistry and Physics 74ᵗʰ edition*   CRC Press, London   (1994).

Liu S.H.   *Fractal Model for the AC Response of a Rough Interface*   Physical Review Letters   **55**   529-532 (1985).

Lyklema J.   *Fundamentals of Interface and Colloid Science Vol II*   Academic Press, London   (1995).

Lyklema J., Rovillard S. and de Coninck J.   *Electrokinetics: The properties of the Stagnant Layer Unraveled*   Lanmguir   **14**   5659-5663 (1998).

Lyklema J., van Leeuwen H.P. and Minor M.   *DLVO-theory, a Dynamic Reinterpretation*   Adv. Coll. Int. Sci   **83**   33-69 (1999).

Müller T., Gerardino A., Schnelle Th., Shirley S.G., Bordoni F., De Gasperis G., Leoni R. and Fuhr G.   *Trapping of Micrometre and Sub-micrometre Particles by High-frequency Electric Fields and Hydrodynamic Forces*   Phys. D: Appl. Phys.   **29**   340-349 (1996).

Pauling L.   *Nature of the Chemical Bond and Structure of Molecules and Crystals 3ʳᵈ edition*   Cornell Uni. Press, Ithaca, New York   (1960).

Pethig R.   *Dielectric and Electric Properties of Biological Materials*   John Wiley & Sons Ltd., Chichester   (1979).

Ramos A., Morgan H., Green N.G. and Castellanos A.   *AC Electrokinetics: A review of forces in microelectrode structures*   J. Phys. D: Appl. Phys.   **31**   2338-2353 (1998).

Ramos A., Morgan H., Green N.G. and Castellanos A.   *AC electric-field-induced fluid flow in microelectrodes*   J. Colloid and Int. Sci.   **217**   420-422 (1999).

Russel W.B., Saville. D.A. and Schowalter W.R.   *Colloidal Dispersions*   Cambridge Uni. Press, Cambridge   (1989).

Schwan H.P.   *Electrical Properties of Tissues and Cells* in Lawrence J.H., Tobias C.A. (Eds.) *Advances in Biological Medical Physics*   **15**   147-209 (1957).

Schwan H.P.   *Determination of Biological Impedance* in   Physical Techniques in Biological Research, Nastuk W.L. (ed.), New York   Academic Press   **6**   323-406 (1963).

Schwan H.P.   *Electrode Polarisation Impedance and Measurements in Biological Materials*   Ann. New York Acad. Sci.   **148**   191-209 (1968).

Schwan H.P.   *Linear and Non-linear Electrode Polarisation of Biological Materials*   Ann Biomed. Eng.   **20**   269-288 (1992).

Sluyters-Rehbach M. and Sluyters J.H. in Bard A.J. (Ed.) *Electroanalytical Chemistry* Dekker, New York   **14**   Chap 1   (1970).

Stern O. *The theory of the electrical double-layer* Z. Elecktrochem **30** 508-516 (1924).

Trau M., Saville D.A. and Aksay I.A. *Assembly of colloidal crystals at electrode interfaces* Langmuir **13** 6375-6381 (1997).

Verwey E.J.W. and Overbeek J. Th. G. *Theory of Stability of Lyophobic Colloids* Elsevier, Amsterdam (1948).

Wien M. Über den spannungseffeckt der elektrolytischen leitfähigkeit in sehr starken feldern Ann. Physik **1** 400-416 (1929).

Yeh S-R., Seul M. and Shraiman B.I. *Assembly of ordered colloidal aggregates by electric-field-induced fluid flow* Nature (London) **386** 57- 59 (1997).

Chapter Seven

Surface conduction and polarisation

In this chapter, we examine in detail the influence of the electrical double layer on particle electrokinetics. When an electric field is applied to a suspension of particles, the ions in the diffuse part of the double layer surrounding the particle experience a Coulomb force and move, *i.e.* the particle polarises. We also know that in the case of a static or DC field, the particle moves, a phenomenon called electrophoresis. To the outside observer, the double layer charge screens any surface charge on the particle so that it is electrically neutral. The question is then, why does the particle move? When the field is applied, the ions move around the particle, and as they do so, they drag the fluid with them *i.e.* the fluid moves under electroosmosis. This movement occurs in a low Reynolds number regime; the fluid is viscous so that it remains motionless and the particle moves in the opposite direction to the fluid flow. In other words, during electrophoresis the particle *appears* to move as if the field were acting on its surface charge.

In an AC field the ions still move, the diffuse double layer ions respond to the harmonically oscillating AC field and move around the particle giving rise to a large induced dipole moment. This ion movement around the particles is responsible for the very large values of the dielectric constant measured for suspensions of colloidal particles at low frequencies.

7.1   Surface conductance

We begin by considering the double layer as a conducting sheet surrounding the particle, and discuss how this gives rise to a particle surface conductance. Consider a planar surface with free charge placed in an electric field. When charge moves along the surface under the influence of the field, a surface current flows as shown in figure 7.1. The surface current density $\mathbf{J}'$ (Am^{-1}) is related to the surface conductance K_S (Siemens) and the electric field tangential to the surface in the same way that the volume current density is related to volume conductivity, *i.e.*

$$\mathbf{J}' = K_s \mathbf{E}_t \qquad (7.1)$$

where $\mathbf{E}_t$ is the component of the electric field tangential to the surface of the particle. Defining a current sheet $\Delta \mathbf{J}(y)$ in a thin layer dy, the total excess current density parallel to the surface is given by the difference between $\Delta \mathbf{J}(y)$ and the

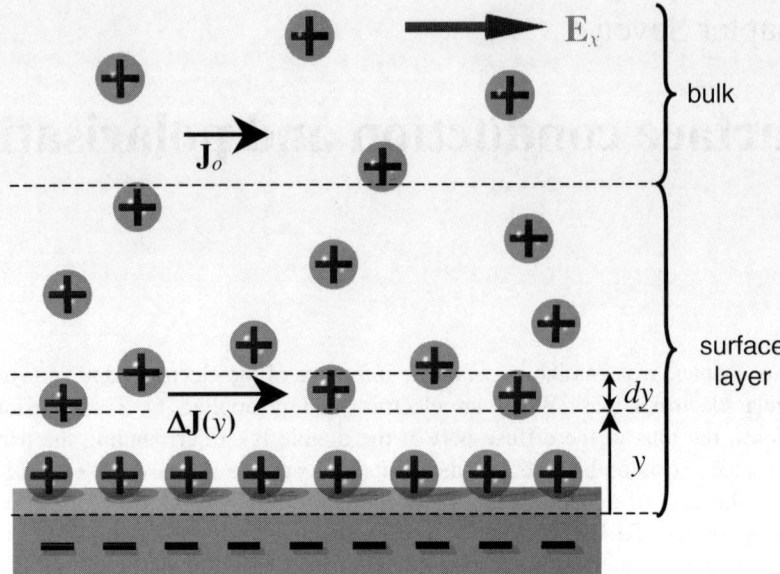

Fig. 7.1 A schematic diagram showing the increase in current density in an electrolyte close to a charged surface. The surface charge produces a double layer at the surface, which in this case has a higher concentration of positive ions. The tangential field gives rise to a current density that depends on height above the surface, y.

bulk current density $\mathbf{J}_o$, integrated from the surface to infinity (Lyklema 1995)

$$\mathbf{J}' = \int_0^\infty (\Delta \mathbf{J}(y) - \mathbf{J}_o) dy \qquad (7.2)$$

From Chapter Six, we know that for each ion of type j, the current density $\mathbf{J}_j = z_j n_j \mu_j \mathbf{E}$, so that the excess current can be re-written as

$$\Delta \mathbf{J}(y) - \mathbf{J}_o = q \sum_j (n_j(y) - n_{o,j}) z_j \mu_j(y) \mathbf{E} \qquad (7.3)$$

and the surface conductance is

$$K_S = q \sum_j z_j \int_o^\infty (n_j(y) - n_{o,j}) \mu_j(y) dy \qquad (7.4)$$

where $n_{o,j}$ is the bulk concentration of ion j and $n_j(y)$ is the concentration in the thin layer at height y. Therefore, the surface conductance can be thought of as the two-dimensional analogue of bulk conduction, with the charge movement confined to the surface.

7.1.1   Diffuse layer and Stern layer conductance

As discussed in the previous chapter, the double layer can be separated into two distinct layers. Considering the differences between the diffuse double layer and the Stern layer, it is also reasonable to separate the surface conductance into two separate components (Lyklema 1995). The total surface conduction becomes the sum of a conduction in the diffuse part of the double layer $K_{S,d}$ and a conduction behind the slip plane in the Stern layer $K_{S,s}$, therefore

$$K_S = K_{S,s} + K_{S,d} \tag{7.5}$$

The conductance in the Stern layer is given by the sum of the product of the surface charge density in this part of the double layer $\sigma_{qs,j}$ and the ion mobility μ_j for all ions j

$$K_{S,s} = \sum_j \sigma_{qs,j} \mu_j \tag{7.6}$$

which, for a layer consisting of only one monovalent ion, is $K_{S,s} = \sigma_{qs}\mu$.

In the diffuse part of the double layer there is an added effect, that of electroosmotic transport of charge carriers, which must be considered along with conduction. The sheet current in this part of the double layer, $\Delta\mathbf{J}(y)$ is the sum of two components, one due to electroosmosis and the other due to conduction. For a symmetrical electrolyte the surface current is (Lyklema 1995)

$$\Delta\mathbf{J}(y) = q(z_- n_-(y) - z_+ n_+(y)) \frac{\varepsilon}{\eta} (\zeta - \phi(y))\mathbf{E} + q(z_- n_-(y) + z_+ n_+(y))\mathbf{E} \tag{7.7}$$

$$\underbrace{\phantom{q(z_- n_-(y) - z_+ n_+(y)) \frac{\varepsilon}{\eta} (\zeta - \phi(y))\mathbf{E}}}_{\text{Electroosmosis}} \quad \underbrace{\phantom{q(z_- n_-(y) + z_+ n_+(y))\mathbf{E}}}_{\text{Conduction}}$$

This expression is substituted into equation (7.2) and integrated to give

$$K_{S,d} = \frac{4F^2 c z^2 D}{RT\kappa} \left(1 + \frac{3m}{z^2}\right) \left(\cosh\left[\frac{zF\zeta}{2RT}\right] - 1\right) \tag{7.8}$$

which, when written for consistency in the units used throughout this book, is

$$K_{S,d} = \frac{4q^2 n_o z^2 D}{10^3 k_B T\kappa} \left(1 + \frac{3m}{z^2}\right) \left(\cosh\left[\frac{zq\zeta}{2k_B T}\right] - 1\right) \tag{7.9}$$

In this expression, ζ is the zeta potential of the particle (the potential at the slip plane) and the contribution from the electroosmotic transport is represented by the dimensionless parameter m, given by (Lyklema 1995)

$$m = \frac{2}{3} \frac{\varepsilon}{\mu\eta} \left(\frac{k_B T}{q} \right)^2 \tag{7.10}$$

7.1.2 Variation in total double layer surface conductance

A plot showing the variation of the diffuse layer surface conductance, $K_{s,d}$, as a function of electrolyte concentration and ζ potential is shown in figure 7.2. At low electrolyte conductivities (molarity) the diffuse layer conductance is small in comparison to the Stern layer conductance, which is typically of the order of 1nS. However, at higher molarities the two are comparable and the diffuse layer conductance can make a significant contribution to the total conductance. This particular analysis, based on the combination of the two conductances, has been used to successfully explain both the dielectric and dielectrophoretic properties of colloidal particles.

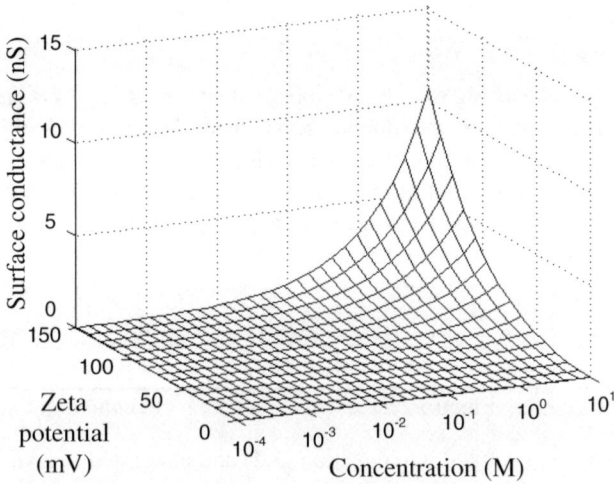

Fig. 7.2 Variation of the surface conductance in the diffuse layer in KCl as a function of zeta potential and electrolyte concentration.

In the remainder of this chapter we explore the influence of the double layer and surface conductance in controlling the polarisability of colloidal particles. We discuss the implications of this for the electrophoretic movement and polarisation of particles.

7.2   Electrophoresis: movement of a charged particle in electric fields

We know that when a particle is suspended in an electrolyte it moves under the influence of an electric field, a motion called electrophoresis. We also know that the double layer forms around the particle in order to maintain electroneutrality so that overall the particle is uncharged. This section explores the relationship between the motion of ions in the double layer and the macroscopic scale movement of the particle. (Further reading on electrophoresis can be found in books such as Hunter 1981; Russel *et al.* 1989; Lyklema 1995.) We begin by defining the zeta potential from a hydrodynamic point of view.

7.2.1   The zeta potential

For a fluid moving across a surface, the boundary condition is that the velocity of the fluid at the surface is zero. How can we interpret this from the microscopic point of view, where the liquid is not a continuous medium but consists of individual molecules? At the surface there exists a stationary phase or stagnant layer, one or two molecular sizes in width, in which the liquid is stationary. Outside this stationary layer, the fluid velocity (relative to the surface) increases with distance from the surface, reaching a maximum value at some distance away from the surface. When a charged surface is immersed in an electrolyte, part of the double layer lies within this stagnant layer. The slip plane is defined to be the outer limit of the stagnant layer, *i.e.* behind the slip plane the ions cannot move with the fluid, whilst outside it they can, as discussed in Chapter Five.   The potential at the slip plane is defined to be the zeta (ζ) potential, a quantity that can be measured from experiment.

Although this description applies to a fluid moving across a surface, the same holds true for a particle moving in a stationary fluid, the electrical potential at the slip plane is the particle's ζ potential. In general, the ζ potential is often equated with the potential of the diffuse double layer ϕ_d, and by implication the stagnant layer is equivalent to the Stern layer (where much of the double layer charge may reside).

7.2.2   Electrophoretic mobility

For the definition of electrophoresis, we refer back to Chapter Three and Four. A charged particle in an electric field experiences a force, given by $\mathbf{F} = q\mathbf{E}$. In a vacuum, a particle accelerates (until relativity comes into play) but in a viscous medium, the presence of friction leads to movement with a constant velocity $\mathbf{v}$. The *electrophoretic mobility* (μ_E) of the particle is a measure of the velocity of a particle in an electric field. It is defined as

$$\mu_E = \frac{|\mathbf{v}|}{|\mathbf{E}|} = \frac{v}{E} \tag{7.11}$$

When a particle is placed in an aqueous suspending medium, the presence of the double layer means that to an observer the particle is electroneutral; the excess charge in the double layer is exactly equal and opposite to the charge on the particle. The particle therefore appears to have zero net charge but when placed in an electric field it moves. Clearly, if the ions in the double layer were fixed to the surface, then the particle would not move. However, in the diffuse part of the double layer the ions have the same mobility as the bulk solution; those in the Stern layer are bound but have a non-zero mobility. Since the counterions have the opposite sign to the surface charge, they move in the direction opposite to that in which the particle would move if it were unscreened. This ion movement gives rise to fluid motion around the particle (electroosmosis). However, we are in the viscous limit for the fluid (the particles are very small) so that on the global scale the fluid is stationary. As a result, the moving ions push the particle in the opposite direction, in other words in the same direction that the particle would move, were it in a vacuum.

The situation is more complicated than this, since co-ions around the particle move ahead of the particle and counterions move in the direction opposite to the particle. As a result, a polarisation field, arising from the displacement of the charge and countercharge of the particle is established. This additional field can go with or against the applied field, either increasing or decreasing the electrophoretic velocity respectively.

Obviously, the electrophoretic mobility of a colloidal particle depends on the properties of the double layer. More particularly, it depends on the thickness of the double layer compared with the size of the particle. This is generally described by the ratio of the particle radius to the Debye length given by κa. There are two limiting cases to consider: (i) when the double layer is thin compared to the particle, $\kappa a \gg 1$, the double layer has very low curvature and can be considered to be flat allowing the planar theory to be used. (ii) when the double layer is thick compared to the particle, $\kappa a \ll 1$; this is exactly the same physical situation as the spherical double layer around a point charge. The electrophoretic mobility varies with κa, between the two limiting values as follows.

(i) *Thin double layer, $\kappa a \gg 1$*
For a double layer which is thin compared with the particle radius and therefore appears flat on the scale of the Debye length, the derivation of the equation describing the motion of the ions and the fluid is identical to the derivation of electroosmotic flow in the previous chapter. However, in this case the surface moves relative to the stationary fluid. The electrophoretic mobility is

$$\boxed{\mu_E = \frac{\varepsilon \zeta}{\eta}}$$

(7.12)

This is referred to as the Helmholtz-Smoluchowski limit.

(ii) *Thick double layer*, $\kappa a \ll 1$

When a particle has a thick double layer (small particles or very dilute electrolytes), the forces acting on the double layer are not felt by the particle. The net force acting on the particle is simply the difference between the Coulombic force and the drag force given by Stokes' law. We can write the force as $F = qE = 6\pi\eta a v$, giving $\mu = v/E = q/6\pi\eta a$. The zeta potential can be identified with the potential of a point charge, *i.e.* $\zeta = q/4\pi\varepsilon a$. Combining these expressions gives the electrophoretic mobility as

$$\mu_E = \frac{2\varepsilon\zeta}{3\eta} \qquad\qquad (7.13)$$

This is referred to as the Hückel-Onsager limit.

The difference between the high and low limits of κa is only a factor of 1.5. Improvements on the theories have been made, notably by Henry (1931) who took into account the distortion of the field by the particle and also the ratio of the particle conductivity to the suspending electrolyte conductivity.

The mobility is written as a general function

$$\mu_E = \frac{2\varepsilon\zeta}{3\eta} f(\kappa a) \qquad\qquad (7.14)$$

Henry (1931) and latterly Oshima (1994) provided an expression for $f(\kappa a)$ for low ζ potential (< 50mV) as

$$f(\kappa a) = 1 + \frac{1}{2}\left[1 + \frac{5}{2\kappa a(1 + 2e^{-\kappa a})}\right]^{-3} \qquad\qquad (7.15)$$

This function is plotted in figure 7.3, showing the variation between 1 and 1.5 as κa varies between 0 and ∞, allowing values of electrophoretic mobility to be more accurately determined between the limiting cases.

Further developments of the theory must take into account the effect of the electric field on the structure of the double layer and the surface conductivity of the particle, for example by including surface conduction behind the slip plane.

7.3   Polarisation

We now return to a theme that is central to the subject of AC electrokinetics, which is the origin of the particle polarisation. Clearly in any process where a charged (or even uncharged) particle is polarised, the double layer is going to play an important role. In particular, the Maxwell-Wagner interfacial polarisation mechanism is central to the concept of a frequency-dependent dielectrophoretic force where the polarisation process arises from the motion of charges at the interface between particle and electrolyte, *i.e.* inside the double layer.

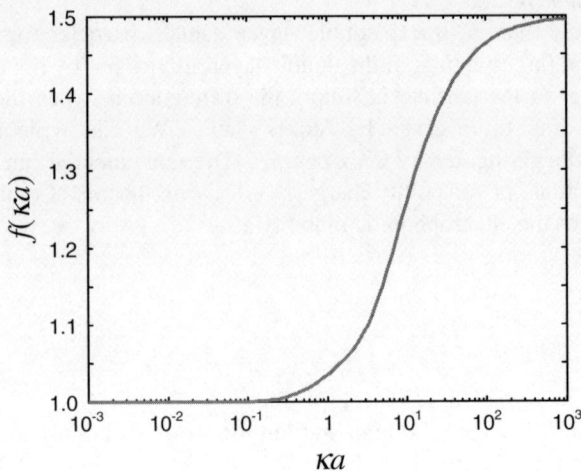

Fig. 7.3 Plot of the approximate value of the function $f(\kappa a)$, from equation (7.15), against κa.

During the first half of the twentieth century, dielectric spectroscopic measurements of suspensions of colloidal particles identified two phenomena that were at the time unexpected and unexplained. The first was that the Maxwell-Wagner relaxation frequency for charged colloids was higher than expected from theory, indicating that the particle conductivity was higher than expected. Miles and Robertson (1932) and Fricke and Curtis (1937) were among the first to propose that the presence of the double layer affects the dielectric properties of the particle. Later, O'Konski (1960) showed that the dielectric properties of colloidal particles in this frequency regime were dominated by surface conductance effects.

The second unexpected result was the observation of a large dispersion in the permittivity of the solution at low frequencies, below the Maxwell-Wagner relaxation (Schwann 1957; 1962; 1963). This relaxation is related to a macroscopic polarisation of the double layer ions around the entire particle. The characteristic relaxation time for this mechanism is the time taken for the ions to travel around the particle. This means that the polarisation mechanism is slow and has a lower relaxation frequency than the interfacial polarisation mechanism.

7.3.1 Interfacial polarisation and surface conductance

The high value of particle conductivity determined from dielectric measurements was explained by the inclusion of a surface conductance component (see section 7.1) in the derivation of the particle's dipole moment (O'Konski 1960). The concept of a particle surface conductance had been around for a number of years and had been used successfully in explaining electroosmosis and electrophoresis

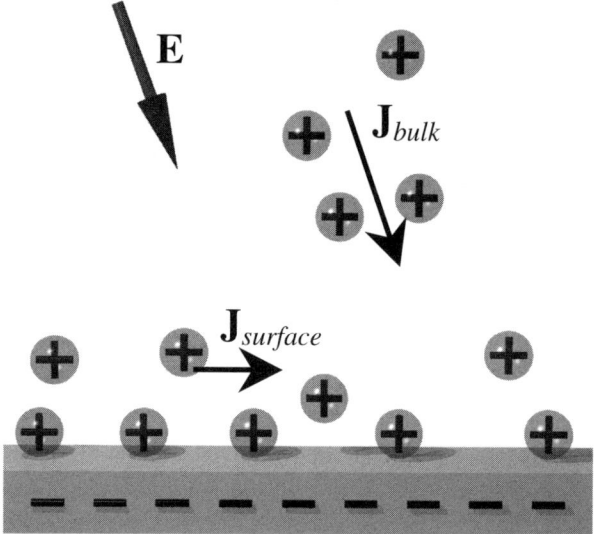

Fig. 7.4 Schematic diagram of the mechanism used by O'Konski to explain how surface conductance affects interfacial polarisation. The electric field produces both a bulk flow of ions and a surface flow around the particle.

measurements (Bickerman, 1933). The model developed by O'Konski is shown in figure 7.4. He assumed that the flux due to the transport of charge carriers, associated with the fixed charge on the particle surface, could be added to the flux due to the transport of bulk charge to and from the surface. Both terms are then included in the charge conservation equation for the interface. With this assumption, he derived the potential around the spherical particle and derived the equation for the dipole moment of a dielectric sphere (equation (3.7)) but with the particle conductivity given by the sum of the bulk conductivity of the particle and a surface conductivity term: $\sigma_p = \sigma_{p,bulk} + \sigma_{p,surface}$. The surface conductivity of the particle is given by

$$\boxed{\sigma_{p,surface} = \frac{2K_s}{a}} \qquad (7.16)$$

where K_s is the **surface conductance**. This approach has been successfully used by many groups to explain the experimental observations of the frequency dependent dielectric and dielectrophoretic behaviour of larger particles (greater than 1μm) at frequencies where Maxwell-Wagner polarisation is dominant (*e.g.* Arnold *et al.* 1987).

However, detailed measurements of the dielectric and electrokinetic behaviour of colloidal particles have demonstrated that the surface conductance does not remain constant with the conductivity of the suspending medium (Green and Morgan 1999; Hughes *et al.* 1999). O'Konski's model can be developed further by invoking a surface conductance that consists of two separate components, one from the diffuse layer and another independent one due to the Stern layer, as discussed in section 6.1. Lyklema and Minor (Lyklema 1995; Minor *et al.* 1998; 1999) have demonstrated that the dielectric properties of colloidal particles can be explained in terms of this model. The same model has been successfully used to explain measurements of the dielectrophoretic behaviour of colloidal particles (see Chapter Eleven).

7.3.2 Relaxation of the double layer

Dielectric measurements of suspensions of cells and colloidal particles at low frequencies (below the Maxwell-Wagner interfacial relaxation frequency) showed the presence of a second large dispersion (Schwan 1957; 1962; 1963). This dispersion is commonly referred to as the α-dispersion in biophysics, or the Low Frequency Dielectric Dispersion (LFDD) in colloidal and physical chemistry. It is related to the macroscopic polarisation of the double layer around the particle as shown in figure 7.5 and has a characteristic relaxation time of the order of

$$\boxed{\tau_\alpha = a^2 / 2D} \tag{7.17}$$

Following on from the work of O'Konski (1960), Schwarz (1962) derived a model to describe the polarisation of the diffuse double layer around the particle. He considered two components of flux around the particle. The first is due to the movement of the ions by the electric field and the second, the diffusion flux which acts against this ion movement, is caused by the accumulation of ions at the poles of the polarised particle. In deriving this model, Schwarz assumed that the ions were relatively free to move around the surface of the particle in a thin highly conducting layer (the double layer) but were bound to it in such a way that they could not leave this layer, as shown in figure 7.5. The field arising from the applied potential was also assumed to be sufficiently small so that the alternating perturbation in charge densities arising from the applied AC field was small. This gives the following expression for the flux of ions in the charged layer (Schwarz 1962)

$$\mathbf{J}'_j = -\frac{\mu_j}{a}\sigma_{qo,j}\mathbf{E}_t - \frac{\mu_j}{a}\frac{k_B T}{q}\frac{\partial \sigma_{qo,j}}{\partial \theta} \tag{7.18}$$

where $\sigma_{qo,j}$ is the surface charge density of ion j and $\mathbf{E}_t$ is the component of the electric field tangential to the surface, θ is defined according to the standard definition of spherical polar co-ordinates (Lorrain *et al.* 1988) and $\mathbf{J}'$ is a surface

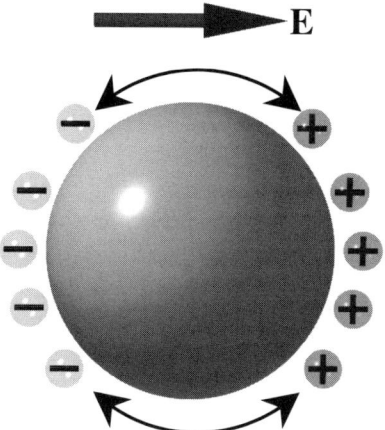

Fig. 7.5 Schematic diagram of the polarisation of an induced
double layer around a particle at low frequencies.

flux. The first term on the right hand side of equation (7.18) is the flux of ions at
the surface and the second term is the diffusion flux acting against this. Assuming
a symmetrical electrolyte, then from equation (3.4) the condition for the charge
density perturbation $\Delta\sigma_q$ at the interface is

$$\frac{\partial\Delta\sigma_{qo}}{\partial t} = \frac{D_s}{a^2 \sin\theta}\frac{\partial}{\partial\theta}\left(\sin\theta\left[\frac{\partial\Delta\sigma_{qo}}{\partial\theta} + \frac{q\sigma_{qo}}{k_B T}\frac{\partial\phi_o}{\partial\theta}\right]\right) \qquad (7.19)$$

where D_s is the surface diffusion constant for the ion. In this case, the solution of
the potential around the sphere is the same as for the dielectric sphere (derived in
Chapter Three) and the O'Konski model, with the surface conductance now given
by a frequency dependent term

$$K_{s\alpha} = \frac{i\omega\tau_\alpha}{1+i\omega\tau_\alpha}\frac{2K_s}{a} \qquad (7.20)$$

where $\tau_\alpha = a^2/2D$ is the relaxation time[†]. The complex permittivity of the particle
is then

$$\tilde{\varepsilon}_p = \varepsilon_p + \frac{2K_s}{a}\frac{\tau_\alpha}{1+\omega^2\tau_\alpha^2} - i\left(\frac{\sigma_p}{\omega} + \frac{2K_s}{a}\frac{\omega\tau_\alpha^2}{1+\omega^2\tau_\alpha^2}\right) \qquad (7.21)$$

[†] Note that in this model the relaxation time τ_α is derived from physical parameters and
not assumed.

It is clear from this expression that as $\omega \rightarrow 0$, the contribution of the surface conductance to the particle conductivity also tends to zero. This is at odds with experimental observations and led Schurr (1964) to develop the following expression for the particle complex permittivity

$$\tilde{\varepsilon}_p = \varepsilon_p + \frac{2K_s}{a}\frac{\tau_\alpha}{1+\omega^2\tau_\alpha^2} - i\left(\frac{\sigma_p}{\omega} + \frac{2K_s}{a}\left[1+\frac{\omega\tau_\alpha^2}{1+\omega^2\tau_\alpha^2}\right]\right) \tag{7.22}$$

Schwarz's model for the polarisation of the double layer, while simplistic compared with currently accepted models for the electrical double layer, was quite successful in a number of ways. Equations (7.21) and (7.22) predict that the relaxation frequency is inversely proportional to the particle radius squared, a fact which has been experimentally proven. The model also predicts high values for the dielectric constant of suspensions of particles at low frequencies, as seen experimentally. Although the measured frequencies of the α-relaxation are not exactly as predicted, Schwarz (1962) states in his paper that the value of the mobility of the ions in the double layer (integral to the calculation of the dipole) will be different from the bulk value.

A major deficiency in Schwarz's model is that the exchange between the suspending electrolyte and the charged layer is neglected. As we have seen, ion movement to and from parts of the double layer is an integral part of double layer polarisation. In addition, contributions to the polarisation arising from deformation of the double layer are ignored. This will be negligible for large particles, such as cells; for sub-micrometre particles, where the double layer is large compared to the particle radius, this could be an important consideration.

Following on from this early pioneering work, many models have been developed and refined. Indeed, understanding the polarisation of the double layer is now a considerable area of research interest (*e.g.* Dukhin and Shilov 1974; Lyklema *et al.* 1983; 1986; Lyklema 1995; Hinch and Sherwood 1984; Chew and Sen 1982; O'Brien 1986; 1988; Fixman 1983; Shubin *et al.* 1993). Current research focuses on both the α-relaxation and high frequency effects where the double layer thickness is of the same order of size as the particle.

7.4   Relaxation of the double layer revisited: current opinion

In his book on dielectrophoresis, Pohl (1978) discussed the theories of O'Konski and Schwarz in terms of their influence on the AC electrokinetic properties of particles. He also introduced the more advanced theoretical work performed by Dukhin, Shilov and co-workers (Dukhin and Shilov 1974). Following on from the work of Dukhin, Lyklema presented several theories for polarisation mechanisms of the double layer. In his book, Lyklema (1995) reviews previous publications in the field (Lyklema 1983a; 1983b; 1986) together with more up-to-date models for the polarisation of the double layer.

Many of the theories for double layer polarisation were developed to explain electrophoresis experiments. The theory of Dukhin and Shilov is restricted to high κa, a limitation that was not present in the numerical work of DeLacey and White (1981). Fixman (1983) published an analytical expression for $\zeta < 200\text{mV}$ and $\kappa a \gg 1$ and Chew and Sen (1982) derived an analytical expansion up to first order in $1/\kappa a$. The different theories, including more recent elaborations, do not differ greatly in the end result. Recently, there have been a number of papers that offer new insights into the physical mechanisms involved in the polarisation of the double layer (*e.g.* Werner *et al.* 1998; Lyklema *et al.* 1998; 1999; Matsumura *et al.* 1999; Grosse *et al.* 1999; Verbich *et al.* 1999; Shilov *et al.* 2000; Simonova *et al.* 2000; 2001; Lyklema 2001).

7.4.1   Polarisation of the Stern or bound layer

Current thinking about the double layer is that the polarisation mechanism arises from a combination of diffuse and bound layer contributions. Lyklema (Lyklema *et al.* 1983, Lyklema 1995) derived a general model for the polarisation of the bound part of the double layer *i.e.* the Stern layer. This is similar to the model of Schwarz in that the ions cannot leave the surface of the particle and exchange with the medium. Lyklema generalised this model to take account of the physical structure of the double layer and included the effect of the diffuse part of the double layer in the model.

There are three categories of charge that must be taken into account:

- fixed surface charge on the particle σ_{qo} (*e.g.* covalently bound ionic groups)
- charge in the bound layer σ_{qs}
- charge in the diffuse layer σ_{qd}

Figure 7.6 is a schematic diagram illustrating how the bound and diffuse layers around the charged dielectric particle polarise under the applied field. At low frequencies, the bound ions move under the influence of an applied electric field to enrich the negative side (to the right of the diagram). However, the diffuse ions outside the bound layer can exchange with the bulk and attempt to maintain electroneutrality by enriching the positive side (on the left).

In this model a fixed charge is defined to be a charge with negligible surface mobility. Counterions with a surface diffusion coefficient smaller than the bulk, $D_s < D$, are considered fixed for frequencies corresponding to $\omega \gg 2D_s a^{-2}$ and mobile if $\omega \ll 2D_s a^{-2}$. The mobility of the ions in the diffuse part of the double layer is greater than the mobility of counterions in the bound layer. This means that for thin double layers, the diffuse ions can redistribute fast enough to be able to almost maintain equilibrium between the bound and diffuse parts of the double layer. Therefore, the bound and surface charges are almost continually screened by the diffuse ions, with the conclusion that the polarisation of the diffuse layer cannot be ignored.

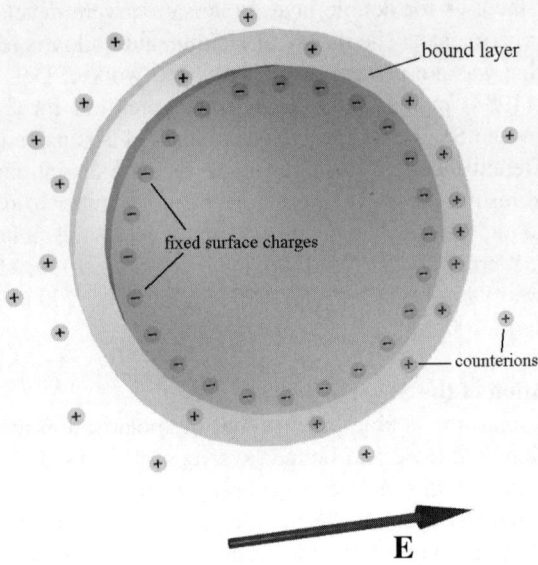

Fig. 7.6 Schematic diagram of the polarisation of the bound layer and the screening action of the diffuse layer proposed by Lyklema (1983).

This diffuse layer polarisation reduces the relaxation time of the bound ions and, as a result raises the relaxation frequency of the α-dispersion above that predicted by the Schwarz bound layer model (equation 7.21). The conservation of charge (continuity) equation for the alternating perturbation in the total charge at the particle surface $\Delta\sigma_q$, due to the applied AC field is

$$\frac{\partial \Delta\sigma_q}{\partial t} = \frac{D_s}{a^2 \sin\theta} M \frac{\partial}{\partial\theta}\left(\sin\theta\left[\frac{\partial \Delta\sigma_q}{\partial\theta} + \frac{q\sigma_q}{k_B T}\frac{1}{M}\frac{\partial\phi_o}{\partial\theta}\right]\right) \tag{7.23}$$

which is similar to equation (7.19) but with the addition of a factor M, which arises from the electroosmotic contribution of the ion flux of the double layer given by

$$M = 1 + \frac{qU_q}{k_B T C_d'} \tag{7.24}$$

where C_d' is the differential (or effective) capacitance of the bound layer given by

$$C_d' = \varepsilon_m \kappa \cosh\left(\frac{q\zeta}{2k_B T}\right) \tag{7.25}$$

Lyklema *et al.* (1983) solved this equation following the method of Schwarz and obtained a similar expression for the complex dipole moment of the particle, but with the relaxation time now given by

$$\tau_{a,L} = \frac{a^2}{2DM} \tag{7.26}$$

Owing to the presence of the factor M, the relaxation time is decreased and the predicted dispersion relaxation frequency increases.

In the same paper (Lyklema *et al.* 1983) the authors also presented an analysis of the polarisation of the diffuse layer taken from Dukhin (1974). In this case the physical mechanism was electroosmotic flow around the particle in the absence of a Stern layer; full details can be found in the papers by these authors.

7.4.2   The relaxation of the whole double layer

Lyklema (1995) presented a comprehensive theory to account for the dielectric dispersion caused by the relaxation of the whole electrical double layer. The theory describes the contribution of charge currents, the surface potential and electroosmosis in the diffuse layer to the surface conduction of the particle. The solution for the perturbation in the potential outside the spherical particle, in the far field is

$$\Delta \phi = \left[-Er\cos\theta + \frac{\tilde{d}_e a^3}{r^2} E\cos\theta \right] e^{i\omega t} \tag{7.27}$$

where E is the magnitude of the electric field which has been assumed to lie along the z-axis and r and θ are defined as for a spherical co-ordinate system. $\tilde{d}_e$ is a term equivalent to the Clausius-Mossotti factor, given by

$$\tilde{d}_e = -\frac{1}{2} + \frac{3Du^d}{2(1+Du^d(\gamma+1))} \tag{7.28}$$

where Du^d is the *Dukhin number* (Lyklema 1995) for the diffuse layer, which for a symmetrical electrolyte is given by

$$Du^d = \frac{2}{\kappa a}\left(1+\frac{3m}{z^2}\right)\left[\cosh\left(\frac{zq\zeta}{2k_BT}\right)-1\right] \tag{7.29}$$

The Dukhin number describes the ratio of the surface conductance to the bulk. The other terms in equation (7.28) are the electroosmotic contribution m, given by equation (7.10), and γ, which is

$$\gamma = 1 - \frac{(\omega\tau)^{3/2} - i\omega\tau}{(1 + \sqrt{\omega\tau})(1 + \omega\tau)} \qquad (7.30)$$

The relaxation time for this mechanism is that given by Schwarz (*i.e* τ_α). From the equation for the potential (7.27), the dipole moment around the particle is

$$\tilde{\mathbf{p}} = 4\pi\varepsilon_m \tilde{d}_e a^3 \mathbf{E} \qquad (7.31)$$

Figure 7.7 shows how the dipole moment varies with frequency and suspending medium conductivity, calculated from equation (7.28) for a sub-micrometre particle.

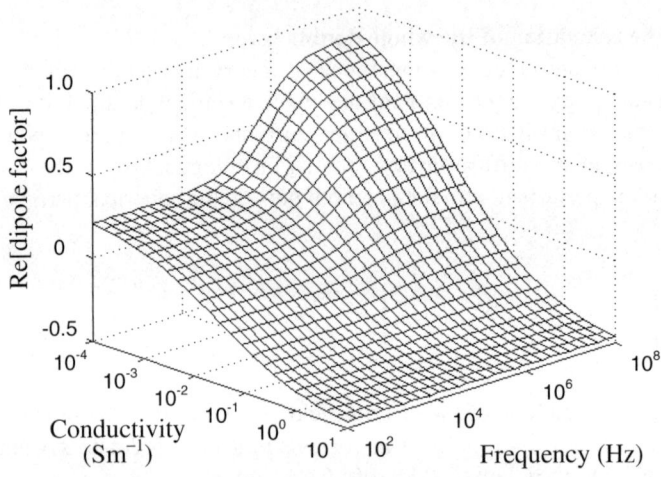

Fig. 7.7 Variation in the dipole factor (equivalent to the Clausius-Mossotti factor) from Lyklema (1995), plotted as a function of medium conductivity and applied field frequency for a 500 nm diameter spherical particle with a zeta potential of 50 mV in KCl.

7.5   Other approaches

Following on from Dukhin and Shilov (1974), several groups developed theories to explain electrophoresis and dielectric experiments. An exhaustive review of the work in the field is beyond the scope of this book, but for a representative sample the reader should refer to Chew and Sen (1982a; 1982b); their collaborative work with Hinch and Sherwood (Hinch *et al.* 1984); Fixman (1983); O'Brien (1986; 1988) and the later work of Rosen and Saville (1991) and Minor and Lyklema (Lyklema *et al.* 1999; Minor *et al.* 1998; 1999). More recently, numerical programs have been developed to determine the dielectric response of suspension of particles (Mangelsdorf and White 1990: 1992; 1997; 1998a; 1998b).

7.5.1   Diffusion beyond the double layer

In Chew and Sen's paper, the mechanism responsible for the polarisation of the particle/double layer was expanded to include a neutral diffusion cloud beyond the Debye length (Chew *et al.* 1982a; 1982b). The principle is that the polarisation mechanism is affected by diffusion, even in the electroneutral region outside the double layer, a mechanism also applied to electrode polarisation (Gunning 1995). The diffusion cloud, although electroneutral, is important in determining the effective polarisation of the particle/double layer system.

7.5.2   Relaxation time

We have already mentioned that the relaxation time derived in Schwarz's model ($\tau_\alpha = a^2 / 2D$) is not completely correct. We also discussed the modification to this expression through the inclusion of the differential capacitance term, equation (7.25). DeLacey and White (1981) also proposed a different relaxation time, in which the time taken for the ions to traverse the double layer perpendicular to the surface must be included. The authors estimated that the relaxation time in this case would be of the order of

$$\tau = \frac{2\pi(a + \kappa^{-1})^2}{D} \tag{7.32}$$

O'Brien (1986) has discussed the relaxation of ions in the double layer at high frequencies. In general, the frequency is assumed to be much lower than the charge relaxation frequency, *i.e.* the reciprocal of the time taken for an ion to diffuse a distance equal to the Debye length and for the double layer to form. O'Brien discussed the relaxation mechanism assuming that this was not the case. The changes in the charge take a finite amount of time to adjust to the perturbation created by the applied field. For frequencies greater than the reciprocal of this time, the polarisation due to the surface conductance does not reach its maximum value, and at the high frequency limit, the surface conductance does not contribute to the particle conductivity. The corresponding relaxation time for the diffuse layer in this case is

$$\tau = 1 / 2\kappa^2 D \tag{7.33}$$

Substituting for the reciprocal Debye length κ and the diffusion coefficient D, this characteristic time is found to be approximately the charge relaxation time of the medium. This time is important from the point of view of AC electrokinetics. Since experiments cover a wide range of suspending medium conductivities, it is possible that the Maxwell-Wagner relaxation time for the particle is higher than the charge relaxation time of the electrolyte. In this case, the surface conductance of the particle cannot be considered to have a constant value.

7.6   Conclusion

In conclusion, this chapter has demonstrated the contribution the double layer makes in controlling the polarisation of particles. From the point of view of sub-micron particles, surface conductance effects are still not fully understood, an issue we shall return to in Chapter Eleven.

7.7   References

Arnold W.M., Schwan H.P. and Zimmermann U. *Surface Conductance and Other Properties of Latex-Particles Measured by Electrorotation*   J. Phys. Chem. **91** 5093-5098 (1987).

Bickerman J.J. *Ionentheorie der Elektrosmose der Strömungsströme unde der Oberflächenleitfähigkeit* Z. Physik. Chem. **A163** 378-394 (1933).

Chew W.C. and Sen P.N. *Potential of a sphere in an ionic solution in thin double layer approximations* J Chem. Phys. **77** 2042-2044 (1982a).

Chew W.C. and Sen P.N. *Dielectric enhancement due to electrochemical double layer: thin double layer approximation* J. Chem. Phys. **77** 4683-4693 (1982b).

DeLacey E.H.B. and White L.R. *Dielectric Response and conductivity of dilute suspension of colloidal particles* J. Chem. Soc. Faraday. Trans. 77 2007-2039 (1981).

Dukhin S.S. and Shilov V.N. *Dielectric Phenomena and the Double Layer in Disperse Systems and Polyelectrolytes* Wiley, New York (1974).

Fixman M. *Thin double layer approximation for electrophoresis and dielectric response* J. Chem. Phys. **78** 1483-1491 (1983).

Fricke H. and Curtis H.J. *The dielectric properties of water-dielectric interphases* J. Phys. Chem. **41** 729-745 (1937).

Green N.G. and Morgan H. *Dielectrophoresis of sub-micrometre latex spheres* J. Phys. Chem. **103** 41-50 (1999).

Grosse C., Pedrosa S. and Shilov V.N. *Calculation of the dielectric increment and characteristic time of the LFDD in colloidal suspensions of spheroidal particles* J. Col. Int. Sci. **220** 31-41 (1999).

Gunning J., Chan D.Y.C. and White L.R. *The impedance of the planar diffuse double layer: An exact low-frequency theory* J. Col. Int. Sci. **170** 522-537 (1995).

Henry D.C. *The cataphoresis of suspended particles. I. The equation of cataphoresis.* Proc. Roy. Lond. A **133** 106-129 (1931).

Hinch E.J., Sherwood J.D., Chew W.C. and Sen P.N. *Dielectric response of a dilute suspension of spheres with thin double layers in an asymmetric electrolyte* J. Chem. Soc. Faraday. Trans. **80** 535-551 (1984).

Hughes M.P., Morgan H. and Flynn M.F. *The dielectrophoretic behaviour of sub-micron latex spheres: influence of surface conductance* J. Colloid Int. Sci. **220** 454-457 (1999).

Hunter R.J. *Zeta Potential in Colloid Science* Academic Press, San Diego (1981).

Lorrain P., Corson D.R. and Lorrain F. *Fundamentals of Electromagnetic Phenomena* W.H. Freeman and Co, New York (2000).

Lyklema, J., Dukhin, S.S. and Shilov V.N. *The relaxation of the double-layer around colloidal particles and the low-frequency dielectric-dispersion. 1 theoretical considerations* J. Electroanalytical Chem. **143** 1-21 (1983).

Lyklema J., Springer M. M., Shilov V. N. and Dukhin S. S. *The relaxation of the double-layer around colloidal particles and the low-frequency dielectric-dispersion 3. application of theory to experiments.* J. Electroanalytical Chem. **198** 9-26 (1986).

Lyklema J. *Fundamentals of Interface and Colloid Science Vol II* Academic Press London (1995).

Lyklema J., Rovillard S. and de Coninck J. *Electrokinetics: The properties of the Stagnant Layer Unraveled* Langmuir **14** 5659-5663 (1998).

Lyklema J., van Leeuwen H.P. and Minor M. *DLVO-theory, a Dynamic Re-interpretation* Adv. Coll. Int. Sci. **83** 33-69 (1999).

Lyklema J. *Surface Conduction.* J. Phys. Condens. Matter **13** 5027-5034 (2001).

Mangelsdorf C.S. and White L.R. *Effects of Stern Layer conductance on electrokinetic transport properties of colloidal particles* J. Chem. Soc. Faraday Trans. **86** 2859-2870 (1990).

Mangelsdorf C.S. and White L.R. *Electrophoretic mobility of a spherical colloidal particle in an oscillating electric field* J. Chem Soc. Faraday Trans. **88** 3567-3581 (1992).

Mangelsdorf C.S. and White L.R. *Dielectric response of a dilute suspension of spherical colloidal particles to an oscillating electric field* J. Chem. Soc. Faraday Trans. **93** 3145-3154 (1997).

Mangelsdorf C.S. and White L.R. *The dynamic double layer: Part 1 Theory of a mobile Stern layer* J. Chem. Soc. Faraday Trans. **94** 2441-2452 (1998a).

Mangelsdorf C.S. and White L.R. *The dynamic double layer: Part 2 Effects of Stern Layer Conductance on the high frequency electrokinetic transport properties.* J. Chem. Soc. Faraday Trans. **94** 2583-2593 (1998b).

Miles J.B. and Robertson H.P. *The dielectric behaviour of a colloidal particle with an electric double-layer* Phys. Rev. **40** 583-591 (1932).

Matsumura H., Verbich S.V. and Dukhin S.S. *Investigation of dynamic Stern layer of liposomes by utilizing the isoelectric point and isoconducting point* Colloid Surf. A-Physicochem. Eng. Asp. **159** 271-276 (1999).

Minor M., van Leeuwen H.P. and Lyklema J. *Low-frequency dielectric response of polystyrene latex dispersions.* J. Coll. Int. Sci. **206** 397-406 (1998).

Minor M., van Leeuwen H.P. and Lyklema J. *Low-frequency dielectric responses, static conductivities, and streaming potentials of polymer-coated latex dispersions and porous fluffs* Langmuir **15** 6677-6685 (1999).

Myers D.F. and Saville D.A. *Dielectric spectroscopy of colloidal dispersions 2: comparisons between experiment and theory* J. Coll. Int. Sci. **131** 461-470 (1989).

O'Brien R.W. *The High-frequency dielectric dispersion of a colloid.* J. Col. Int. Sci. **113** 81-93 (1986).

O'Brien R.W. and Ward D.N. *Electrophoresis of a spheroid with a thin double layer.* J Coll. Int Sci. **121** 402-413 (1988).

O'Konski C.T. *Electric properties of macromolecules V: theory of ionic polarization in polyelectrolytes* J. Phys. Chem. **64** 605-619 (1960).

Oshima H. *A simple expression for Henry's function for the retardation effect in electrophoresis of spherical colloidal particles* J.Coll. Int. Sci **168** 269-271 (1994).

Pohl H.A. *Dielectrophoresis* Cambridge University Press, Cambridge UK (1978).

Rosen L.A. and Saville D.A. *Dielectric spectroscopy of colloidal dispersions: comparisons between experiment and theory* Langmuir **7** 36-42 (1991).

Russel W.B., Saville. D.A. and Schowalter W.R. *Colloidal Dispersions* Cambridge Uni. Press, Cambridge (1989).

Schurr J.M. *On the theory of the dielectric dispersion of spherical colloidal particles in electrolyte solution* J. Phys. Chem. **68** 2407-2413 (1964).

Schwan H.P. *Electrical properties of tissues and cells* in Lawrence J.H., Tobias C.A. (Eds.) *Advances in Biological Medical Physics* **15** 147-209 (1957).

Schwan H.P., Schwarz G., Maczuk J. and Pauly H. *On the low-frequency dielectric dispersion of colloidal particles in electrolyte solution* J. Phys. Chem. **66** 2626-2635 (1962).

Schwan H.P. *Determination of Biological Impedance* in Physical Techniques in Biologicial Research, Nastuk W.L. (ed.), New York Academic Press **6** 323-406 (1963).

Schwarz G. *A theory of the low-frequency dielectric dispersion of colloidal particles in electrolyte solution* J. Phys. Chem. **66** 2636-2642 (1962).

Shilov V.N., Delgado A.V., Gonzalez-Caballero E., Horno J., Lopez-Garcia J.J. and Grosse C. *Polarization of the electrical double layer. Time evolution after application of an electric field* J. Coll. Int. Sci. **232** 141-148 (2000).

Shubin V.E., Hunter R.J. and O'Brien R.W. *Electroacoustic a*J. Coll. Int. Sci. **159** 174-183 (1993).

Simonova T.S., Shilov V.N. and Shramko O.A. *Electrophoresis of a spherical particle with an arbitrarily thick electrical double layer in a solution of macromolecules: The effect of the depletion layer* Colloid J. **62** 206-210 (2000).

Simonova T.S., Shilov V.N. and Shramko O.A. *Low-frequency dielectrophoresis and the polarization interaction of uncharged spherical particles with an induced Debye atmosphere of arbitrary thickness* Colloid J. **63** 108-115 (2001).

Verbich S.V., Dukhin S.S. and Matsumura H. *Investigation of dynamic stern layer of liposomes by measurements of conductivity and electrophoresis.* J. Dispersion Sci. Technol. **20** 83-104 (1999).

Werner C., Korber H., Zimmermann R., Dukhin S. and Jacobasch, H.J. *Extended electrokinetic characterization of flat solid surfaces* J. Coll. Int. Sci. **208** 329-346 (1998).

Chapter Eight

Electrohydrodynamics

In this chapter we outline the mechanisms of AC electric field-induced fluid flow and illustrate the behaviour of fluid flow in microsystems with experimental observations and measurements. Order of magnitude calculations and numerical simulations of the field-driven fluid flows are presented, allowing the reader to estimate the relative contribution of different electrohydrodynamic forces to the motion of particles and fluids in microsystems.

In electrokinetic systems, the electric field acts on charges and dipoles in the fluid, as well as on the particles, giving rise to a force on the fluid and therefore flow. In a DC field, this gives rise to the familiar phenomenon of electroosmosis, as discussed in Chapter Six, section 6.5; the magnitude of this flow depends predominantly on the applied field and the charge density. In AC fields, fluid can be driven in different directions and at different velocities, depending on a number of inter-related factors: frequency; voltage; fluid conductivity, permittivity, viscosity, and importantly the scale of the system.

In Chapters Five and Six, we saw that two types of electric field-driven fluid flow can occur in electrolytes when the fields are large and the scale of the system is small and this is summarised in figure 8.1. At low frequencies (<100 kHz), the dominant fluid flow is AC electroosmosis. At higher frequencies the magnitude of the AC electroosmotic flow decreases, and the dominant effect becomes an electrothermally induced fluid flow. The behaviour of the fluid and the resulting induced motion of particles are distinct and differentiable for each of the two

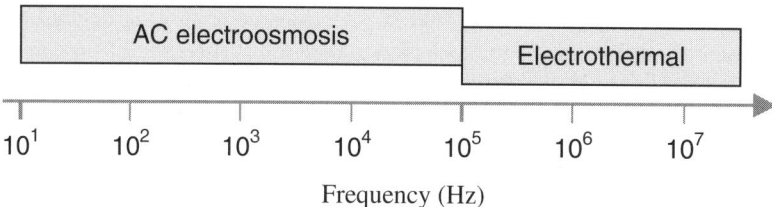

Fig. 8.1 An outline of the frequency ranges in which the different types of electric field-induced fluid flow are observed in AC electrokinetic microsystems. AC electroosmosis dominates in the lower frequency range (<100 kHz) and electrothermal fluid flow at higher frequencies.

mechanisms. In addition to field-driven flow, natural convection can also drive fluid if the conductivity is sufficiently high so that Joule heating produces significant temperature gradients and, therefore, gradients in the density of the fluid. However, in most practical cases, these effects are very small by comparison with other fluid effects and can be ignored.

The AC electrokinetic movement of the fluid is, in itself, an extremely useful effect. It can be used for particle manipulation and separation, either by itself or in concert with other electrokinetic forces. More importantly, the interaction of the electric field with the fluid can be used for pumping fluid through microchannels and microdevices. Significantly, micropumps based on AC electrokinetic methods are solid state, *i.e.* they have no moving parts. As long as the volume of fluid being pumped is small (typically in microfluidic systems $< 1 - 10 \text{ mm}^3 \text{ s}^{-1}$), they will also work with very low applied voltages (< 10 Volts) compared with conventional DC electroosmotic systems.

We begin with a general discussion of the relative magnitudes of different flows for different experimental conditions.

8.1   Mapping frequency & conductivity dependence of field-induced flow

AC electroosmotic fluid flow is governed by the charge induced by the electric field at the electrode/electrolyte interface. Since the amount of charge is frequency dependent, the magnitude of AC electroosmosis is also frequency dependent. As shown in Chapter Five, the flow occurs mainly at low frequencies (of the order of Hz to tens of kHz). In the low frequency limit, the magnitude of the AC electroosmotic flow tends to zero since there is no field in the bulk electrolyte. At high frequencies the AC electroosmotic flow again tends to zero, because there is insufficient time for the induced charge to form in the double layer. Using a simple model, the velocity profile can be plotted as a function of frequency, as shown in Chapter Six, figure 6.15. The characteristic frequency at which the maximum fluid velocity occurs increases with rising electrolyte conductivity. However, as will be discussed in the following section, the magnitude of the fluid flow decreases with increasing electrolyte conductivity. Therefore, although the frequency window over which AC electroosmosis occurs increases with increasing conductivity, the flow is nevertheless not observed above ~100 kHz.

At higher frequencies (>100 kHz), AC electroosmotic flow is negligible and the dominant fluid flow is due to electrothermal effects. This type of flow requires temperature gradients in the fluid, and these can be generated both by internal and/or external sources. The internal source is Joule heating, where the electric field causes power dissipation in the fluid, and the corresponding temperature rise diffuses through the system. This gives rise to gradients in the conductivity and permittivity; the electric field acts on these gradients to give a body force on the fluid and consequently a flow. In this case, the velocity of the flow is proportional to the temperature rise in the fluid, which is in turn proportional to the conductivity of the electrolyte and the magnitude of the electric field. As a result, we expect this type of fluid flow to occur mainly at high electrolyte conductivities.

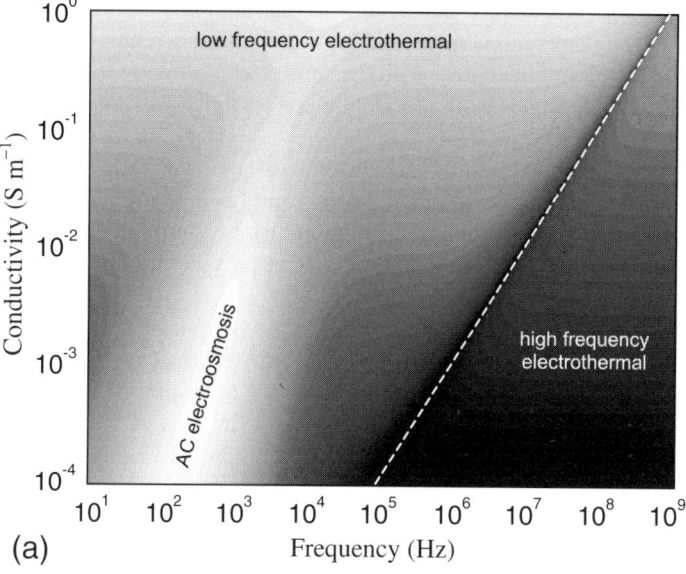

Fig. 8.2 Schematic frequency/conductivity maps of the two types of electric field-driven fluid flow: AC electroosmosis and electrothermal. The white dotted line indicates the frequency at which electrothermal flow changes direction. The velocities are plotted on a log greyscale and were determined for the same position (10 µm from the edge of the electrode) and applied signals: (a) 1 Volt and (b) 10 Volts on each electrode. Note that the ratio of AC electroosmotic flow to electrothermal flow is much less at 10 Volts.

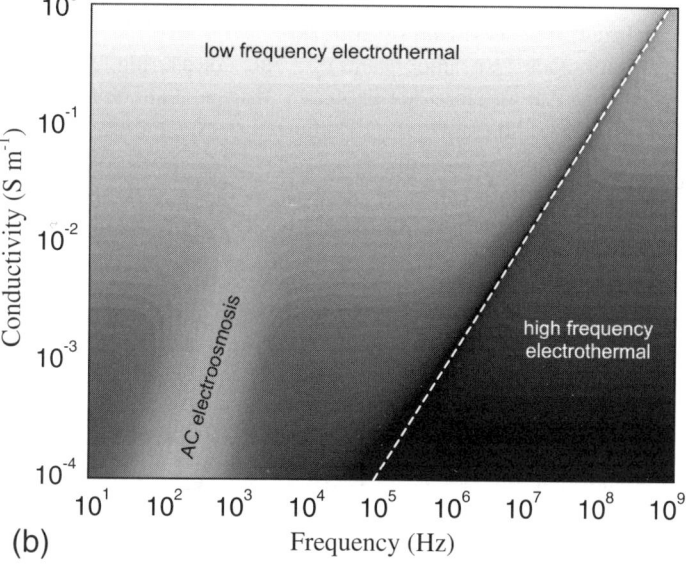

Alternatively, temperature gradients can be imposed externally using heat sources and heat sinks to control the geometry of the temperature field (for example a heater placed in contact with the substrate). In this case, the temperature gradient does not depend on the electrolyte conductivity but the flow velocity is still a function of the electric field.

Apart from the conductivity dependence of the electrothermal effect, the magnitude of the flow is also frequency dependent (as discussed in Chapter Five). The ratio of low to high frequency limiting values is approximately 10:1 and the flow changes magnitude *and* direction at a frequency of the order of the charge relaxation frequency, $\sim\sigma/\varepsilon$, of the medium.

The change in the velocity of the fluid as a function of the two parameters; frequency of the applied field and conductivity of the electrolyte, is complicated and is summarised in figure 8.2. In these plots the logarithm of the magnitude of the flow velocity is indicated by a grey scale as a function of the conductivity of the electrolyte and frequency of the field, for two applied voltages. Both figures demonstrate the regions in which strong fluid flows occur and also which regions might be "safe" from fluid motion. Figure 8.2(a), shows the fluid flow calculated for an applied voltage of 1Volt and a characteristic electrode gap of 10 μm. It clearly shows how electroosmotic flow dominates at low frequencies, and that the magnitude of the electrothermal flow is much smaller except at high conductivities. Note the diagonal dotted line corresponding to the frequency at which the net electrothermal flow is zero (approximately the charge relaxation frequency of the electrolyte). Contrast this figure with 8.2(b), which is the same calculation but with an applied potential of 10 Volts. Now Joule heating is significant and since the electrothermal flow scales with ϕ^4, this flow dominates the behaviour of the system and the relative magnitude of the AC electroosmotic flow is much less.

8.2   Experimental observation of fluid flow

The simplest approach to understanding and measuring fluid flow on microelectrodes is to use an electrode geometry such as that shown in figure 8.3. This type of electrode has several advantages: it is easy to fabricate, it is translationally symmetric and the long axis (z-axis) can be assumed to be infinite, *i.e* during calculations the z-axis can be ignored. With this electrode, it is possible to derive analytical approximations for the electric field and field related forces. In addition, numerical simulations of the system can be simplified to two dimensions. Indeed, experimentally the fluid flow is found to be two-dimensional along the majority of the length of the device (ignoring the ends of the electrodes). Away from the electrode ends, any observation of the fluid motion made in a vertical plane will fully describe the flow. An experimental setup for the observation of fluid flow has been described in Green *et al.* (2000a) and is shown in figure 8.4. This arrangement allows simultaneous observation of the fluid flow profiles in the xz-plane (horizontal) and the xy-plane (cross-section of the electrodes). Fluid flow is measured by observing the movement of small (200 to 500 nm diameter) fluorescent latex particles seeded into the fluid. Images of the particles can be

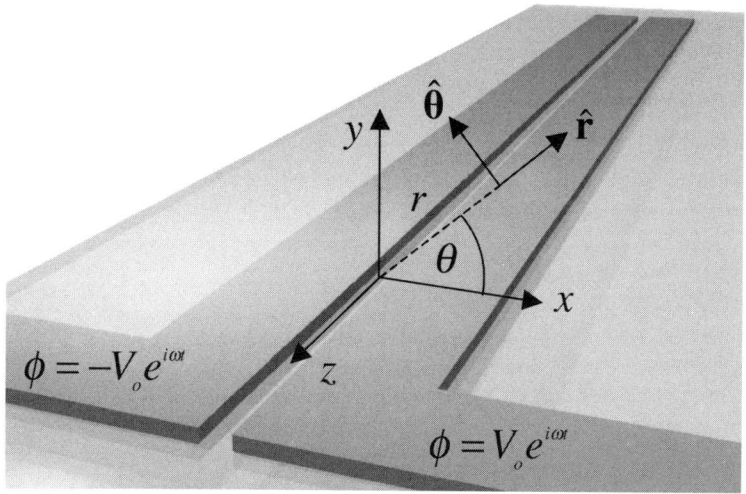

Fig. 8.3 A schematic diagram of the electrode design used to observe fluid flow, consisting of two long thin parallel electrodes. Also shown are the Cartesian co-ordinates used for reference in this chapter and the polar co-ordinates in the xy-plane. The electrodes used for experimental measurements have a gap of 25 µm and widths of 100 µm or 500 µm. Two potential signals of opposite phase and amplitude V_o are applied to the electrodes as shown.

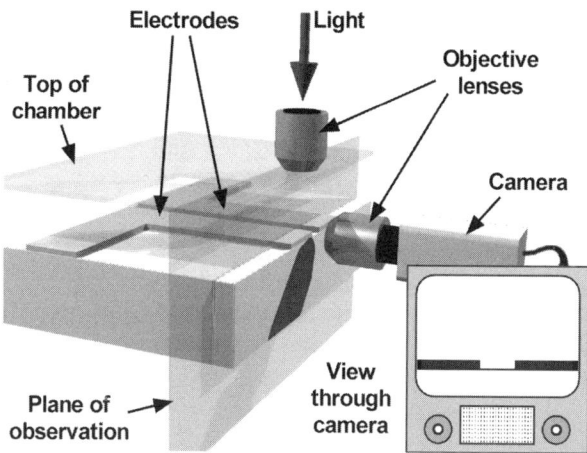

Fig. 8.4 Experimental setup for observation of fluid flow in the microelectrode structure shown in figure 8.3. The motion of tracer particles can be observed in the xz-plane through the microscope and in the xy-plane using the horizontally placed camera and objective. (Reprinted with permission from Green *et al.* 2000a, copyright IOP Publishing Ltd.)

recorded on video and analysed to produce plots of fluid velocity as a function of position in the chamber.

8.3   AC electroosmosis

When fluid moves by AC electroosmosis, particles are pushed from the electrode edge, away from the high field region onto the surface of the electrode. This effect is shown in figure 8.5, where particles can be seen moving away from the edge and

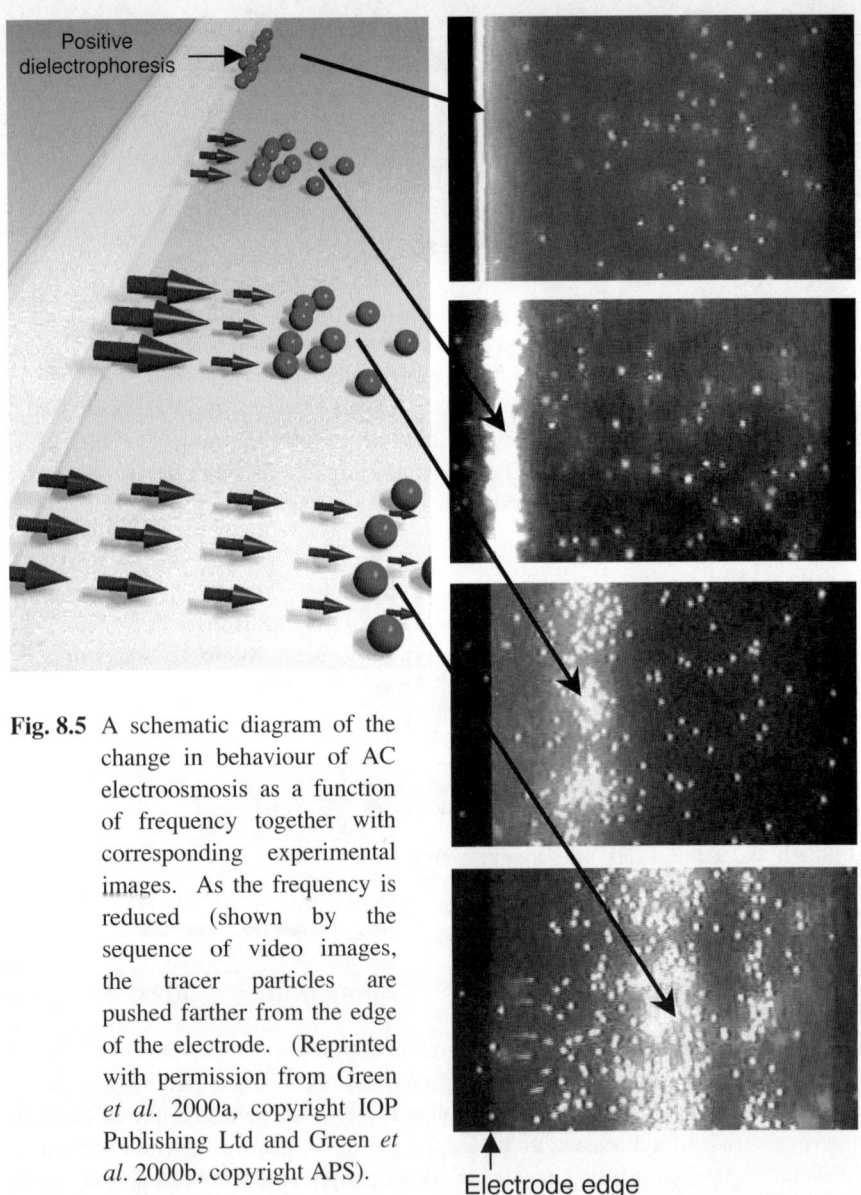

Fig. 8.5 A schematic diagram of the change in behaviour of AC electroosmosis as a function of frequency together with corresponding experimental images. As the frequency is reduced (shown by the sequence of video images, the tracer particles are pushed farther from the edge of the electrode. (Reprinted with permission from Green *et al.* 2000a, copyright IOP Publishing Ltd and Green *et al.* 2000b, copyright APS).

over the electrodes due to the action of fluid movement. When there is no induced flow, particles are initially trapped by positive DEP on the electrode edge. As the frequency of the applied field is reduced, the AC electroosmotic flow increases and the particle collection band moves further and further away from the electrode edge. Measurement of the fluid flow as a function of distance from the electrode edge shows the reason for this effect. Figure 8.6 shows the velocity of fluid flow measured as a function of distance across the electrodes for three different frequencies. The velocity is not a constant value across the electrode but is always a maximum at the edge where the field is highest. At high frequencies, the velocity decreases rapidly with distance, from a peak at the electrode edge, so that

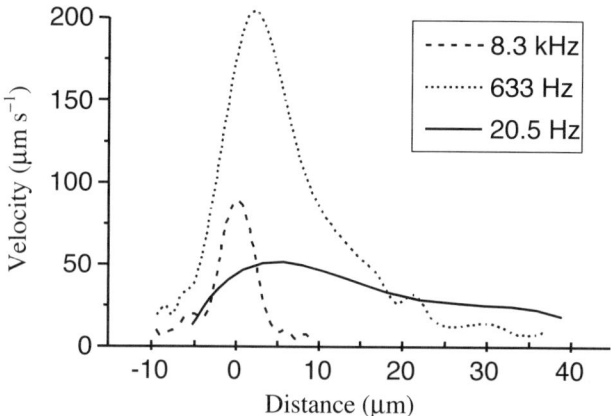

Fig. 8.6 Experimental graph of AC electroosmotic velocity as a function of distance from the edge of a 100 μm wide electrode. The graph shows the velocity determined from position vs time data for applied signals of amplitude 0.5 Volts at the three frequencies shown in the legend, in an electrolyte of conductivity 2.1 mS m^{-1}. (Reprinted with permission from Green *et al.* 2000b, copyright APS.)

in this case particles collect in a band close to the edge. At lower frequencies, although the velocity is still a maximum close to the edge, the decrease with distance is much less pronounced. There is still a significant fluid flow on top of the electrode, which pushes the particles across the surface towards the centre of the electrode, as shown in figure 8.5.

AC electroosmosis is in some respects analogous to electroosmosis in a DC field, where a bulk flow is driven by a surface effect. In closed microsystems, the fluid moves in two symmetrical continuous rolls over the two electrode edges. This flow profile is illustrated in figure 8.7(a) and (b) which shows particle trajectories (*i.e.* the fluid flow streamlines) over the electrodes in the *xy*-plane, produced by the superimposition of successive video frames. The two figures show data taken at different frequencies, 100 Hz and 1 kHz respectively, and demonstrate the change

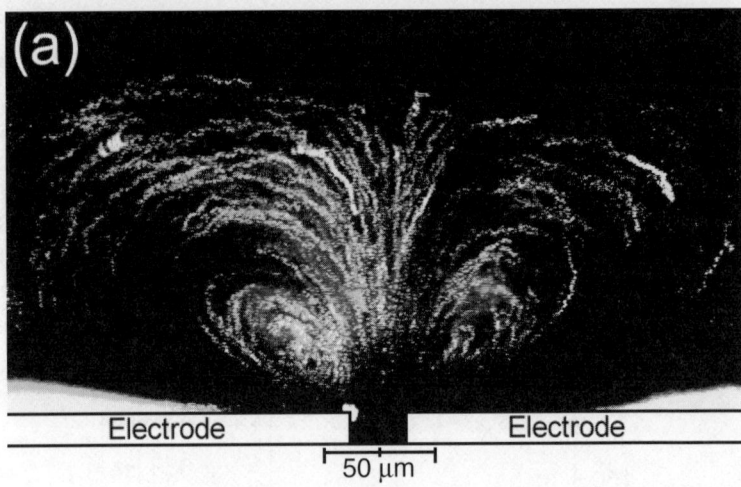

Fig. 8.7 Images of experimental fluid streamlines obtained by superimposing successive frames of a video of particle motion. The electrolyte conductivity was 2.1 mS m^{-1}. At 100 Hz (a), the fluid motion extends much farther into the bulk electrolyte than at 1 kHz (b). (Reprinted with permission from Green *et al.* 2002, copyright APS.)

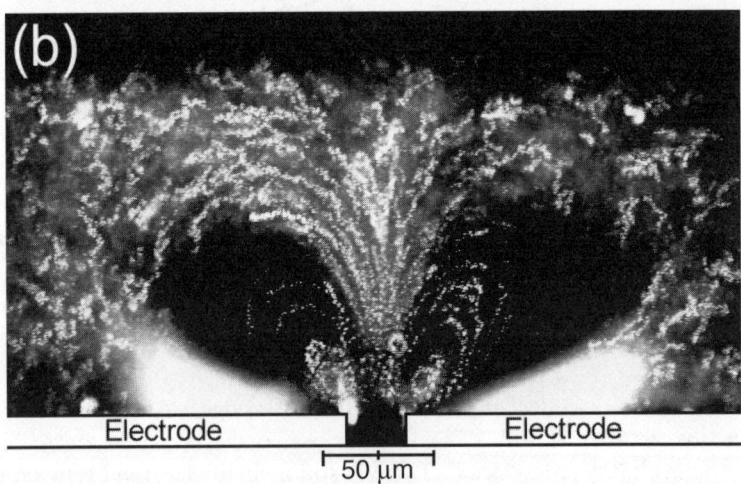

in the fluid velocity profile with frequency. At the higher frequency, figure 8.7(b), the flow is small over much of the electrode surface and high at the electrode edge where the fluid roll is restricted to the region just above the electrode edge. At the lower frequency, figure 8.7(a), the surface flow extends to greater distances and the fluid flow in the bulk extends much farther outwards, with a clear boundary at the upper surface of the chamber. The bright collections of particles at the bottom of the images, show the distance that the particles are thrown by the fluid moving across the top of the electrode, as shown in figure 8.5.

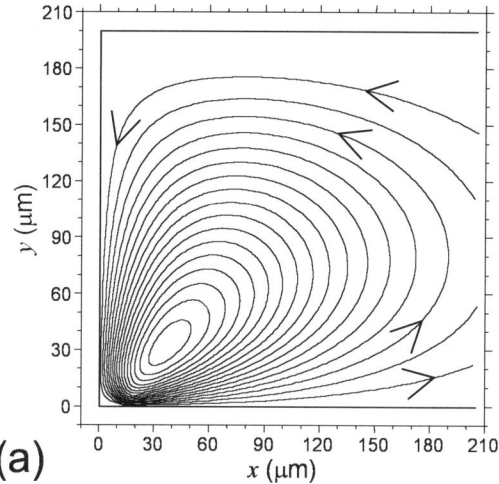

(a)

Fig. 8.8 Streamlines calculated numerically from theory for AC electroosmosis for the same conditions as the experimental images shown in figure 8.7. (Reprinted with permission from Green *et al.* 2002, copyright APS.)

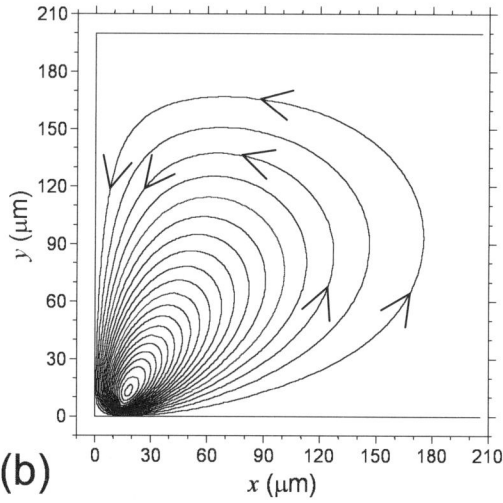

(b)

The bulk fluid flow driven by AC electroosmosis at the surface of the electrodes can be calculated by numerical simulation of the Navier-Stokes equation. The model calculates the fluid flow using the experimentally measured impedance of the double layer to simulate the electrode polarisation as a function of frequency. The numerical streamlines, matching the experimental images are shown in figure 8.8. Comparison of the corresponding figures demonstrates good agreement between experimental and numerically simulated fluid flow profiles.

8.3.1   AC electroosmosis *vs.* distance and dielectrophoresis

When plotted as a function of frequency, the magnitude of the AC electroosmotic fluid flow (measured at a given point on the electrode) has a Gaussian-shaped profile, as shown in figure 8.9. The velocity is almost zero at high and low frequencies and the Gaussian profile extends over two to three decades. The maximum velocity is close to the electrode edge, and for an applied signal of 1.25 Volts (with a 25 μm gap electrode), is of the order of 300 μm s^{-1}. A comparison of the numerically calculated AC electroosmotic fluid flow and the experimental velocity is also shown in figure 8.9. There is good agreement in the frequency dependence and on the dependence with distance from the edge. Numerically calculated values of the magnitude of the flow do not agree with experimental data and a scaling factor has been applied in the figure.

We can compare these fluid velocities with the velocity of a particle undergoing dielectrophoresis under the same experimental conditions for the electrode design shown in figure 8.3. An estimate of the typical *maximum* value of $\nabla|\mathbf{E}|^2$ is 10^{16} V^2 m^{-3}. For a 1 μm diameter particle the steady-state DEP velocity can then be calculated as ~30 μm s^{-1}. This would be 100 times smaller (300 nm s^{-1}) for a 100 nm diameter particle. Comparing these values with the data plotted in figure 8.9 shows that it is at least an order of magnitude less than the maximum AC electroosmotic velocity for the same conditions. Therefore, at the frequencies where AC electroosmosis is strong, the manipulation of sub-micrometre particles by DEP will be difficult. In addition, at these frequencies, a large component of the applied potential is dropped across the double layer (see Chapter Six) and consequently the dielectrophoretic velocity would be even smaller than calculated above. By contrast, for larger particles such as cells, which are of the order of $5 - 10$ μm in diameter, the DEP velocity can be an order of magnitude higher than the AC electroosmotic fluid velocity. However, the fluid can still produce significant particle movement if the particles are some distance from the electrode edge and the DEP force is weak.

This comparison serves to illustrate that in the 10 Hz – 10 kHz frequency range, AC electroosmotic fluid flow can be a dominant force in AC electrokinetic microsystems. Therefore the dielectrophoretic manipulation of sub-micrometre particles in this frequency regime is non-trivial, and particle motion will almost certainly be dominated by the fluid flow.

It is interesting to consider the effect of reducing the electrode dimensions on the behaviour of particles in this regime. The DEP force on a particle is proportional to the gradient of the field magnitude squared, which has units of V^2 m^{-3}. To a first approximation, reducing all characteristic distances by a factor of 10 increases the DEP force by a factor of 1000, for the same applied voltage. Dimensional analysis of the expression for the AC electroosmotic flow (equation (6.40)) shows that the flow velocity also scales as V^2 m^{-3}. Therefore, reducing the electrode dimensions or varying the applied potential does not alter the balance between AC electroosmosis and dielectrophoresis. Of course, AC electroosmotic flow is frequency dependent, so that altering the frequency of the applied field may allow the balance of forces to be reversed.

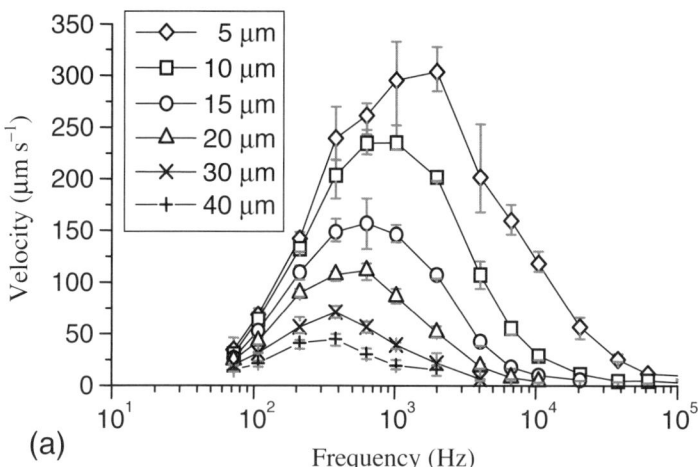

(a)

Fig. 8.9 Experimental data of AC electroosmotic fluid flow plotted as a function of frequency for several distances from the edge of the electrode. (a) electrolyte conductivity of 8.6 mS m^{-1}, applied signal of 1.25 Volts. (b) electrolyte conductivity of 2.1 mS m^{-1}, applied signal 0.25 Volts. The numerical match for the same conditions is also shown, corrected by a scaling factor of 0.24. (Reprinted with permission from Green *et al.* 2000b, copyright APS.)

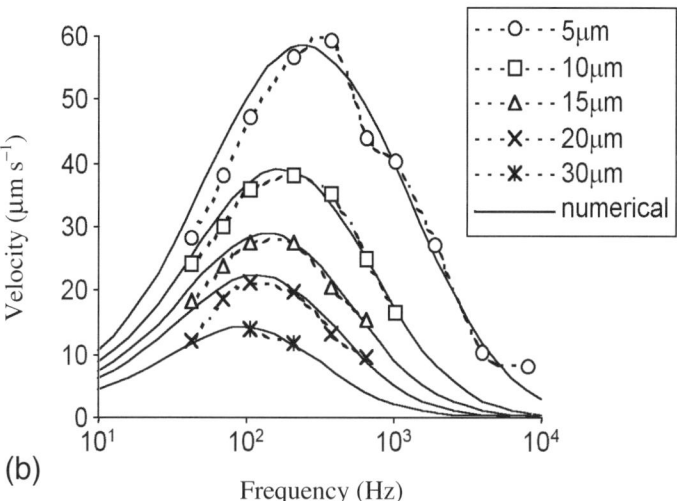

(b)

8.3.2   AC electroosmosis *vs.* conductivity

The picture discussed previously illustrates the variation of AC electroosmotic flow for a single conductivity. However, as shown by figure 8.10, the AC electroosmotic velocity varies with the frequency and the conductivity in a non-linear manner. For the same voltage and electrode geometry, at the lowest conductivity the maximum velocity is ~600 μm s^{-1}, but this falls to a maximum velocity of ~100 μm s^{-1} at the higher conductivity. The frequency at which the maximum velocity occurs is also not constant, increasing with conductivity.

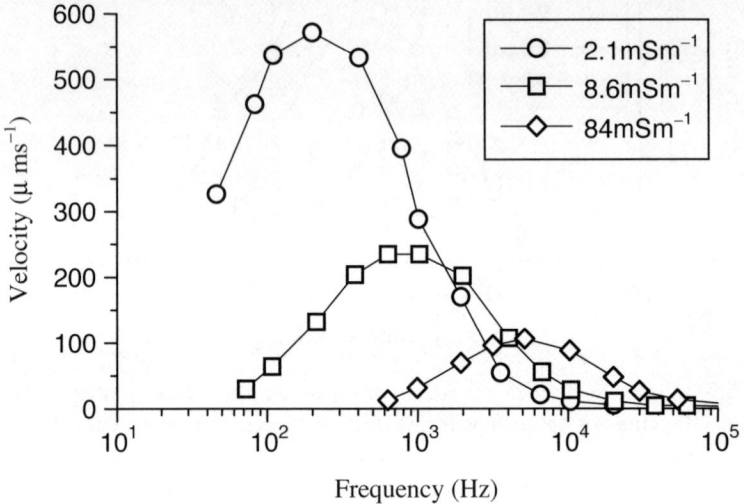

Fig. 8.10 Experimental data of AC electroosmosis at 10 μm from the edge of a 100 μm wide electrode for three different conductivities and applied potentials of 1.25 Volts. As the conductivity increases, the magnitude of the fluid flow decreases. (Reprinted with permission from Green *et al.* 2000b, copyright APS.)

This effect can, in part, be explained by including the Stern layer into the model that describes the system. Recall from Chapter Six that the fluid is driven by the action of the tangential field (at the level of the electrodes) on the induced charge in the diffuse part of the double layer. The Stern layer can be thought of as a fixed thickness capacitor, in between the electrode and the diffuse layer. The thickness of the diffuse part of the double layer is given by the Debye length (equation (6.21)) and varies with the inverse square root of the conductivity. At higher conductivities, the diffuse part of the double layer is very thin and comparable with the dimensions of the Stern layer. As a result, a large percentage of the applied potential is dropped across the Stern layer. Since the fluid velocity depends on the free charge in the diffuse layer, which in turn depends on the potential dropped across the diffuse layer, the AC electroosmotic velocity should decrease with increasing conductivity, as seen experimentally.

8.3.3   Summary

To summarise, the frequency and conductivity dependence of AC electroosmotic flow can be visualised in terms of a grey-scale map shown in figure 8.11. Three maps are shown for three particle sizes: 10 µm, 1 µm and 100 nm diameter. The region where AC electroosmosis dominates is shown in grey. The boundary delineating the two regions is defined by the point at which the fluid velocity and the DEP-induced velocity of the particle are equal. Also shown in this figure is the line indicating the frequency/conductivity at which the potential at the electrode/electrolyte interface is 50% of its maximum value. In the region to the left of this line, AC electrokinetic manipulation of particles becomes increasingly difficult since the potential across the electrolyte, and therefore the field magnitude in the bulk, reduces with decreasing frequency. In the region to the right of this line, the potential across the fluid and AC electrokinetic forces approach their maximum possible values. Note that this figure is qualitative and representative of the electrode array shown in figure 8.3. The exact positions and dependency of the lines depend on the electrode geometry being used. Since the magnitude of the AC electroosmotic flow decreases with increasing conductivity, for sub-micrometre particles and at high conductivities AC electroosmosis is not necessarily the dominant effect. In high conductivity electrolytes Joule heating becomes a much greater problem.

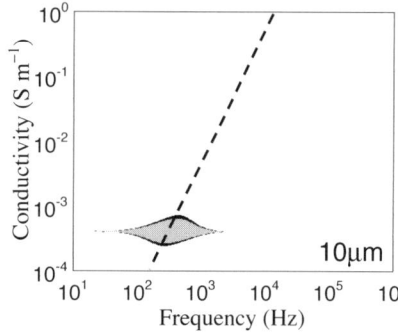

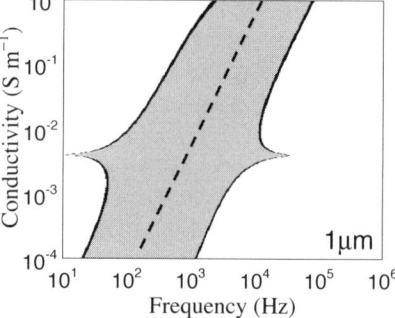

Fig. 8.11 Conductivity/frequency maps of the regions where dielectrophoresis dominates (white) as opposed to the regions where AC electroosmosis dominates (grey), for three different sizes of particle diameter. The dashed lines represent the characteristic frequency of electrode polarisation (taken from equation (6.32)).

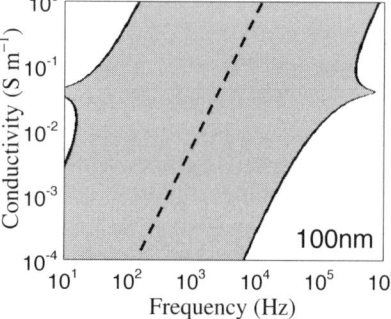

8.4   Electrothermally driven fluid flow

As the frequency of the applied field increases, the magnitude of the AC electroosmotic flow decreases significantly. However, this does not mean that fluid motion ceases. As described by figure 8.2, there is a second type of fluid flow that occurs for all frequencies but generally has a lower magnitude than AC electroosmosis. This is electrothermal fluid flow, the theoretical foundations of which were laid in Chapter Five.

In this section we describe experimental observations of electrothermal fluid flow using the same two-electrode device described in figure 8.3. We begin by deriving a simple analytical approximation for the flow velocity that can be used to make order of magnitude calculations and illustrate the practical ramifications of fluid flow in microstructures. We then review experimental observations of electrothermal fluid flow and with the aid of numerical calculations, illustrate the way in which different heat sources can radically affect the fluid flow patterns.

8.4.1   Analytical approximation to the electrode geometry

We start with an analytical expression for the electric field for two coplanar electrodes, as shown in figure 8.3. Assuming that the gap between the electrodes is much smaller than the width of the electrodes, the electric field can be calculated from the applied AC potential $\phi = V_o e^{i\omega t}$ and is (Zahn 1979)

$$\mathbf{E}(r,\theta) = \frac{2}{\pi} \frac{V_o}{r} \hat{\boldsymbol{\theta}} \tag{8.1}$$

This electric field heats the electrolyte of conductivity σ, giving the time-averaged power density dissipated in the system

$$\langle W \rangle = \frac{1}{2}(\sigma \mathbf{E} \cdot \mathbf{E}^*) = \frac{2}{\pi^2} \frac{\sigma V_o^2}{r^2} \tag{8.2}$$

The temperature field can be calculated using the diffusion equation (equation (5.29)) together with equation (8.2) for the temperature source. Following the argument presented in Ramos *et al.* (1998) the temperature gradient is

$$\nabla T = \frac{1}{\pi k} \frac{\sigma V_o^2}{r}\left(1 - \frac{2\theta}{\pi}\right)\hat{\boldsymbol{\theta}} \tag{8.3}$$

Substituting this expression into equation (5.27) gives the electrothermal body force on the fluid:

$$\langle f_e \rangle = \frac{2}{\pi^3 k} \frac{\varepsilon \sigma V_o^4}{r^3} \Pi\left(1 - \frac{2\theta}{\pi}\right)\hat{\boldsymbol{\theta}} \tag{8.4}$$

where k is the thermal conductivity of the fluid. In this equation, the factor Π is given by

$$\Pi = \left[\frac{\alpha - \beta}{1 + (\omega \tau_q)^2} - \frac{\alpha}{2} \right] \tag{8.5}$$

where $\alpha = (1/\varepsilon)(\partial \varepsilon / \partial T)$, $\beta = (1/\sigma)(\partial \sigma / \partial T)$ and τ_q is the charge relaxation time (ε / σ).

Equation (8.4) shows that the force has both an angular dependence and a distance dependence; it increases to a maximum at the electrode surface, *i.e.* as θ tends to 0 and π and also as r tends to zero. Note also the dependence on voltage to the power 4, which comes from the product of the electrical power, the $\mathbf{E} \cdot \mathbf{E}^*$ term in equation (8.2) together with the $\mathbf{E} \cdot \mathbf{E}^*$ and $|\mathbf{E}|^2$ terms in the body force expression. The force also scales linearly with the conductivity (through the power dissipation term, equation (8.2)).

The frequency dependence of the fluid flow is described by the factor Π, equation (8.5), which tends to a value of $(\alpha/2 - \beta)$ at low frequencies and $-\alpha/2$ at high frequencies. For values typical of aqueous electrolytes these limits are -0.022 K^{-1} and 0.002 K^{-1} respectively. The frequency dependence of fluid flow can be illustrated with reference to figure 8.12. In figure 8.12(a), the factor Π has been calculated as a function of frequency and conductivity. The figure shows that there is a transition in the sign of Π at a frequency slightly higher than the charge relaxation frequency. Π is negative at high frequencies, indicating that the direction of the flow changes. Figure 8.12(b) shows the magnitude of the fluid flow calculated using the above equations. It is significant at high conductivities and is greatest at low frequencies where it is dominated by the induced charge or the Coulomb force. At high frequencies, the dielectric force dominates and the fluid flow is reduced. The ratio of the high to low frequency fluid velocity is approximately 10:1 and the magnitude of the flow is zero at a frequency that is approximately the charge relaxation frequency of the system.

The flow pattern is illustrated in figure 8.13, which shows the streamlines calculated from the analytical approximation (Ramos *et al.* 1998). The streamlines show how the fluid moves from the bulk down to the surface in the gap between the two electrodes and then out over the electrode. At low frequencies the direction of the flow is as shown in the figure and it is the obverse at high frequencies. An order of magnitude calculation for the typical fluid flow can be made from

$$|\mathbf{u}| \sim |f_{\mathrm{E}}| \frac{l_o^2}{\eta} \tag{8.6}$$

where l_o is a distance characterising the microelectrode structure *e.g.* $10 - 100 \ \mu\text{m}$ (Ramos *et al.* 1998).

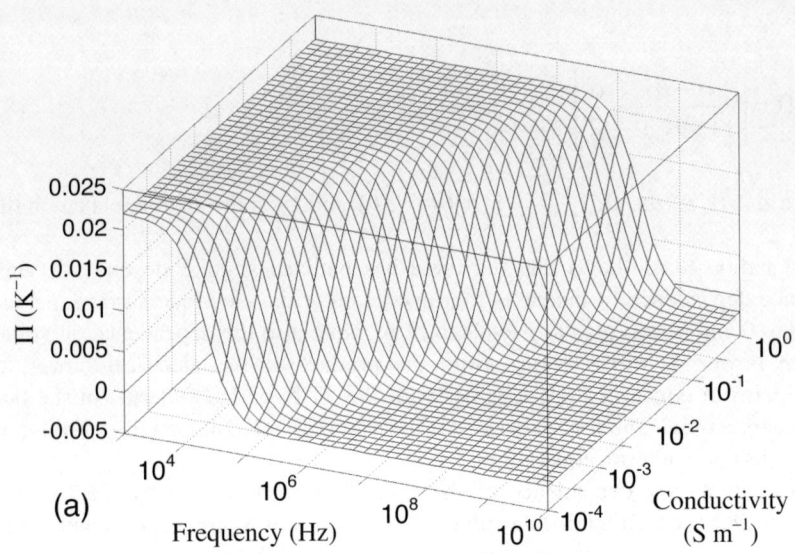

(a)

Fig. 8.12 A plot of the factor Π (a) and the electrothermal fluid velocity (b) *vs* applied signal frequency and electrolyte conductivity. The amplitude of the applied signal was 1 Volt and the velocity was calculated for a position 10 μm from the edge of the electrode.

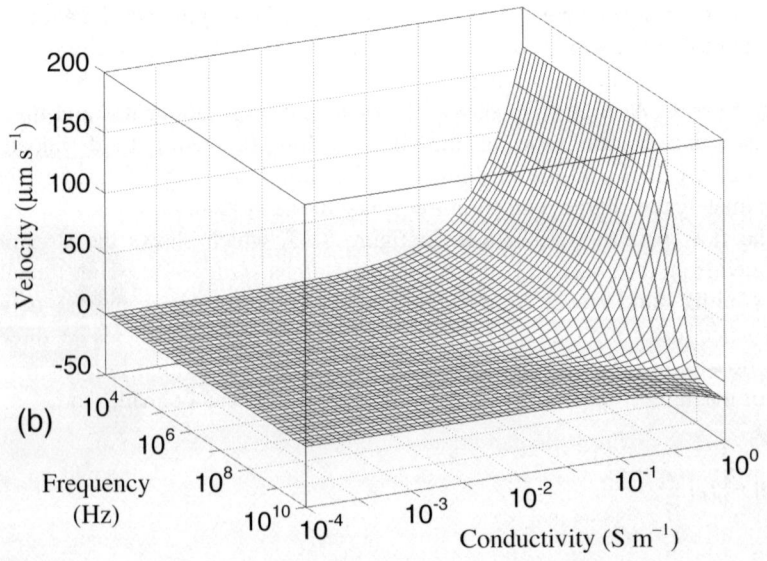

(b)

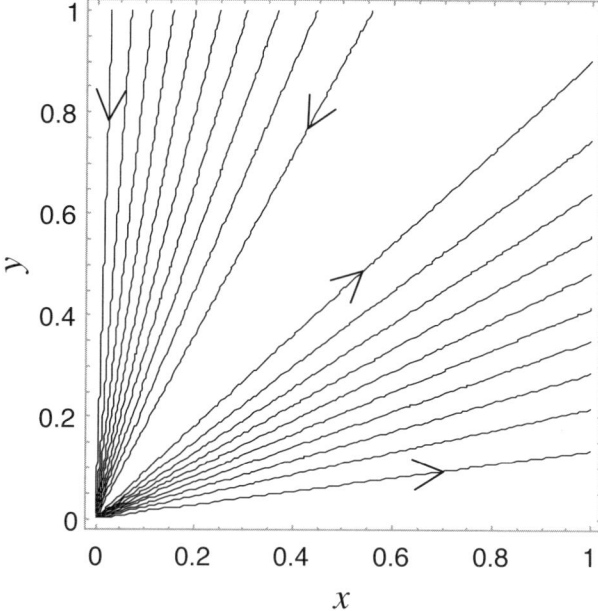

Fig. 8.13 Streamlines of the electrothermal fluid flow calculated from the analytical approximation. The fluid moves down to the gap at $x = 0$ and out over the electrode surface.

In practical terms, the magnitude of the fluid flow generated by Joule heating is small at low conductivities and low applied potentials, as shown by figure 8.12. As the electrolyte conductivity increases, the magnitude of the flow increases to the point where it can be stronger than dielectrophoresis. Figure 8.14 shows a conductivity/frequency map of the regions where electrothermal flow dominates (at higher conductivities) or dielectrophoresis dominates. This was calculated for two particle sizes, 1 μm (figure 8.14(a) and (c)) and 100 nm (figure 8.14(b)) for an applied potential of 1 Volt and 10 Volt, using equations (8.4) and (8.6). For particles bigger than 1 μm, the DEP velocity is always greater than electrothermal fluid flow at this potential. However, as can be seen from equation (8.4), the force on the fluid scales both with conductivity and the fourth power of the applied potential but the DEP velocity scales only with ϕ^2. Therefore, if the applied potential is increased, the fluid flow becomes more important as shown in figure 8.14(c).

Although the effects of Joule heating are relatively small, if alternative sources of heat are present then the fluid flow can be much greater. For example, the fluid can be heated by the source of illumination (Green *et al.* 2000a) or by a combination of heat sources and sinks. Assuming, for example, that the source of

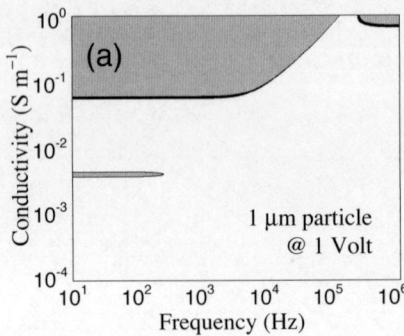

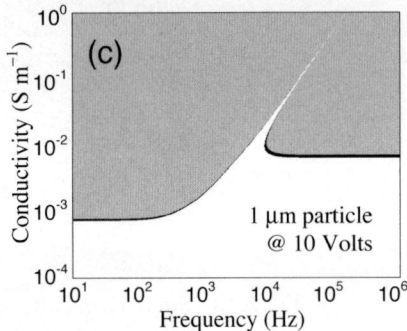

Fig. 8.14   Conductivity/frequency maps of the regions where dielectrophoresis dominates (white) as opposed to the regions where electrothermal fluid flow dominates (grey) for the conditions shown. The particle and fluid velocities were calculated for a position 10 μm in from the electrode edge.

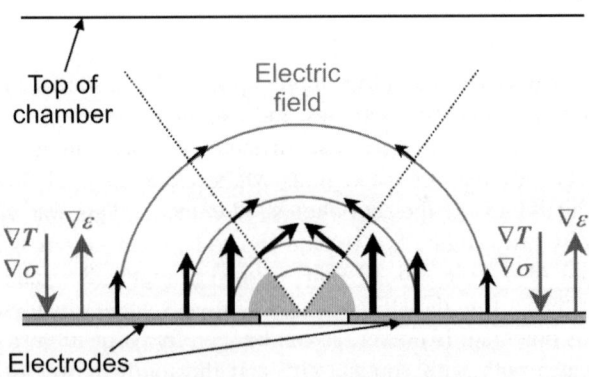

Fig. 8.15   A schematic diagram of the temperature gradient generated by a heated substrate beneath the electrodes and the resulting gradients in conductivity and permittivity. Also shown are the electric field lines and an approximate representation of the electrothermal force at low frequencies (bold arrows). (Reprinted with permission from Green *et al.* 2000a, copyright IOP Publishing Ltd.)

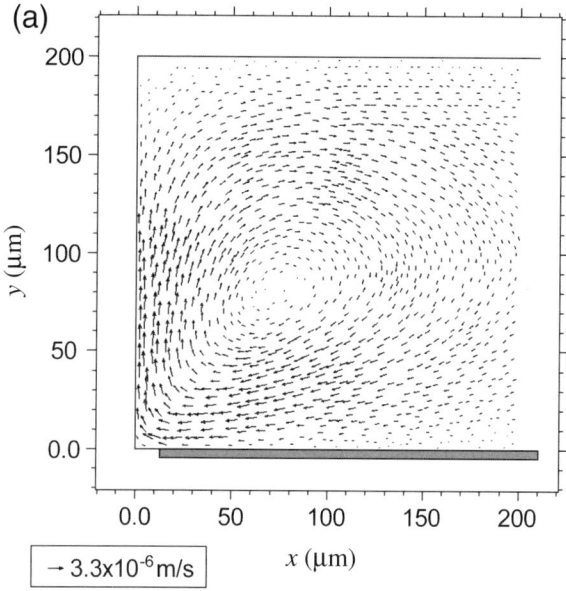

Fig. 8.16 Numerical simulations of the electrothermal fluid flow for one half
of the two electrode system and two different heat sources: (a) a
heated substrate and (b) Joule heating. Both solutions are for low
frequencies. The electrolyte conductivity was 2.1 mS m^{-1} and the
maximum velocity is shown in the boxes for each case. (Reprinted
from Green *et al.* 2001, with permission from Elsevier Science.)

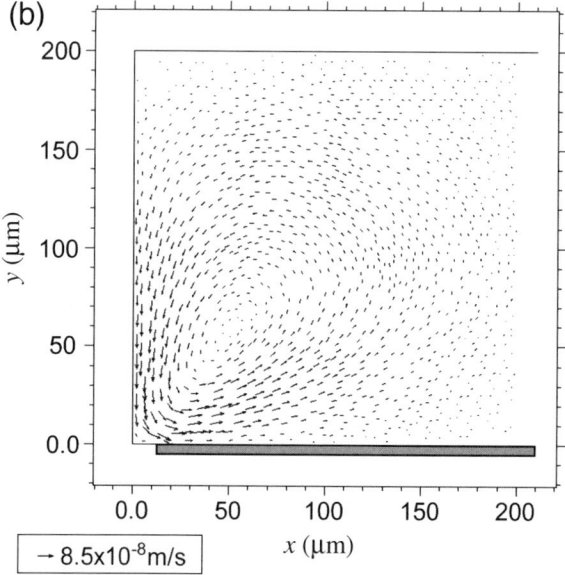

heat is external to the system and at the base (for example a heated substrate), a temperature gradient will be established, as shown in figure 8.15. Here the heat radiates upwards through the electrolyte to produce gradients in temperature, conductivity and permittivity as shown in the figure. When a potential is applied to the electrodes, fluid is driven to produce the flow pattern shown in figure 8.16(a). This is a numerical simulation calculated at low frequencies for the two electrode system illustrated in figure 8.3 in a closed chamber 200 μm high, with a temperature differential of 0.7 K (*i.e.* a temperature gradient of 3.5×10^3 K m^{-1}). Contrast this figure with figure 8.16(b) which shows the numerically calculated flow pattern for the same applied signal, but where the temperature gradients is caused by Joule heating (Green *et al.* 2001). Two things are immediately apparent. First, the direction of the fluid flow is reversed and second, the magnitude of the flow from Joule heating is approximately 40 times smaller, showing that even a moderate external temperature gradient can produce significant fluid flow. Note also that the centre of the fluid roll is in a different location.

When the temperature gradient is created by an external source, an analytical approximation for the force on the fluid can be derived. For low frequencies, ignoring the permittivity gradient, and assuming the simple circular electric field (equation (8.1)) this is

$$\langle f_e \rangle = \frac{2}{\pi^2} \frac{\varepsilon V_o^2}{r^2} \left(\frac{1}{\sigma} \frac{d\sigma}{dT} \right) |\nabla T| \cos\theta \, \hat{} \qquad (8.7)$$

This expression can be used to estimate flow magnitudes for arbitrary applied temperature gradients and potentials. Note that in this case, the velocity is proportional to the *square* of the applied potential compared with ϕ^4 for Joule heating.

8.4.2 Observations of electrothermal fluid flow

Figure 8.17 shows an experimental image of electrothermal fluid flow above a two-electrode array, produced by superimposing successive video frames. As can be seen there are two rolls in the fluid, one over each electrode, with the centre of the rolls approximately midway between the electrodes and the top of the chamber. Contrast this with AC electroosmosis (figure 8.7), which is surface, rather than bulk-driven and has rolls with centres close to the electrodes. The heating mechanism responsible for the fluid flow in this case was heating of the electrode by the illumination source (Green *et al.* 2001).

8.4.3 Summary

To summarise electrothermal fluid flow we refer back to figure 8.2. Electrothermal fluid flow, as we have seen, is produced in the bulk fluid in the region near the electrodes and has noticeable conductivity and voltage dependence. The flow velocity changes direction at a frequency close to the charge relaxation frequency ($\sim\sigma/\varepsilon$) of the medium. There is a linear dependence on conductivity,

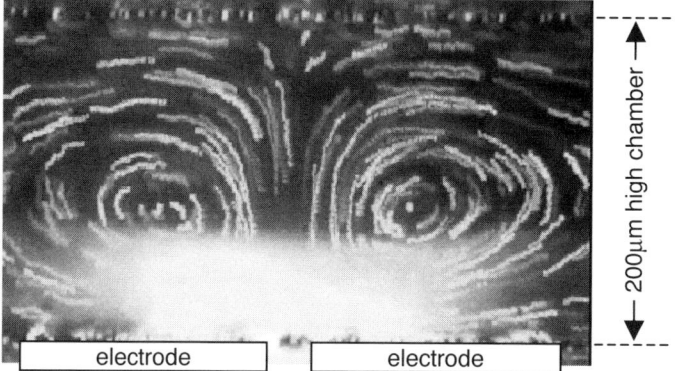

Fig. 8.17   An image of experimental electrothermally driven fluid flow streamlines obtained by superimposing successive frames of a video sequence.

with electrothermal fluid flow becoming the dominant effect at high conductivities. There is also a substantial voltage dependence: ϕ^4 for Joule heating and ϕ^2 for an external heat source. Electrothermal fluid can therefore be a significant problem in AC electrokinetic microsystems, particularly at high conductivities.

8.5   Technological applications of electrohydrodynamics in microsystems

We conclude this chapter by reviewing the application of electrohydrodynamics (EHD) in the development of electric field-driven micropumps. By definition, such pumps would be solid-state with no moving parts and operate on the principle of the interaction of an AC field with the fluid. There are two categories of pump: those driven by AC electroosmosis and those driven by electrothermal principles.

8.5.1   4-phase travelling wave electrothermal pumps

The concept of exploiting the electrothermal force to pump liquid has been around for nearly 25 years. The principle of such a pump was first demonstrated by Melcher (1966) and Melcher and Firebaugh (1967) who showed that a travelling electric field could move induced charges in an insulating liquid to give unidirectional fluid motion. Maximum fluid flow occurred at a frequency corresponding to the charge relaxation time of the system (Melcher and Firebaugh 1967; Crowley 1980). The underlying mechanism of a travelling wave EHD pump is rooted in the electrothermal body force on the fluid. An array of electrodes is energised to produce a travelling electric field using, for example, a four-phase AC voltage as described in Chapter Three for twDEP. Because of Joule heating the electrodes provide a heat source which creates a gradient in temperature and therefore in conductivity and permittivity. Alternatively larger temperature gradients can be created using an external temperature source and/or sink. A

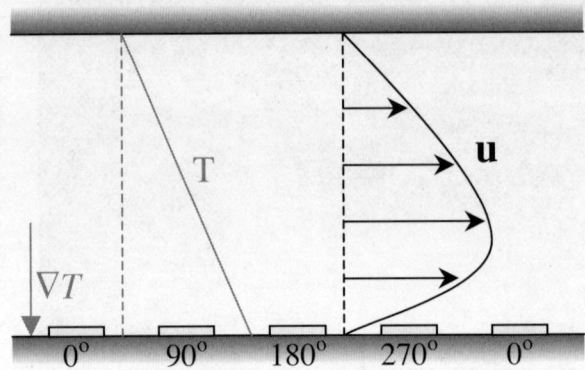

Fig. 8.18　Schematic diagram of a four phase travelling wave electrohydrodynamic pump. A temperature gradient is applied across the chamber by a heater. Four potential signals, successively 90° phase shifted, are applied consecutively to the bar electrodes on the bottom substrate to create the travelling electric field.

schematic diagram of a travelling wave EHD pump is shown in figure 8.18. In the presence of an electric field, the inhomogeneities in the fluid lead to the creation of a space charge in the bulk. At low frequencies the induced charge has sufficient time to follow the field vector so that the net flow along the electrode array is zero. However, local flow due to electrothermal effects can still occur above the electrodes, but there is no net fluid transport. At high frequencies only the dielectric force exists and there is no net transport. At frequencies around the charge relaxation time of the medium, there is a phase lag between the field and the induced charge, so that, as the field moves, the charge experiences a force and the liquid moves along the electrode array. Maximum movement occurs when the period of the electric field corresponds to the charge relaxation time of the medium. Müller *et al.* (1993) conducted a thorough investigation of the properties of 4-phase travelling wave pumps with conducting media. It was shown that using an electrode of 35 μm width and gap, the maximum average fluid velocity was 100 μm s^{-1} (generated with an applied voltage of 35 Volts and a 100 μm channel height at a conductivity of 4 mS m^{-1}). The velocity profile was Gaussian in shape with a maximum occurring at the charge relaxation time. Theoretically, it is predicted that the fluid flow profile is maximum at some small distance above the electrodes and zero at both the electrode surface and top surface of the device. Contrast this profile with Poiseuille flow in a micro channel, where the maximum velocity occurs in the channel centre. Maximum fluid velocity occurs at the charge relaxation time of the electrolyte, and in order to produce sufficient temperature gradients from Joule heating, relatively high conductivities and voltages are

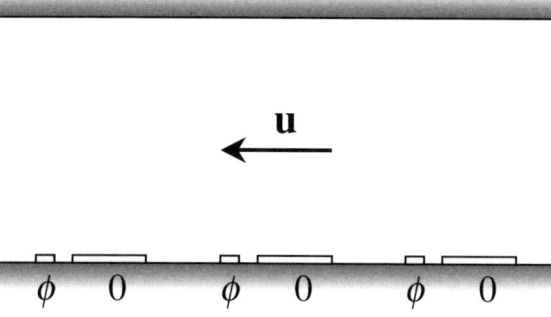

Fig. 8.19 Schematic diagram of the asymmetric electrodes used for AC electroosmotic pumping. A potential is applied to every second electrode, with the others earthed and the fluid moves in the direction shown.

required. For example, with a conductivity of 10 mS m^{-1}, maximum flow occurs at 2.3 MHz.

Preliminary observations of the low-frequency behaviour of fluids in travelling electric fields indicate that at low frequencies an additional fluid flow occurs. The magnitude of this flow is much higher than the electrothermal flow, and is probably due to AC electroosmotic effects (Cui *et al.* 2001).

8.5.2 Asymmetric electrode arrays for pumping

Travelling wave micropumps are difficult to fabricate because each electrode has to be independently connected in sequence to the correct phase of the supply. This means that multilayer fabrication methods must be used (see Koch *et al.* 2000, Morgan *et al.* 1997). It would be much simpler if single level lithography could be used to fabricate devices so that only one or two AC signals are required. Interestingly, fluids can be pumped using such arrays, provided the electrode geometry is *asymmetric*. Ajdari (2000) first suggested that large-scale global motion of fluid could be generated provided the fluid is in a locally asymmetric environment. Brown *et al.* (2000) showed how an asymmetric electrode array, driven by a single-phase voltage could be used to pump liquids unidirectionally using the principle of AC electroosmosis. This design of asymmetric electrode structure is shown in figure 8.19. It consists of a small electrode separated from a larger electrode by a small gap. Pairs of electrodes are separated by a large repeat distance and alternate electrodes are connected to an AC signal, as shown in the figure. Experiments showed that unidirectional fluid motion occurs in the direction shown in the figure. Experimental velocity *vs* frequency plots were analysed using a previously developed model (Ramos *et al.* 1999) and good correlation was found between theory and experiment. Velocities approaching 100 μm s^{-1} were measured using voltages of 1.25 Volts (*rms*). These low voltage pumps have great potential for integration into microdevices since not only do they operate at very low voltages compared with electrothermal pumps, but they are simple to make and the velocity of the fluid is comparable.

8.6   Summary

In this chapter we have addressed the subject of AC electric field-induced fluid flow in microsystems. The two main types of fluid flow, AC electroosmosis and electrothermal, are well-defined functions of experimental parameters which we have discussed in this chapter. For particle electrokinetics, the fluid flow is an important parameter, and should be included in experimental design and analysis of AC electrokinetic manipulation and separation systems. Through careful and innovative design of new electrode geometries, AC electric fields have great potential as the basis of a new generation of solid-state micropumps.

8.7   References

Ajdari A.   *Pumping liquids using asymmetric electrode arrays*   Phys. Rev. E   **61** R45-R48   (2000).

Bart S.F., Tavrov L.S., Mehregany M. and Lang J.H.   *Microfabricated electrohydrodynamic pumps*   Sensors and Actuators   **A21-23**   193-197   (1990).

Brown A.B.D., Smith C.G and Rennie A.R.   *Pumping of water with ac electric fields applied to asymmetric pairs of microelectrodes*   Phys. Rev. E.   **63**   art. 016305   (2000).

Crowley J.M.   *The efficiency of electrohydrodynamic pumps in the diffraction mode*   J. Electrostatics   **8**   171-181   (1980).

Cui L., Holmes D. and Morgan H.   *The dielectrophoretic levitation and separation of latex beads in microchips*   Electrophoresis   **22**   3893-3901   (2001).

Green N.G., Ramos A., González A., Castellanos A. and Morgan H.   *Electric field induced fluid flow on microelectrodes: the effect of illumination*   J. Phys. D: Appl. Phys.   **33**   L13-17   (2000a).

Green N.G., Ramos A., Gonzalez A., Morgan H. and Castellanos A.   *Fluid flow induced by non-uniform AC electric fields in electrolytic solutions on micro-electrodes. Part I: Experimental measurements*   Phys. Rev. E   **61**   4011-4018   (2000b).

Green N.G., Ramos A., González A., Castellanos A. and Morgan H.   *Electrothermally induced fluid flow on microelectrodes*   J. Electrostatics   **53**   71-87   (2001).

Green N.G., Ramos A., Gonzalez A., Morgan H. and Castellanos A.   *Fluid flow induced by non-uniform AC electric fields in electrolytes on micro-electrodes. Part III: Observation of streamlines and numerical simulation*   Phys. Rev. E   (submitted) (2002).

Koch M., Evans A. and Brunnschweiler A.   *Microfluidic Technology and Applications*   Research Studies Press, Herts, England   (2000).

Melcher J.R.   *Traveling wave induced electroconvection*   Phys. Fluids   **9**   1548-1555 (1966).

Melcher J.R. and Firebaugh M.S.   *Traveling-wave bulk electroconvection induced across a temperature gradient*   Phys. Fluids   **10**   1178-1185   (1967).

Morgan H., Green N.G., Hughes M.P., Monaghan W. and Tan T.C. *Large area travelling-wave dielectrophoresis particle separator* J. Micromech. Microeng. **7** 65-70 (1997).

Müller T., Arnold W.M., Schnelle T., Hagedorn R., Fuhr G. and Zimmermann U., *A travelling-wave micropump for aqueous solutions: comparisons of 1 g and μg results* Electrophoresis **14** 764-772 (1993).

Ramos A., Morgan H., Green N.G. and Castellanos A. *AC Electrokinetics: A review of forces in microelectrode structures* J. Phys. D: Appl. Phys. **31** 2338-2353 (1998).

Ramos A., Morgan H., Green N.G. and Castellanos A. *AC electric-field-induced fluid flow in microelectrodes* J. Coll. Int. Sci. **217** 420-422 (1999).

Richter A., Plettner A., Hofmann K.A. and Sandmaier H. *A micromachined electrohydrodynamic pump* Sensors and Actuators **A29** 159-168 (1991).

Zahn M. *Electromagnetic Field Theory: A Problem Solving Approach* Wiley, New York (1979).

Chapter Nine

Colloidal particle dynamics

In this chapter we will examine the forces and mechanisms involved in controlling the dynamics of sub-micrometre particles. We begin with a brief introduction to the effect that a deterministic force (gravity) has on a particle. We then discuss the effect of Brownian motion, the random force that arises from the thermal energy of the system. All particles experience this to a degree, but for particles smaller than a micrometre, Brownian motion can cause significant particle displacements and velocities compared with typical deterministic forces, such as DEP or gravity.

The random or stochastic force caused by Brownian motion can be significant for very small particles when compared with other deterministic forces. Therefore, in order to produce controlled movement of particles smaller than a micrometre, the applied force must be of sufficient magnitude so that the random force is of secondary importance. In his book, Pohl (1978) stated that the DEP-induced deterministic movement of sub-micrometre particles could not be achieved at realisable electric field strengths. This assumption was based on the argument that the force required to move sub-micrometre particles had to overcome a diffusion barrier. However, we now know that colloidal particles in the range 5 nm to 1 μm can be manipulated by dielectrophoresis (*e.g.* Washizu 1994; Müller *et al.* 1995; 1996; Fuhr *et al.* 1995; Schnelle *et al.* 1996; Morishima 1996; Green and Morgan 1997; Morgan and Green 1997; Morgan *et al.* 1999; Green *et al.* 2000) provided that the required high electric field gradients can be generated, usually with microelectrodes.

We begin by examining the two natural forces that exist in microsystems, the deterministic force gravity and the stochastic force which produces Brownian motion.

9.1 Gravity and sedimentation

When a particle sits in a fluid medium, it displaces an amount of fluid equal to its volume. If we define the mass density of the particle as $\rho_{m,p}$ and the suspending medium as $\rho_{m,m}$, then the effective mass of a sphere (of radius a) is equal to the volume (v) times the difference in mass densities: $\Delta m = (4/3)\pi a^3(\rho_{m,p} - \rho_{m,m})$. In a gravitational field, the sphere experiences a force given by $\mathbf{F}_{Bouyancy} = \Delta m\mathbf{g}$, where $\mathbf{g}$ is the acceleration due to gravity (9.81 m s^{-2}). This force, which is downward if the density of the particle is greater than the medium, is counteracted by the Stoke's drag acting on the particle. For a sphere this is $-6\pi\eta a\mathbf{v}$, where $\mathbf{v}$ is the particle velocity. The sedimentation velocity of the particle is then

$$|\mathbf{v}| = \frac{\Delta m|\mathbf{g}|}{6\pi\eta a} = \frac{2}{9}\frac{a^2(\rho_{m,p} - \rho_{m,m})|\mathbf{g}|}{\eta} \tag{9.1}$$

We see from this expression that the sedimentation velocity of a sphere depends on its surface area. As an example, for a 500 nm latex sphere of mass density 1050 kg m^{-3}, the sedimentation velocity is 7 nm s^{-1}, *i.e.* the particle moves a distance of the order of its diameter in a minute. This sedimentation velocity can be compared to that of a cell, which sediments at a rate of between 1 and 10 μm s^{-1} in water. Although the gravitational forces acting on sub-micron particles are small, they are not insignificant and cannot always be ignored.

In section 9.4.1, we shall see that a suspension of sub-micrometre particles does not sediment to the bottom of a container. This is because, when we have more than one particle, *diffusion* must be considered. Diffusion acts against gravity so that not all the particles reach the bottom, and a gradient in particle concentration is established.

9.2   Brownian motion

Particles in solution experience a random force due to the thermal energy of the system, causing them to move in a random manner. The physical origin of this force is the collisions between the vibrating molecules of the solution and the much larger particles. Robert Brown first observed this random motion in 1827, studying the movement of pollen in liquid. Perrin (1909) carried out a detailed analysis of particle trajectories. Einstein published the theory describing Brownian motion in 1905 (Einstein 1905; 1956), giving the distribution in probability of finding a single particle at any point after a given time. The theory assumes that the impacts of the molecules are statistically independent and extremely frequent, and also that the molecules are much smaller than the particle, allowing the fluid to be treated as a continuous medium.

Brownian motion is a stochastic process, *i.e.* the random movement of the particle does not depend on past history. Assuming that the particle starts from a position $\mathbf{x_0}$ at time $t = 0$, and at some time later has moved to position $\mathbf{x}$. Then, if the displacement of the particle is averaged over a large number of steps, the average displacement is zero, *i.e.* $\langle \mathbf{x} - \mathbf{x_0} \rangle = 0$. However, the *root mean square* (rms) of the displacement, Δ, is not. This is

$$\Delta = \langle |\mathbf{x} - \mathbf{x_0}|^2 \rangle^{\frac{1}{2}} \tag{9.2}$$

Einstein gave the value of Δ as $\sqrt{2Dt}$ per degree of freedom, so that the *rms* displacement is $\sqrt{6Dt}$ for movement in three dimensions. In these expressions, D is the diffusion constant for the *particle*, given by

$$D = \frac{kT}{f} \tag{9.3}$$

where f is the friction factor. The probability distribution of the position of the particle after time t has a Gaussian profile.

To put this into context, let us calculate how long it would take for an oxygen molecule to diffuse into a bottle of wine. The diffusion constant of oxygen is approximately 2×10^{-9} m^2 s^{-1}, so that it diffuses a distance of the order of 1 μm in a time given by $t \approx x^2/2D = 0.25$ ms. In one second the particle will have moved 60 μm. In the absence of any stirring or convection, we would have to wait four months before the oxygen molecule has diffused to the bottom of the bottle of wine after opening it!

The force which gives rise to Brownian motion can also be represented by a Gaussian distribution with zero mean. The particles have a thermal energy of $\frac{1}{2}kT$ per degree of freedom, giving $\frac{3}{2}kT$ in three dimensions. Assuming that the thermal energy is translated entirely into kinetic energy, the *rms* velocity of the particle is then

$$\left\langle |\mathbf{v}|^2 \right\rangle^{\frac{1}{2}} = \sqrt{\frac{3kT}{m}} \tag{9.4}$$

in three dimensions.

To illustrate the probabilistic nature of the process, imagine the particle moving in one dimension and taking a small step of length l randomly in either direction every time interval δt seconds. The particle starts from $x = 0$ at time $t = 0$ and has taken a very large number of steps so that the probability of finding a particle between distance x and dx, at some time t is given by the Gaussian probability distribution

$$\boxed{P(x)dx = \frac{1}{\sqrt{4\pi Dt}} \exp\left(-\frac{x^2}{4Dt}\right) dx} \tag{9.5}$$

The variance of this distribution is $2Dt$ and the standard deviation is $\sqrt{2Dt}$, which means that 66.7% of the particles will be found within a distance $\sqrt{2Dt}$ from the centre.

To illustrate the consequences of diffusion and random walks, consider how long it would take a small molecule, such as an odour molecule to travel across a room. If the molecule has a molecular weight of 100, and has a mass of 1.7×10^{-22} g, then from equation (9.4), the *rms* velocity can be calculated as 270 m s^{-1}! The molecule should cross the room in a hundredth of a second. However, this molecule is not in a vacuum; it sits in a soup made up of many billions of other molecules. The molecule continuously bumps into other molecules and does not move in a straight line but wanders around in a random way. We can never be sure of finding it at the other side of the room; to be absolutely certain, we would have to wait an infinite amount of time. However, we can assign a high probability to the likelihood of finding it. The diffusion constant of this molecule can be estimated to be approximately $5 \times 10^{-10} \, \text{m}^2\text{s}^{-1}$. In order to be fairly certain of finding it we can say that it must be within one standard deviation of the Gaussian probability distribution, then the time needed is $x^2 / 6D \approx 10^{11} \text{s} = 3000 \, \text{years}$! Clearly diffusion is not responsible for moving molecules around our environment.

The manipulation of sub-micron particles by DEP requires a force sufficient to overcome Brownian motion, but how large does this need to be to produce deterministic movement? The dielectrophoretic velocity (or force) experienced by particles greater than 1 μm can be measured relatively easily and Brownian motion is of little consequence. For smaller particles, however, Brownian motion can dominate particle dynamics and a probabilistic method of calculating forces is often preferable. In the next section we shall derive an expression that can be used to estimate the force required to produce "observable" movement of a sub-micron particle against the randomising Brownian motion.

9.2.1   An observable deterministic force

Consider the simplest system, a single particle suspended in a volume sufficiently large that the boundary has no effect on the particle over the time interval of the experiment. At time $t = 0$ the particle is at point $\mathbf{x} = 0$. In the absence of any deterministic force the particle experiences only Brownian motion. After a time interval δt, the position of the particle is represented by a normal distribution with standard deviation Δ, as shown in figure 9.1(a). After the time interval δt there is a 67% probability of finding the particle within a distance Δ of $\mathbf{x} = 0$. Similarly there is a 95% probability of finding the particle within 2Δ and so on.

Now the particle is subjected to an arbitrary uniform force $\mathbf{F}$. The force has been applied long enough prior to $t = 0$ for the system to be in steady state. As discussed in section 5.4, for sub-micrometre particles the time required to reach steady state is very small, so that the particle moves at a steady state velocity given by $\mathbf{v} = \mathbf{F} / f$. Over the time interval δt, the particle displacement $\delta \mathbf{x} = \mathbf{v} \delta t$, and substituting for $\mathbf{v}$ gives

$$= \frac{\mathbf{F}}{f} \delta t \qquad (9.6)$$

In a typical experiment, the particle is observed at time $t = 0$ ($\mathbf{x} = 0$) and subsequently after the time interval δt. Over this time interval, the particle moves under the influence of both the deterministic force $\mathbf{F}$, and Brownian motion. The probability of finding a particle at the time of the second observation ($t + \delta t$) is a Gaussian distribution around the point $\mathbf{x} = \delta \mathbf{x}$, with standard deviation Δ.

This is illustrated in figure 9.1(b), which shows the probability distribution of a particle moving under the influence of the force $\mathbf{F}$ for $\delta x = \Delta$, 2Δ, 3Δ, i.e. one, two

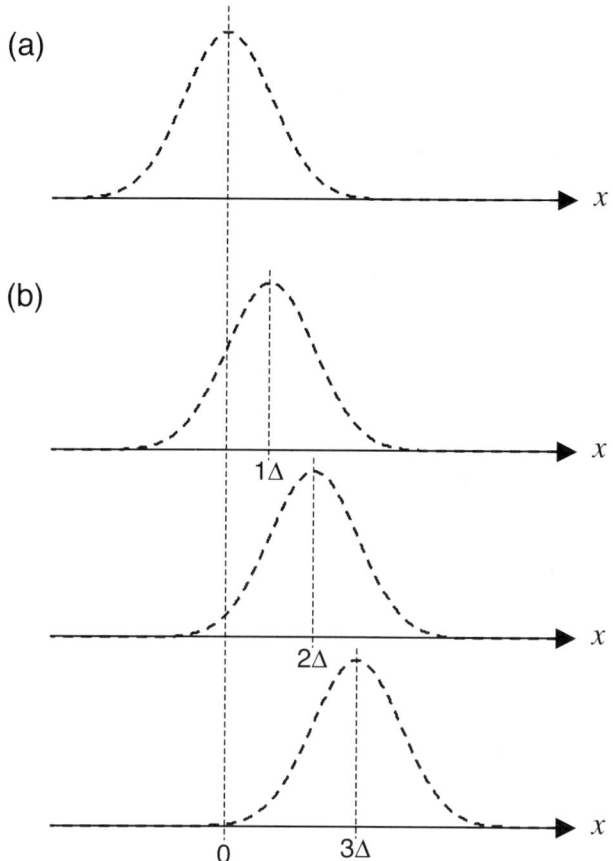

Fig. 9.1 Schematic diagram of the distribution in position in one dimension
after a time δt for a particle experiencing Brownian motion. (a) no
applied force, where the distribution is symmetrical about the
starting value $x = 0$. (b) distribution in position, where the applied
force produces displacements of 1, 2 and 3 times the standard
deviation Δ after time δt.

and three times the standard deviation. The displacement δx is governed by the
magnitude of the force, so that for the three displacements, $|\mathbf{F}| = f\Delta/\delta t$, $2 f\Delta/\delta t$ and
$3 f\Delta/\delta t$. As the magnitude of the applied force and the displacement increases, the
probability that the observed displacement is due to the applied force rather than
Brownian motion increases.

The question of what is and what is not a significant force can now be
addressed. We can define an *observably deterministic threshold force* as the force
required to displace the particle by three times the standard deviation (3Δ) over the

time interval δt. Substituting 3Δ for the displacement δx in equation (9.6) and substituting for Δ and D gives

$$3\sqrt{2\frac{kT}{f}\delta t} = \frac{|\mathbf{F}|}{f}\delta t \qquad (9.7)$$

Rearranging this in terms of force and time we have

$$|\mathbf{F}|\sqrt{\delta t} = \sqrt{18kTf} \qquad (9.8)$$

This expression gives the relationship between the magnitude of the force $\mathbf{F}$, the time interval δt between observations and the friction factor for the particle. The last depends on the size, shape and orientation of the particle as discussed in Chapter Five. In figure 9.2 the threshold force for a spherical particle is plotted as a function of δt and the radius of the particle. It can be seen that for a given δt, as the radius of the particle decreases the force required to produce deterministic movement also decreases. This is due to the fact that although the random displacement increases, the drag on the particle decreases in proportion to the radius. Increasing the time interval between observations also leads to a decrease in the required force, since the force is inversely proportional to the square root of the time interval.

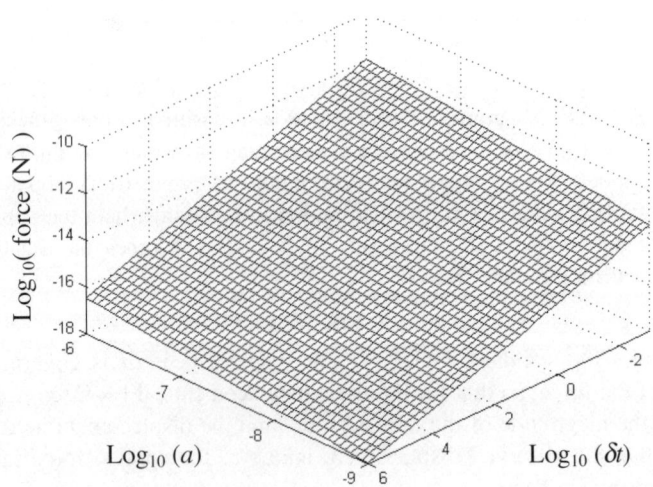

Fig. 9.2 Plot of the threshold force as a function of the observation time δt in seconds and the particle radius a in metres.

Using this definition, the threshold dielectrophoretic force necessary to produce an observable movement of a particle against the random effects of Brownian motion can be calculated, bearing in mind that the DEP force is proportional to the particle volume.

9.2.2 Experimental determination of threshold force

An estimate of the force required to initiate movement of sub-micrometre particles by DEP can be determined by direct observation of the behaviour of particles on microelectrodes. As an example, the threshold force required to produce movement of the Tobacco Mosaic Virus (TMV) was measured by observing the movement of particles undergoing positive DEP as a function of the applied potential (Morgan and Green 1997). This virus is a long thin rod, and approximating this to a prolate ellipsoid, an expression for the dielectrophoretic force is given by

$$\langle \mathbf{F}_{DEP} \rangle = \pi \varepsilon_m a_1 a_2 a_3 \, \text{Re}[\tilde{K}_n] \nabla |\mathbf{E}|^2 \tag{9.9}$$

where a_1, a_2, a_3 are the half lengths of the axes and $\tilde{K}_n$ is the frequency dependent factor for axis n (equation (3.13)). In the experiments, the lowest value of the applied voltage for which dielectrophoresis was observed at one specific position was recorded as a function of $\tilde{K}_n$. The electric field term in equation (9.9), $\nabla |\mathbf{E}|^2$, was calculated numerically for each point on the electrodes where the motion of particles were observed. By inference, for each value of χ, the value of the force at which dielectrophoresis is observed should be equal to the threshold force. Therefore, a plot of $\nabla |\mathbf{E}|^2$ against $1 / \text{Re}[\tilde{K}_n]$ should be a straight line with gradient

$$\frac{|\mathbf{F}_{thr}|}{\pi a_1 a_2 a_3 \varepsilon_m} \tag{9.10}$$

where $\mathbf{F}_{thr}$ is the threshold force required for movement. Figure 9.3 shows experimental data for TMV, reproduced from Morgan and Green (1997). It can be seen that the data fits a straight line, and the threshold force determined from this plot was $|\mathbf{F}_{thr}| = 3.16 \times 10^{-15}$ N. The assumptions made in determining this value are that multipole effects can be ignored and that particle-particle interactions do not lead to the formation of aggregates. Figure 9.4 shows a plot of the variation of the predicted deterministic threshold force against observation time calculated from equation (9.8) for one, two and three standard deviations. This plot shows that if the full 3-standard deviations criteria is adhered to, then the force required to produce motion is of the order of 1×10^{-14} N and 4×10^{-15} N for time intervals of 1 and 10 seconds respectively, concurring with the experimentally derived values.

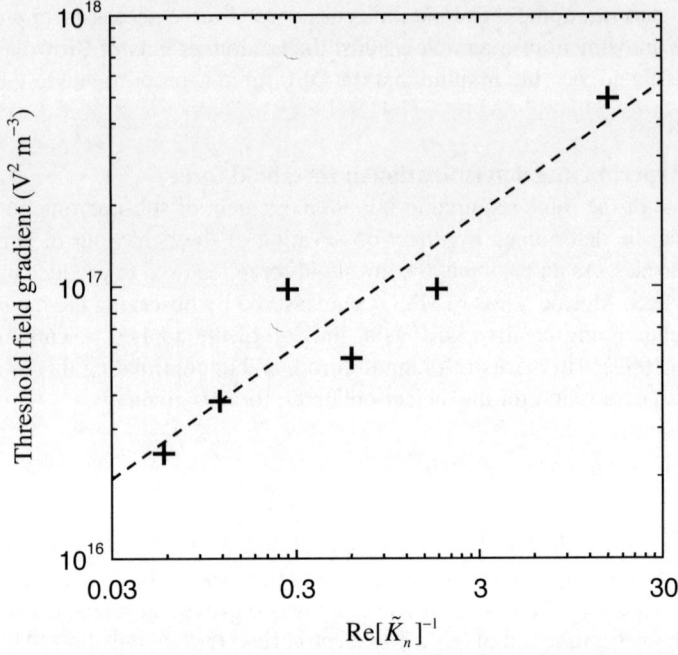

Fig. 9.3 Plot of the experimentally determined threshold field gradient versus
the reciprocal of the particle susceptibility. (Reprinted from Morgan
and Green (1997) with permission from Elsevier Science.)

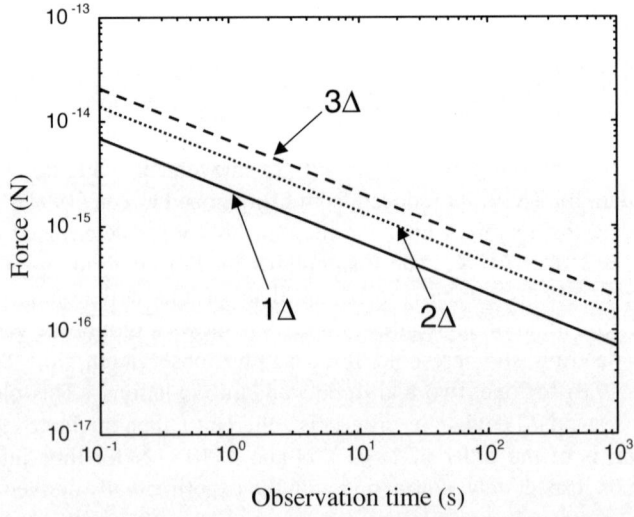

Fig. 9.4 Plot of the calculated threshold force *vs.* observation time, assuming
displacements of 1, 2 and 3 times the standard deviation.

9.2.3   Single particle trapping

One particular application of AC electrokinetics is the non-contact trapping of single particles. The trapping of single sub-micrometre particles by negative dielectrophoresis in quadrupole microelectrode structures has been demonstrated experimentally (Hughes and Morgan 1998). For example, figure 9.5 shows a photograph of a 93 nm diameter latex sphere trapped in a planar quadrupole trap.   Trapping in this manner is of particular interest, since it allows single particles to be isolated without resorting to invasive physical or chemical methods.

In the quadrupole trap, the dielectrophoretic force pushes the particle towards the centre. However, because of Brownian motion, the particle does not stay trapped at a single point in space. Whilst it is held in the trap, the particle is bombarded by impacts from the solvent molecules. Because these impacts are random in size, there is always a certain probability that the particle will experience an impact strong enough to knock it out of the trap.   Holding a particle for an indefinite time is impossible, but an estimate can be made of the force required to hold a particle for a reasonable length of time in a trap.

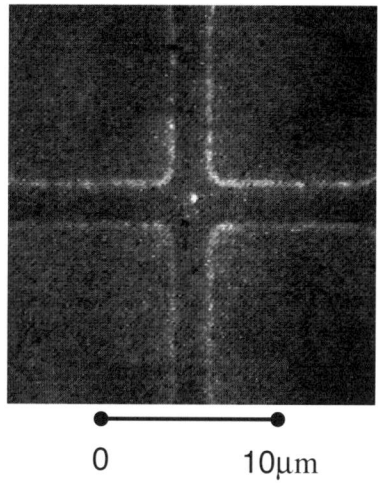

0   10µm

Fig. 9.5 Experimental image of a single 93nm diameter latex sphere trapped in the centre of four electrodes by dielectrophoresis.

Figure 9.6 shows the potential energy and magnitude of the DEP force plotted for a quadrupole electrode array of the type that has been used to trap a single particle (see Chapter Ten). The electrodes are planar, so that the dielectrophoretic trap has an open top and gravity is responsible for the downward force holding the particle on or near the surface.   Of course, particles with near neutral buoyancy, such as latex spheres, are less likely to be held in the trap by gravity.   For these particles, octopole electrodes have been used to create stable three-dimensional DEP traps (Schnelle *et al.* 1993; 1996; 2000; Fuhr *et al.* 1995).

Rule-of-thumb guides to the magnitude of the electric fields required to trap particles can be obtained from electrostatic energy considerations. However, rather than comparing the thermal energy of a particle with the dielectrophoretic energy, the trapping limit can also be computed by comparing the induced velocities of a particle.   For the quadrupole trap, both particle movement through the "sides" of the trap in a plane parallel to the electrodes (the *xy*-plane) and movement in a vertical direction out of the top of the trap can be considered separately.

(a)

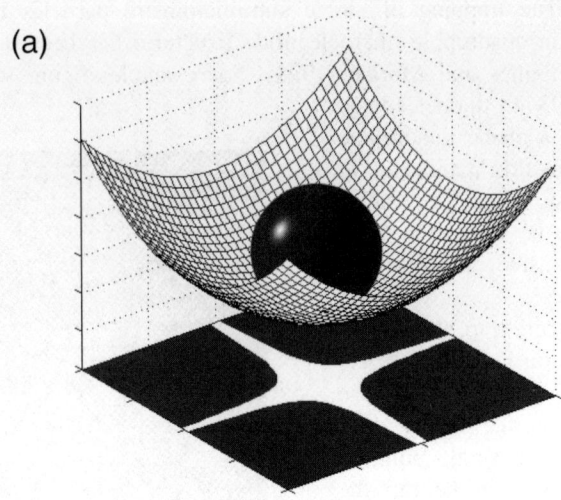

Fig. 9.6 Schematic diagram of the trapping of a particle in the centre of a
four electrode (quadrupole) array, showing the potential energy
(a) and the force magnitude (b). The plot was calculated from
the analytical solution for electric field for a polynomial
electrode given in Chapter Ten.

(b)

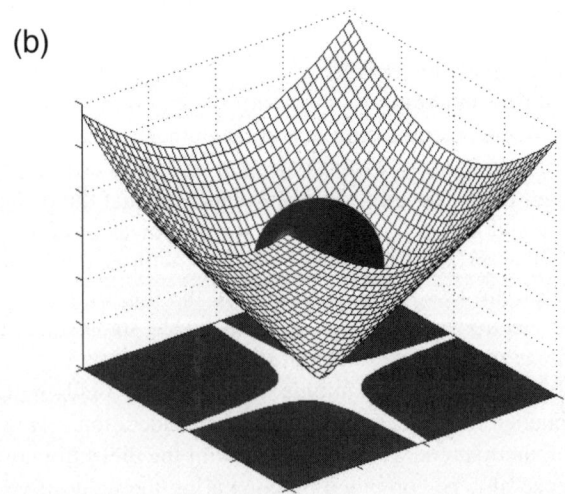

(i) *Movement upwards, in the z-direction*
For the quadrupole electrodes the trap is open topped, negative dielectrophoresis acts upwards and gravity acts downwards through buoyancy. If the particles are denser than water, there is a balance between DEP and buoyancy, (as well as Brownian motion), and particles move randomly around a mean stable levitation position. Of course, if the buoyancy force is stronger than the DEP force everywhere, then the particles sink to the surface of the electrodes. At the levitation position, the DEP force is balanced against buoyancy and the time averaged force is zero, *i.e.*

$$\langle \mathbf{F}_{Total} \rangle = \langle \mathbf{F}_{DEP} \rangle + \mathbf{F}_{Buoyancy} = 0 \tag{9.11}$$

Since the particles are not trapped directly in the centre (owing to Brownian motion, they move about at random), the dipole approximation can be used for the DEP force. Equating the DEP force for a spherical particle with the buoyancy force gives the electric field gradient for the observed levitation height as

$$\nabla |\mathbf{E}|^2 \bigg|_z = -\frac{4(\rho_{m,p} - \rho_{m,m})g}{3\varepsilon_m \, \mathrm{Re}[\tilde{f}_{CM}]} \tag{9.12}$$

This equation defines the condition that must be satisfied in order to levitate a particle to a stable height within a trap, dictated by the trap depth and field magnitude. Obviously particles with density equal to (or less than) that of the suspending medium cannot be trapped in the z-direction.

(ii) *Movement in the horizontal, xy-plane*
When the particle is in a stable levitation position, the efficiency of the trap in the horizontal plane can be examined. We make a simple order of magnitude calculation by assuming that the particle experiences a negative DEP force $\mathbf{F}_{DEP}$ pushing it into the centre of the trap. For simplicity, we further assume that the particle experiences only dipole forces, an assumption which is not strictly true at the centre of the trap where the field strength is minimum but will allow us to make an order of magnitude calculation of trap efficiency. The particle moves at terminal velocity $\mathbf{v}_{DEP} = \mathbf{F}_{DEP} / f$, and the time taken to traverse a small distance Δd over which the force (and $\nabla |\mathbf{E}|^2$) is constant, is $t_{DEP} = \Delta d / |\mathbf{v}_{DEP}|$ or

$$t_{DEP} = \frac{f \Delta d}{|\mathbf{F}_{DEP}|} \tag{9.13}$$

Considering Brownian motion, from equation (9.2) and (9.3) the mean time $\langle t_B \rangle$ taken for a particle to move a distance Δd in one dimension can be derived as

$$\langle t_B \rangle = \frac{f(\Delta d)^2}{2kT} \tag{9.14}$$

So that stable trapping can occur, the time taken for the particle to move from the edge to the centre of the trap should be significantly less than the time taken for the particle to escape under Brownian motion. In other words $t_{DEP}/\langle t_B \rangle$ must be smaller than e.g. 0.1 so that

$$\frac{f\Delta d}{|\mathbf{F}_{DEP}|} < \frac{1}{10}\left(\frac{f(\Delta d)^2}{2kT} \right) \tag{9.15}$$

and

$$\Delta d > \frac{20kT}{|\mathbf{F}_{DEP}|} \tag{9.16}$$

Taking the expression for the dielectrophoretic force from Chapter Four (equation (4.8)), we can re-write equation (9.16) in terms of the minimum particle radius that can be trapped for a given applied electric field

$$a > \sqrt[3]{\frac{20kT}{\pi\varepsilon_m \operatorname{Re}[\tilde{f}_{CM}]\Delta d \left|\nabla|\mathbf{E}|^2\right|}} \tag{9.17}$$

In cases where $\nabla|\mathbf{E}|^2$ varies as a function of distance, the trapping efficiency should be determined using the maximum value of the function ($\Delta d \left|\nabla|\mathbf{E}|^2\right|$). This function can be calculated for different electrode geometries and applied potentials.

Figure 9.7 shows a plot of the sum of the x and y-components of $\nabla|\mathbf{E}|^2$ in a plane across a quadrupole trap electrode array at a height of 7 μm above the surface. The electrode was of the hyperbolic polynomial design (see Chapter Ten) and the data was calculated for an applied voltage of 2.5 V and for an electrode with a central diameter of 6 μm and nearest electrodes distance of 2 μm. The trap efficiency is governed by the smallest distance (along a radius) a particle has to travel in order to escape, and it can be seen from these plots that $\Delta d\nabla|\mathbf{E}|^2$ reaches a broad maximum when Δd is in the range 2 to 4 μm and the corresponding value of $\nabla|\mathbf{E}|^2$ is approximately $2\times10^{17}\,\mathrm{V^2m^{-3}}$. The minimum diameter of particle that can be stably trapped with these particular parameters can be calculated to be 66 nm. Obviously other electrode geometries and field strengths would give different results. These simple calculations show that sub-micrometre particles can in theory be trapped in field cages, as proven experimentally. This technique can be thought of as an electrostatic equivalent to the optical trap (Ashkin 1984) and has potential for non-invasive manipulation and characterisation of single particles.

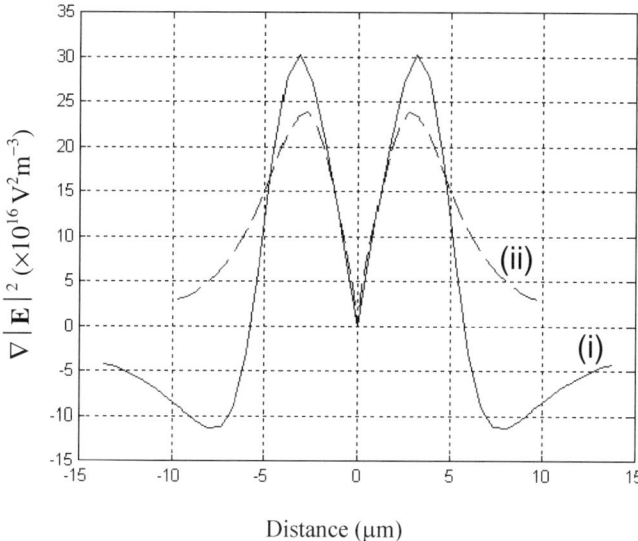

Fig. 9.7 Plot of the sum of the x and y components of the field term in the dielectrophoretic force for the polynomial electrodes, plotted along symmetry lines running across the electrodes (i) and along the gap (ii) determined using numerical methods. (Reprinted with permission from Hughes *et al.* (1998), copyright IOP Publishing Limited.)

9.3   Stochastic processes: the Langevin equation

The equation of motion for a sub-micrometre particle includes the externally applied deterministic forces **F**, such as DEP or gravity, together with the random force due to Brownian motion $\mathbf{F}_{Rand}$ and the frictional force ($f\mathbf{v}$)

$$m\frac{d\mathbf{v}}{dt} = \mathbf{F}_{Deterministic} + \mathbf{F}_{Rand} - f\mathbf{v} \tag{9.18}$$

This equation is referred to as Langevin's equation (Langevin 1905) and because of the presence of the Brownian motion term, it is a stochastic differential equation. We will now examine some aspects of this equation and how it can be used to simulate colloidal particle dynamics.

Analytical solutions for the Langevin equation are difficult to obtain in many situations. The dielectrophoretic force depends on the position of the particle and, unless the electrode shape is particularly simple, does not have an analytical representation. In addition, expressions for other forces such as the random force and particle-particle interactions must be obtained. Consequently, numerical methods are used to solve this equation.

9.3.1   Numerical simulation of the Langevin equation

The behaviour of sub-micron particles can be simulated by stepwise integration of the Langevin equation. The applied or deterministic forces, $\mathbf{F}$ in equation (9.18), are those such as dielectrophoresis and gravity but also, if a collection of particles is being considered, the interaction of the particles with each other.

Llamas *et al.* (1998) performed numerical simulation of the Langevin equation for a large collection of particles, in a uniform rather than a non-uniform electric field. This was used to determine the dynamic evolution of particle pearl chaining in two dimensions. In the simulation, the authors considered particle-particle interactions as a force term, invoking a hard sphere model. In this model the repulsive interaction force between particles rapidly increases as the distance between particles tends to zero. For a spherical particle of radius a, this force is given by (Klingenberg 1989).

$$\mathbf{F}_{Interaction} = \frac{3|\alpha|^2 |\mathbf{E}|^2}{64\pi\varepsilon_m a^4} e^{-100\left(\frac{r}{2a}-1\right)} \hat{\mathbf{r}} \qquad (9.19)$$

where α is the particle polarisability. In this model, the repulsive force prevents particles from overlapping and cancels dipolar attraction when $r = 2a$. In the simulation, the higher order multipoles were ignored and electrostatic interaction between particles and electrodes were considered using the method of image dipoles.

The random force due to Brownian motion ($\mathbf{F}_{Rand}$) can be modelled as white noise (Honerkamp 1994). For the purpose of numerical simulation, the force can be represented by a random number generated by a computer in a Gaussian set of unity standard deviation multiplied by a constant derived from the thermal energy of the system, $f\sqrt{kT/m}$. Llamas *et al.* (1998) simulated a collection of particles as a function of particle polarisability and applied electric field. For biological particles, the authors found good correlation between the simulation and experiment. The authors point to the fact that measurement of the average length of particle chains can be used for estimating the dielectric parameters of the particles. The simulations were performed with large particles (7 μm diameter) so that Brownian motion was of secondary importance in the overall scale of forces. The situation with colloidal particles is somewhat different since the effects of Brownian motion are much more pronounced.

The trajectories of single particles can be simulated with relative ease using programs written on personal computers. This has been performed for a number of simple systems to illustrate the relative effects of deterministic and stochastic forces on particle trajectories. The simplest case is that of a particle moving within a force field such as gravity, where the force is independent of position. Figure 9.8 shows an example of a random path followed by a 100 nm diameter particle (with a density of 1050 kg m^{-3}) in the presence of a gravitational field. It can be seen

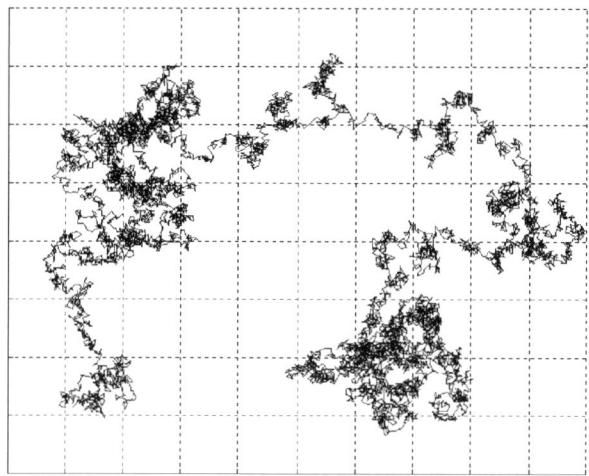

Fig. 9.8 Plot of the numerically determined random path of a 100 nm diameter particle moving under the influence of Brownian motion and a gravitational field acting downwards.

from the figure that the particle follows a random path, and that there is no sedimentation during the time course of the "experiment". For this particle, a typical random impulse is 10^{-10} N compared with a gravitational force of 10^{-18} N. This demonstrates that, in this case, gravity is an insignificant force compared with Brownian motion.

Of more interest is the simulation of the same particle subjected to a well-defined DEP force, such as that produced by an array of parallel planar interdigitated electrodes. As discussed in Chapter Ten, in the far field, the DEP force from this type of array can be described by a simple exponential function, with the electric field gradient term in the DEP force equation given by

$$\nabla |\mathbf{E}|^2 = A_{DEP} \frac{V^2}{d^3} e^{-\frac{\pi y}{d}} \tag{9.20}$$

with d is the electrode width and gap, y the distance above the electrode place, V the applied voltage and A_{DEP} is a geometric constant. Particle trajectories for four different electric field strengths ($V = 50$ V, 25 V, 10 V and 5 V, with $d = 10$ μm) are shown in figure 9.9(a), (b), (c) and (d). In these simulations, the DEP force increases in the ratio of 1 : 4 : 25 : 100, and it is obvious that this has a significant influence on the particle trajectory. Even though the DEP force increases exponentially with distance as the particle approaches the electrodes, for the lowest voltage the particle spends a large amount of time "searching" the space close to

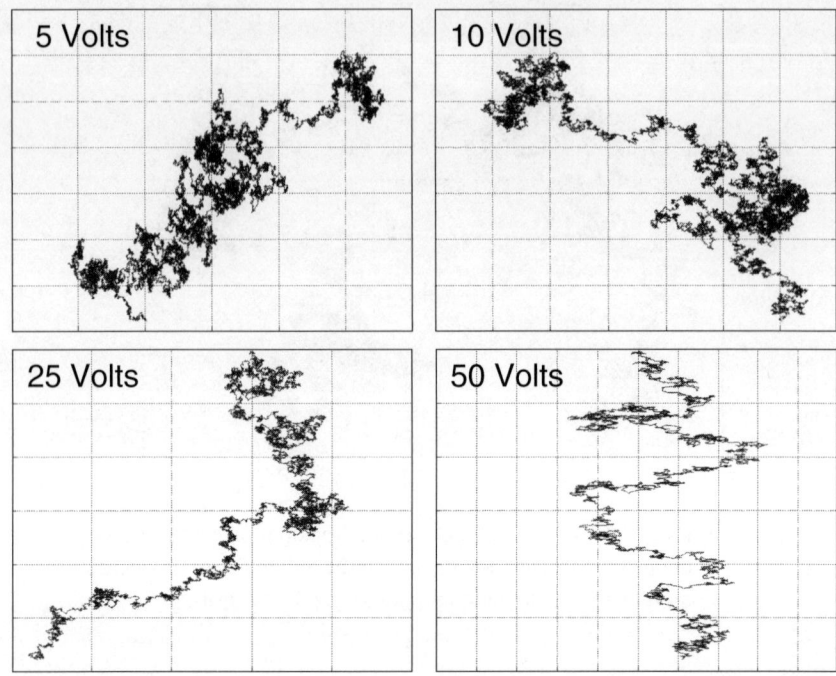

Fig. 9.9 Plots of the tracks of particles undergoing Brownian motion in a exponential DEP force field, for four different applied potentials. The electrodes are at the bottom of the space and the DEP force acts towards the electrodes, vertically downwards in this case.

the electrode before "accelerating" to the surface (at $y = 0$). Although these graphs represent single simulations, calculations show that the ratios of the average velocities of the particles depend on the square of the voltage (as expected). More accurate solutions for the DEP force can also be used. For example, when a 2-dimensional solution for the DEP force is used, the particle can be seen to move directly towards the electrode edge where the DEP force is maximum, as shown in figure 9.10. Although these simulations show that the particle always moves towards the electrode in a short time interval, this might not always be so since an impulse from Brownian motion may be sufficient to send a particle away from the electrode.

9.4   Diffusion

So far we have considered stochastic processes, limited to single particles or small numbers of non-interacting particles. In a system consisting of a large number of particles, thermodynamics requires that the entropy increases until the energy density is uniform, *i.e.* the system is in a stable equilibrium when the particles are

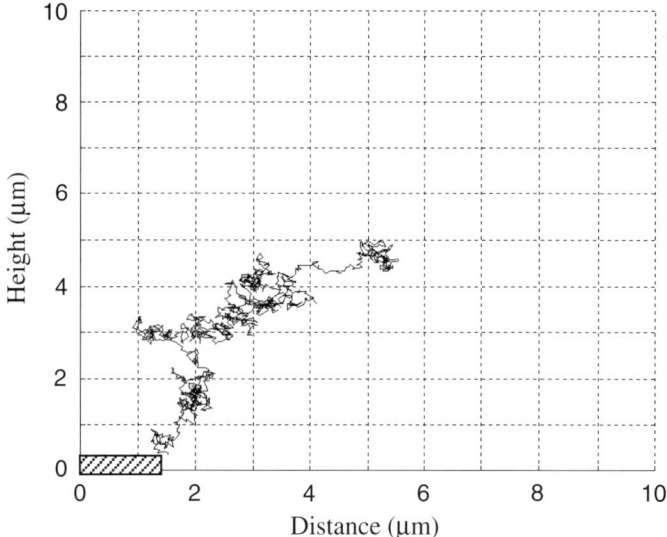

Fig. 9.10   Plot of the tracks of a particle undergoing Brownian motion in a
2D electric field, with the electrode edge shown by the shaded
box. The positive DEP force pulls the particle down to the edge
where it remains held.

uniformly, randomly distributed. Application of an external force causes the
distribution to become non-uniform; particles move under the action of a force
from one place to another, *i.e.* there is a flux of particles. The establishment of this
non-uniform distribution gives rise to an *opposing* flux (the diffusion flux) that acts
to restore uniformity.

Diffusion is the transport of molecules or particles from regions of high to low
concentration, where the driving mechanism is the stochastic Brownian motion of
the individual particles. For a suspension of colloidal particles in an electrolyte,
Fick's first law gives the diffusion flux in terms of the gradient in the concentration
of particles

$$\mathbf{J}_{Diff} = -D\nabla n \tag{9.21}$$

which in one dimension reduces to

$$J_{Diff} = -D\frac{\partial n}{\partial y} \tag{9.22}$$

The time derivative of the concentration gives Fick's second law

$$\frac{\partial n}{\partial t} = D\nabla^2 n \tag{9.23}$$

which in one dimension reduces to

$$\frac{\partial n}{\partial t} = D\frac{\partial^2 n}{\partial y^2} \tag{9.24}$$

The solution to this differential equation (in one dimension) gives the particle concentration in time and space. To illustrate this equation, consider the diffusion of particles, from the semi-infinite region $y < 0$ containing particles at concentration n_o, into a region $y > 0$, which is empty of particles. In this case, equation (9.25) has the solution given by equation (9.5)

$$n(y,t) = \frac{n_o}{\sqrt{4\pi D}}\frac{1}{\sqrt{t}}e^{-\frac{y^2}{4Dt}} \tag{9.25}$$

The solution is plotted in figure 9.11(a) as a function of distance for constant Dt, where Dt has the five values indicated in the graph. The plot shows that for a constant D, the concentration profile quickly reaches a uniform value. Figure 9.11(b) shows the variation of the concentration as a function of time, for a constant D (10^{-9} m^2 s^{-1}) and for the three values of distance y indicated in the graph.

In order to determine the motion of the particles in an ensemble, the flux of particles must be considered (Kittel 1965; Green *et al.* 2000). The total flux in a system consists of the sum of the fluxes due to the different deterministic forces, together with diffusion and fluid motion

$$\mathbf{J}_{Total} = \mathbf{J}_{Diff} + \mathbf{J}_{Buoyancy} + \mathbf{J}_{DEP} + \mathbf{J}_{Fluid} \tag{9.26}$$

where $\mathbf{J}_{Diff}$ is given by equation (9.21) and the other fluxes are determined using the general expression for a flux: $\mathbf{J} = n\mathbf{v}$ where $\mathbf{v}$ is the velocity. The other flux terms in equation (9.26) are: $\mathbf{J}_{Buoyancy} = n(\mathbf{F}_{Buoyancy}/f)$, $\mathbf{J}_{DEP} = n(\mathbf{F}_{DEP}/f)$ and $\mathbf{J}_{Fluid} = n\mathbf{v}_{Fluid}$. Incorporation of these fluxes into the particle conservation equation and solving gives the steady-state particle concentration profile.

9.4.1 Sedimentation

When particle movement is governed only by gravity and diffusion (no external forces and no fluid motion), particles experience a buoyancy force which leads to sedimentation. Starting from a uniform particle distribution, sedimentation leads to a gradual increase in particle concentration at the bottom of a vessel. This creates a diffusion flux upwards, counteracting the effect of the gravitational or buoyancy

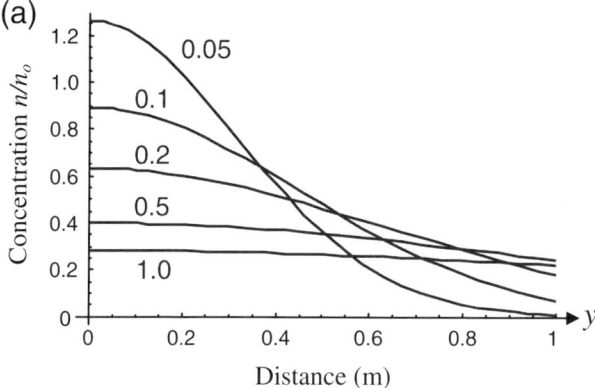

Fig. 9.11 Plot of the variation in relative concentration for diffusion as a function of distance (a) and time (b).

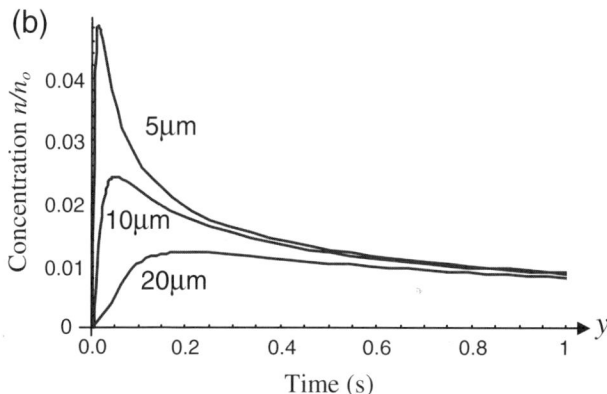

force (Hiemenz 1986). It should be noted that, while this description is obvious for negatively buoyant particles, the situation for positively buoyant particles is the same only reversed, with the particles moving upwards and experiencing diffusion downwards. Eventually the system reaches a steady state either with particles suspended in the bulk with a fixed distribution profile, or complete sedimentation takes place when diffusion is negligible compared to gravity.

For a suspension of particles, the particle conservation equation can be written

$$\frac{\partial n}{\partial t} + \mathbf{v} \cdot \nabla n = -\nabla \cdot \mathbf{J}_{Total} \tag{9.27}$$

Under steady-state conditions the net flux is zero, *i.e.* $\mathbf{J}_{Total} = 0$ and $\mathbf{J}_{Diff} = \mathbf{J}_{Buoyancy}$. If we have a simple planar system, we can consider the forces to be acting in the vertical (y) direction only. At steady state, we have the equilibrium condition

$$D\frac{\partial n}{\partial y} = n\frac{F_{Buoyancy}}{f}$$

(9.28)

where $\mathbf{F}_{Buoyancy}$ acts only in the y direction. This expression can be written as

$$D\frac{\partial n}{\partial y} = \frac{n}{f}\frac{\partial U}{\partial y} \quad \text{or} \quad \frac{dn}{n} = \frac{1}{Df}dU$$

(9.29)

where the gravitational force is defined as the gradient of the gravitational potential U. Integrating this equation gives

$$\frac{n_y}{n_o} = e^{-\frac{(U_y - U_o)}{Df}}$$

(9.30)

The difference in gravitational potential energies between a particle at height y and zero is the potential energy (mgy) so that equation (9.30) can be written as

$$\frac{n_y}{n_o} = e^{-\frac{\upsilon\Delta\rho_m gy}{kT}}$$

(9.31)

where υ is the particle volume and $\Delta\rho_m$ is the differential mass density between particle and fluid. From this equation we see that the particle concentration decreases exponentially with height by a factor e in a distance of $kT/\upsilon\Delta\rho_m g$. Using this equation the particle distribution can be easily calculated. For example, figure 9.12 shows the distribution of 1 μm and 0.5 μm diameter latex particles in a gravitational field.

9.5   The Fokker-Planck Equation (FPE)

As discussed in section 9.3, the Langevin equation can be solved numerically to give a representative trajectory for a single particle. For an identical particle, each time the equation is solved, the trajectory will be different. When considering more than one particle, the interaction force should be included. The computation becomes impractical in describing large numbers of particles in terms of position in space with respect to time and velocity, so that a method capable of describing the probability distribution of particles in space and time is required. This can be done using the Fokker-Planck equation, also known as the Smoluchowski, Kolmogorov, or continuity equation (Gardiner 1985; Doi and Edwards 1986; Risken 1989).

The Fokker-Planck equation can be used to describe the behaviour, in space and time, of the concentration $n(\mathbf{x}, t)$ of particles in a suspending medium when subjected to an arbitrary external force. Under the influence of a small time-

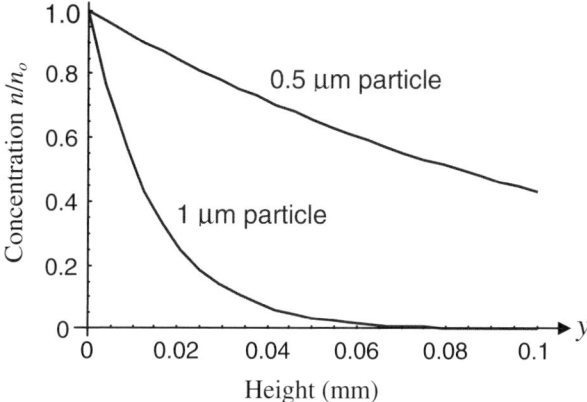

Fig. 9.12 Plot of the relative concentration as a function of distance from the surface for two sizes of latex particle in a gravitational field.

average DEP force $\mathbf{F}_{DEP}(\mathbf{x})$ then for non-interacting particles, the concentration $n(\mathbf{x}, t)$ is related to the particle flux $\mathbf{J}_{Total}(\mathbf{x},t)$ (with $\mathbf{J}_{DEP} = (n/f)\mathbf{F}_{DEP}$), by

$$\frac{\partial n(\mathbf{x},t)}{\partial t} = -\nabla \cdot \mathbf{J}_{Total}(\mathbf{x},t) = -\frac{1}{f}\nabla(n(\mathbf{x},t)\mathbf{F}_{DEP}(\mathbf{x}) + D\nabla \cdot (\nabla n(\mathbf{x},t)) \qquad (9.32)$$

This expression can be solved (numerically or analytically) to give a particle concentration profile, provided that all the constants are known.

The Fokker-Planck equation can be used to model the flux of particles arriving at an electrode array (Bakewell and Morgan 2000). Figure 9.13 shows the numerically simulated "collection rate" of sub-micrometre particles (216 nm diameter) arriving at an interdigitated electrode array as a function of time and frequency of the applied field. The simulation shows that at low frequencies, where the particle polarisability is greatest, the rate of arrival is greatest. The steady-state solution of the FPE ($t \to \infty$) gives the situation described in the next section, where a diffusion flux balances the DEP driven flux.

9.6   Dielectrophoresis versus diffusion

For sub-micrometre particles, the dielectrophoretic collection of particles (for example onto a plane) occurs through the action of a position-dependent DEP force, which often acts together with the gravitational force but is opposed by diffusion. For a suspension of particles in equilibrium, the application of a DEP force (as a step function) disrupts the equilibrium. Particles are either repelled from the electrodes by negative dielectrophoresis or attracted to the electrodes by positive dielectrophoresis. Taking the case of particles collecting on an array of interdigitated electrodes, we can visualise three realistic scenarios.

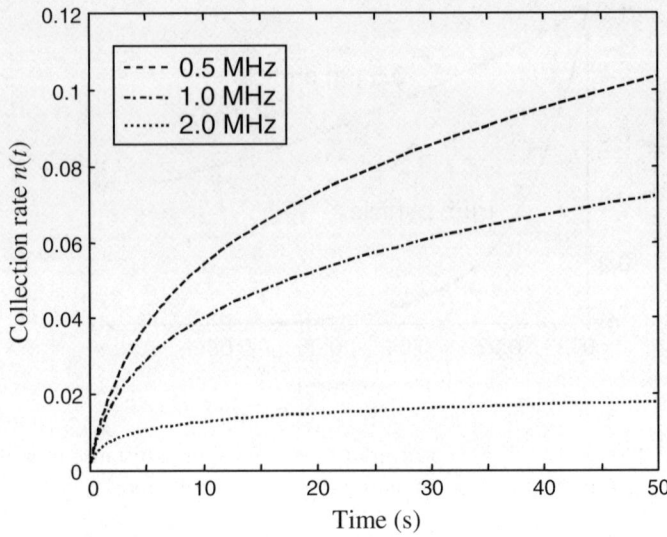

Fig. 9.13   Theoretical plot of the variation of the particle collection rate versus time for three different frequencies for 216 nm diameter latex particles collecting on an interdigitated electrode array, at three different applied frequencies.

(i) *Negative dielectrophoresis*

The particles experience negative dielectrophoresis and are repelled from the electrodes upwards into the bulk. The chamber holding the sample of particles is not infinitely high, so the particles cannot move beyond the upper barrier. The increase in concentration of the particles above the electrodes results in diffusion down towards the electrodes. After sufficient time the result is a steady-state distribution, with zero at the electrode and maximum in the bulk.

(ii) *Weak positive dielectrophoresis*

If the particles experience a weak positive DEP force of the same order of magnitude as the force associated with the random thermal motion, they are attracted to the electrodes but are not immobilised there. They can still receive a sufficiently large impulse from Brownian motion to knock them back into the bulk. This situation is similar to case (i) except that the particles are attracted downwards with diffusion acting upwards. The particles will assume a steady-state concentration profile with maximum at the surface, tending to zero as $y \to \infty$. The temporal and spatial distribution of this particle concentration can be calculated from the Fokker-Planck equation. Under steady state the DEP and diffusion fluxes exactly balance, *i.e.*

$$D\frac{\partial n}{\partial y} = n\frac{F_{DEP}}{f} \qquad (9.33)$$

This equation can be integrated to give the steady-state particle distribution (see Chapter Twelve).

(iii) *Strong positive dielectrophoresis*
If the particles experience a strong DEP force, they move quickly towards the electrode edges and can be considered to be immobilised there since there is a very low probability of them receiving a strong push from Brownian motion. As a result, there is an immediate decrease in the concentration of the particles above the electrodes, creating a local concentration gradient above the electrodes. Diffusion in this case will then act downwards towards the electrodes, in concert with DEP. In theory, this process will occur indefinitely as long as particles continue to be immobilised on the electrodes (an infinite sink), resulting in collection on the electrodes. The dielectrophoretic collection of sub-micrometre particles will be discussed further in Chapter Twelve in relation to DEP field flow fractionation systems.

9.7   Summary
In this chapter we have explained how Brownian motion affects the behaviour of particles in microsystems. We have discussed the physics of stochastic forces versus deterministic forces, such as dielectrophoresis and gravity. We have shown that single sub-micrometre particles can be manipulated experimentally and their behaviour analysed using the Langevin equation. When dealing with large numbers of particles, the system must be analysed in terms of particle flux and the distribution of the particles in time and space is described by the Fokker-Planck equation. These subjects will be further explored in Chapter Eleven on particle manipulation and in Chapter Twelve on the subject of separation using Brownian ratchets.

9.8   References
Ashkin A.   *Stable radiation-pressure particle traps using alternating light beams* Opt. Lett. **10**   454-456   (1984).
Bakewell D. and Morgan H.   *Measuring the frequency dependent polarisability of colloidal particles from dielectrophoretic collection data*   IEEE Trans. Dielectrics & Electrical Insulation **8**   566-571   (2001).
Doi M. and Edwards S.F.   *The theory of polymer dynamics*   Oxford Uni. Press, Oxford   (1986).
Einstein A.   *On the movement of small particles suspended in a stationary liquid demanded by the molecular kinetics theory of heat* Ann. Phys. **17**   549-560   (1905).
Einstein A.   *Investigations on the Theory of Brownian Movement*   New York: Dover (1956).

Fuhr G., Schnelle Th., Hagedorn R. and Shirley S.G.   *Dielectrophoretic field cages: techniques for cell, virus and macromolecule handling*   Cellular Engineering **1** 47-57 (1995).

Gardiner C.W.   *Handbook of stochastic methods 2nd edition*   Springer-Verlag, Heidelberg (1985).

Green N.G. and Morgan H.   *Dielectrophoretic Separation of Nano-particles*   J. Phys. D: Appl. Phys. **30** L41-L44 (1997).

Green N.G., Morgan H. and Milner J.J.   *Manipulation and Trapping of Nanoscale Bioparticle using Dielectrophoresis*   J. Biochem. Biophys. Methods **35** 89-102 (1997).

Green N.G., Ramos A. and Morgan H.   *AC electrokinetics: A survey of sub-micrometre particle dynamics*   J. Phys. D: Appl. Phys. **33** 632-641 (2000).

Hiemenz P.C.   *Principles of Colloid and Surface Chemistry*   Marcel Dekker, New York (1986).

Honerkamp J.   *Stochastic Dynamical Systems*   VCH, New York (1994).

Hughes M.P. and Morgan H.   *Dielectrophoretic Trapping of single sub-micron scale bioparticles*   J. Phys. D: Appl. Phys. **31** 2205-2210 (1998).

Kittel C.   *Elementary Statistical Physics*   Wiley, New York (1965).

Klingenberg D.J., van Swol F. and Zukoski C.F.   *Dynamic Simulations of Electrorheological Suspension*   J. Chem. Phys. **91** 7888-7895 (1989).

Langevin P.   *A theory of Brownian motion*   C.R. Acad. Sci. **146** 530-533 (1908).

Llamas M., Giner V. and Sancho M.   *The dynamic evolution of cell chaining in a biological suspension induced by an electrical field*   J. Phys. D: Appl. Phys. **31** 3160-3167 (1998).

Morgan H. and Green N.G.   *Dielectrophoretic Manipulation of Rod-shaped Viral Particles*   J. Electrostatics **42** 279-293 (1997).

Morgan H., Hughes M.P. and Green N.G.   *Separation of sub-micron bio-particles by dielectrophoresis*   Biophys. J. **77** 516-525 (1999).

Morishima K., Fukuda T., Arai F., Matsuura H. and Yoshikawa K.   *Non-contact transportation of DNA molecule by dielectrophoretic force for micro DNA flow system.*   Proc. IEEE Int. Conf. Robotics & Automation   2214-2219 (1996).

Müller T., Gerardino A.M., Schnelle Th., Shirley S.G., Fuhr G., De Gasperis G., Leoni R. and Bordoni F.   *High-frequency electric-field trap for micron and sub-micron particles*   Il Nuovo Cimento **17** 425-432 (1995).

Müller T., Gerardino A.M., Schnelle Th., Shirley S.G., Bordoni F., De Gasperis G., Leoni R. and Fuhr G.   *Trapping of micrometre and sub-micrometre particles by high-frequency electric fields and hydrodynamic forces*   J. Phys. D: Appl. Phys. **29** 340-349 (1996).

Perrin J.   *Mouvement Brownien et réalité molécularie*   Ann. Chim. Phys. **18** 1-114 (1909).

Pohl H.A.   *Dielectrophoresis*   Cambridge University Press, Cambridge UK (1978).

Risken H.   *The Fokker-Planck Equation*   Springer-Verlag, Heidelberg (1989).

Schnelle Th., Hagedorn R., Fuhr G., Fiedler S. and Müller B. *Three dimensional electric field traps for manipulation of cell – calculation and experimental verification.* Biochim. Biophys. Acta **1157** 127-140 (1993).

Schnelle Th., Müller B., Fiedler S., Shirely S.G., Ludwig K., Hermann A. and Fuhr G. *Trapping of viruses in high-frequency electric field cages* Naturwissenschaften **83** 172-176 (1996).

Schnelle Th., Müller T. and Fuhr G. *Trapping in AC Octode field cages* J. Electrostatics **50** 17-29 (2000).

Washizu M., Suzuki S., Kurosawa O., Nishizaka T. and Shinohara T. *Molecular Dielectrophoresis of Biopolymers* IEEE Trans. Ind. Appl. **30** 835-843 (1994).

Chapter Ten

Electrode design and electric field simulation

In order to move sub-micrometre particles by dielectrophoresis, high electric field strengths generated by well-defined electrode geometries, are required. To move a cell of the order of 10 μm diameter, requires an electric field strength of 10^4 V m^{-1}, which can be generated by two electrodes 1 mm apart with a potential difference of 10 Volts. Generating field strengths high enough to control the movement of a 100 nm particle using the same electrodes, would require a potential difference of $10^3 - 10^4$ Volts. Producing such high field strengths over a wide range of frequencies is not a simple matter and an alternative method of increasing the field strength must be found.

Advances in micro and nano fabrication techniques means that small electrodes of the order of 1 μm in size can now be manufactured with relative ease. This means, for example, that two electrodes separated by a gap of 1 μm with an applied potential difference of 10 Volts can generate a field strength of 10^7 V m^{-1}. It is clear, therefore, that microelectrode structures provide the route by which sufficient field strengths can be generated in order to move sub-micrometre particles without requiring high voltage signal generators.

This chapter outlines the design of electrode structures capable of generating optimum non-uniform electric fields for particle manipulation. The geometry of the electric field depends on the shape of the generating electrodes and a number of different electrode designs have been described in the literature, each with a specific objective, in terms of particle manipulation characterisation, trapping or separation. In order to understand the relationship between field geometries and particle behaviour, we will restrict ourselves to three widely-used electrode designs, namely the polynomial, castellated and interdigitated electrode arrays. The electric field and the electrokinetic forces generated by these electrode designs will be calculated. Several methods have been described in the literature for this purpose, both analytical and numerical: point charge; charge density; finite difference; integral equation methods; Green's functions and Fourier series (Schnelle *et al.* 1993; Wang *et al.* 1993; Martinez and Sancho 1983; Pethig *et al.* 1992; Wang *et al.* 1996; Clague *et al.* 2001; Masuda *et al.* 1987; Morgan *et al.* 2001a; 2001b). In this book, unless stated otherwise, the numerical solutions for the electric field have been calculated using commercial finite element method software (FlexPDE®). The software and examples of the equations solved, and approaches used are described in Appendix A.

10.1   Hyperbolic "polynomial" electrodes

The polynomial electrode design is shown in figure 10.1. It consists of four electrodes arranged symmetrically around a point (Huang and Pethig 1991). The three points marked (i), (ii) and (iii) will be used for reference in this chapter.

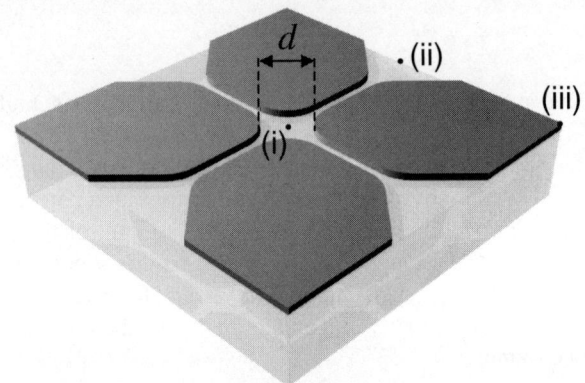

Fig. 10.1 Schematic drawing of the centre part of a set of hyperbolic electrodes.   The curve in the middle is defined by the hyperbola given in equation (10.1).

Several different shapes for the ends of the electrodes have been investigated but the two most commonly used are circles and hyperbolae. The latter is based upon the "isomotive" electrode design described by Pohl (1978). The general case consists of 2n electrodes, but the simplest useful case is the four-electrode system shown in figure 10.1, commonly referred to as a quadrupole. The curved part of the electrode in the centre is defined by the hyperbolic function

$$xy = \frac{d^2}{8} \tag{10.1}$$

where d is the separation of the opposing electrodes measured diagonally across the centre, and the focus of the hyperbola is at the centre of the four electrodes. Practically, the curve is cut off at an arbitrary distance dictated by the capabilities of the manufacturing procedure. Beyond this point the edges of adjacent electrodes are mutually parallel and for sub-micrometre particles, typical electrode dimensions are $5 - 10 \, \mu m$ across the centre (d), with a $2 - 3 \, \mu m$ gap between adjacent electrodes.

10.1.1   Simulation of the electric field

A numerical field simulation for the polynomial electrode array is shown in figure 10.2 for two cross-sectional planes. Figure 10.2(a) shows the field in a plane 100 nm above the surface and figure 10.2(b) a cross section along lines joining points (i) to (iii). There are high field strength regions along the electrode edges (points labelled **B**), particularly between adjacent electrodes and around the central region defining the potential minima for positive dielectrophoresis. The symmetry of the design means that there is also a deep potential minimum for negative

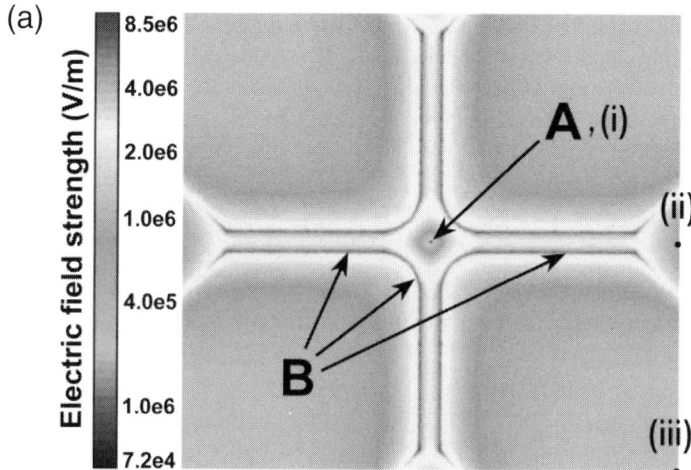

Fig. 10.2   Plot of the numerically calculated electric field magnitude for the polynomial electrode design for two planes: (a) a horizontal plane 100 nm above the upper surface of the electrodes and (b) a vertical plane along the symmetry axis through the electrodes. The reference points are given in figure 10.1. The physical parameters were: central electrode spacing $d = 6\,\mu m$, adjacent electrode parallel edge spacing $2\,\mu m$ and applied potential ± 5 Volts.

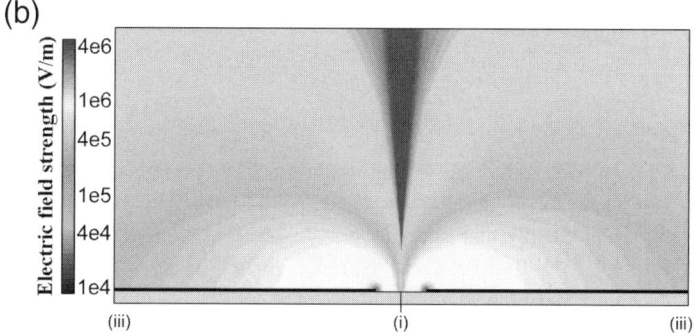

dielectrophoresis in the centre of the four electrodes that can trap and hold particles (point **A**). If d is of the order of the particle size, single particles can be trapped in the centre. If d is much larger than this, these electrodes can be used to measure particle velocities over the central region for both positive and negative dielectrophoresis.

Polynomial electrodes have been used for this purpose since there is an analytical approximation to the electric field in the central area. This is shown in figure 10.3, where the field magnitude is plotted for several horizontal lines across the electrode and parallel to a line joining reference points (i) and (iii) (see figure 10.1). These are plotted at the centre height of the electrodes (50 nm above the substrate), the surface of the electrode and 100 nm above the electrode surface. It can be seen that the electric field magnitude has a constant gradient over two thirds of the diameter between the electrodes. Also shown is the electric field calculated analytically from the expression given by Huang and Pethig (1991)

$$|\mathbf{E}| = \frac{4(V_2 - V_1)}{d^2} r \qquad (10.2)$$

where r is the radial distance. As the figure shows, there is a consistent error (30%) between the numerical calculation and the analytical approximation.

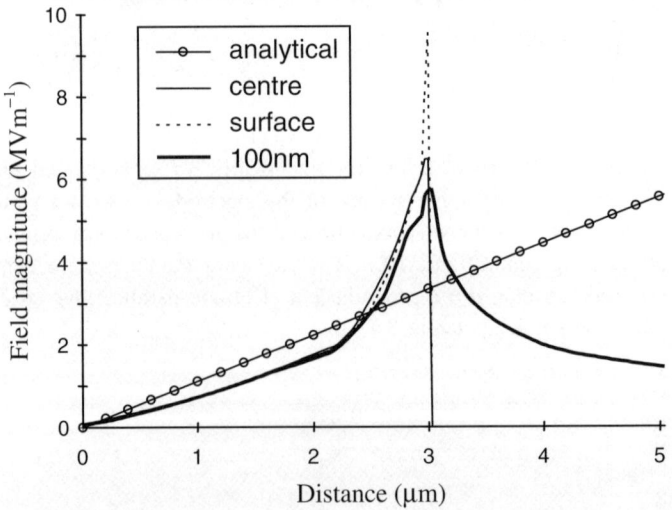

Fig. 10.3 Plot of the radial field magnitude for polynomial electrodes with $d = 6\,\mu\text{m}$ for a number of cases: analytical field calculated from equation (10.2) and the numerical solution along horizontal lines parallel to a line between reference points (i) and (iii) (see figure 10.1), at a number of heights: centre of the electrodes vertically, surface of the electrodes and 100 nm above the electrode surface.

10.1.2   Calculation of the DEP force

The numerical solution of the field gradient term $\nabla|\mathbf{E}|^2$ in the expression for the DEP force for these electrodes is shown in figure 10.4. The arrows show the direction in which a particle experiencing a positive DEP force would move. The arrows are plotted in three dimensions with a fixed length, so that over the electrodes the arrows point downwards and are therefore shorter. The figure also shows the magnitude of $\nabla|\mathbf{E}|^2$ plotted on a logarithmic scale. The solution is a mirror image across the diagonals (i)–(iii) and (i)–(ii) plotted for a height of 100 nm above the surface of the electrodes.

Under positive DEP, a particle just above the electrode surface is pulled towards the electrode edge by a force which increases rapidly in magnitude by over three decades over a short distance. At significant heights above the electrodes, there is a vertical force downwards everywhere. Particles experiencing negative DEP, experience the forces in the opposite direction and will be pushed up and/or into the centre of the array, away from the electrode edges. The analytical expression for the electric field given by equation (10.2) can be used to derive an analytical expression for the electric field gradient term in the dielectrophoretic force giving

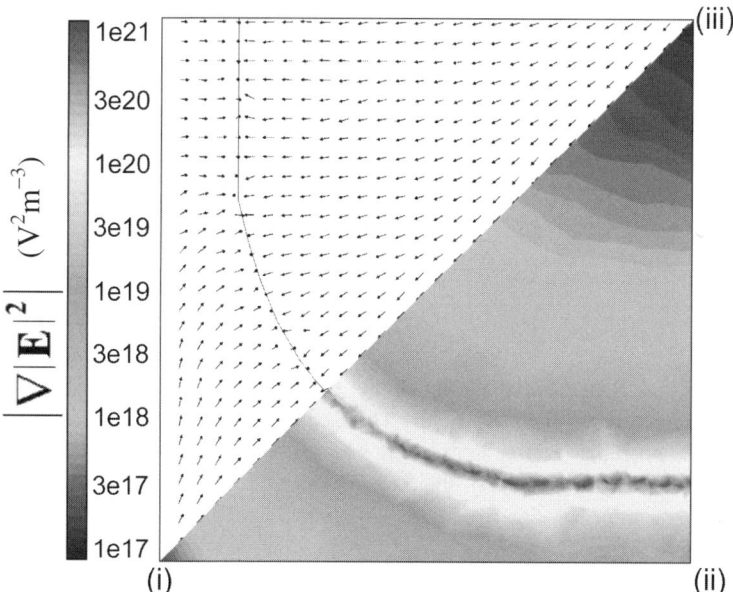

Fig. 10.4 Plot of the vector quantity in the dielectrophoretic force as separate directional arrows and magnitude in a horizontal plane 100 nm above the electrodes. This expression was calculated from the solution for the electric field shown in figure 10.2. In this plane the arrows point towards the electrode edge with the magnitude of the vector increasing rapidly with decreasing distance from the edge.

$$\nabla|\mathbf{E}|^2 = 32\left(\frac{V_2 - V_1}{d^2}\right)^2 r \qquad\qquad (10.3)$$

The quadrupole negative DEP trap in the centre of the array can clearly be seen in the centre marked (i) in figure 10.2. However, this is not a closed DEP trap since there is no downward force in the centre, apart from gravity. Schnelle *et al.* (1993; 2000) have demonstrated how a closed trap can be made using two polynomial electrodes placed one above the other, to produce an octopole. Figure 10.5(a) shows a diagram of the electrode arrangement required to produce an octopole trap. Figure 10.5(b) shows a plot of a surface of constant DEP force potential for the same electrode arrangement, demonstrating the closed nature of the trap.

Planar hyperbolic polynomial electrodes are also widely used for electrorotation (ROT) experiments, with each of the four electrodes sequentially energised with one of four phases of a quadrature AC signal. In order to avoid deleterious DEP effects from unwanted field gradients, the feature sizes for ROT electrodes are much larger, typically with $d = 500$ μm.

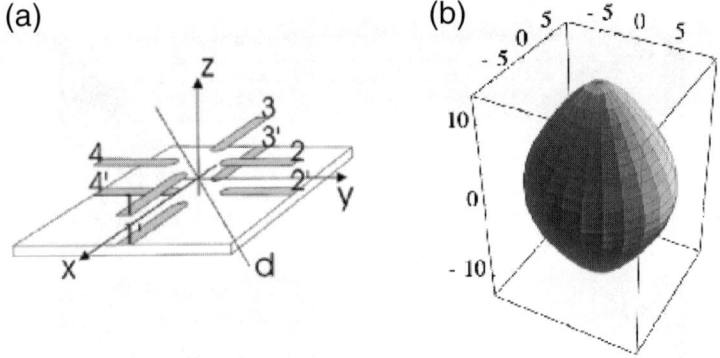

Fig. 10.5 (a) Schematic diagram of the electrode arrangement used for the octopole "field cage". (b) The calculated surface of constant force potential (2×10^{-16} N m) in the centre of the cage, demonstrating the 3D closed nature of the trap. (Reprinted from Schnelle *et al.* 2000 with permission from Elsevier Science.)

10.2   Castellated electrodes

The castellated electrode array consists of square features on parallel wires, as shown in figure 10.6. The line (i)–(ii) will be used for reference in this section. This electrode was first used by Pethig *et al.* (1992) to simultaneously demonstrate positive and negative DEP of bacteria. It was subsequently used in flow-through devices as the basis of a DEP-based separator (Markx *et al.* 1996).

Fig. 10.6 Schematic diagram of the castellated electrode design, which consists of square features on parallel interdigitated electrodes.

Figure 10.7(a) shows the electric field strength in a plane just above the electrode for a castellated electrode array of feature and gap size 5 μm. The high field points are located along the electrode edges, particularly at the tips of the castellations (**B**) but also in the back of the bay (**C**). Well-defined low strength electric field regions occur in each of the bays, as shown by the triangle shape in the figure (**A**). This low field region is closed (locally) as shown by figure 10.7(b), which is a plot of the field in a vertical plane along the line (i) – (ii) indicated in figure 10.6 and 10.7(a). The trap is enclosed vertically, producing a dielectrophoretic cage. For sub-micrometre particles, for which gravity is less important, free-standing electrodes of this design would be able to trap particles in free space by negative DEP.

Figure 10.8(a) shows $\nabla|\mathbf{E}|^2$ plotted in a horizontal plane 100 nm above the surface of the electrodes. Again, the vectors indicate the path of a particle experiencing positive DEP and the path for negative DEP will be in the opposite direction. Particles in the plane of the electrodes are pulled towards the electrode edge by positive DEP, and over most of the gap between the electrodes there is a tendency for particles to move towards the tips of the castellations. The magnitude of the vector increases rapidly towards the corner, indicating that this is a good positive DEP trapping point. In the bay region between the electrodes, the vector points towards the electrode edge at the side and the back of the bay. This behaviour is clearly observed during dielectrophoretic experiments performed using sub-micrometre latex spheres (Green *et al.* 2001). Far above the electrodes the DEP force increases exponentially towards the surface, a behaviour which will be discussed in more detail for the interdigitated electrodes in section 10.3. Close to the surface of the electrodes, particles experiencing negative DEP are pushed towards the bays between the castellations, into an approximately triangular shaped trap. Figure 10.8(b) shows the vector direction for negative DEP in the vertical symmetry plane through line (i)–(ii).

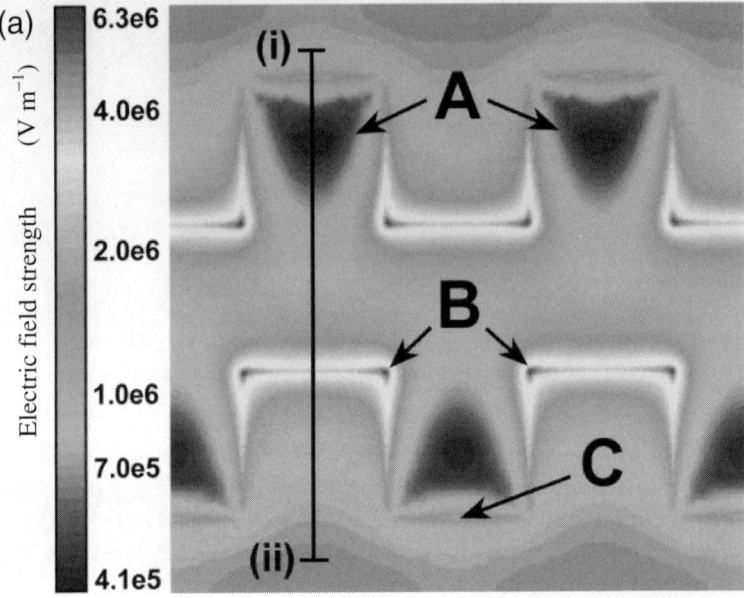

Fig. 10.7   Plot of the numerically calculated electric field magnitude for the castellated electrode design for two planes: (a) a horizontal plane 100 nm above the upper surface of the electrodes and (b) a vertical plane through the line (i)–(ii) shown in figure 10.6. The physical parameters were: feature size and interelectrode gap = 5 μm, applied potential = ±5 Volts.

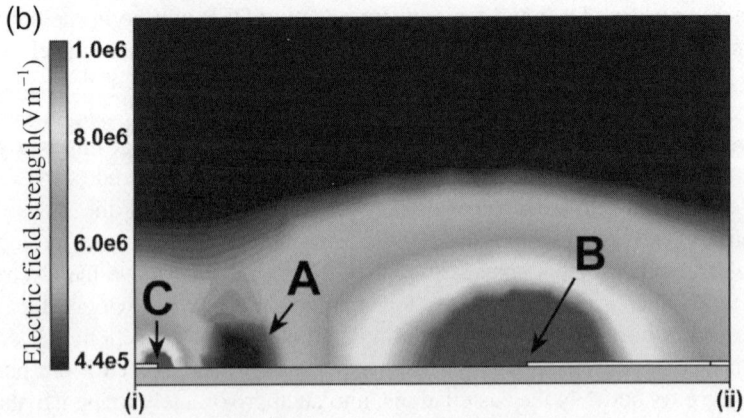

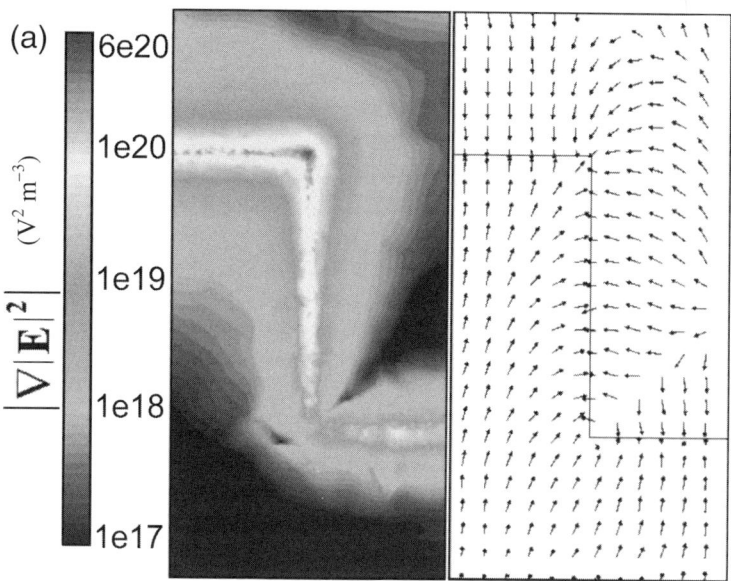

Fig. 10.8 Plot of the vector $\nabla|\mathbf{E}|^2$ for a section of the horizontal plane 100 nm above the surface of the electrodes. In (a) the vectors point towards the electrode edges with a rapidly increasing magnitude. (b) shows a plot of the negative direction in the vertical plane through the line (i)–(ii) shown in figure 10.6. These vectors indicate the direction for negative dielectrophoresis, clearly showing the closed trap in the triangular low field region in the bay.

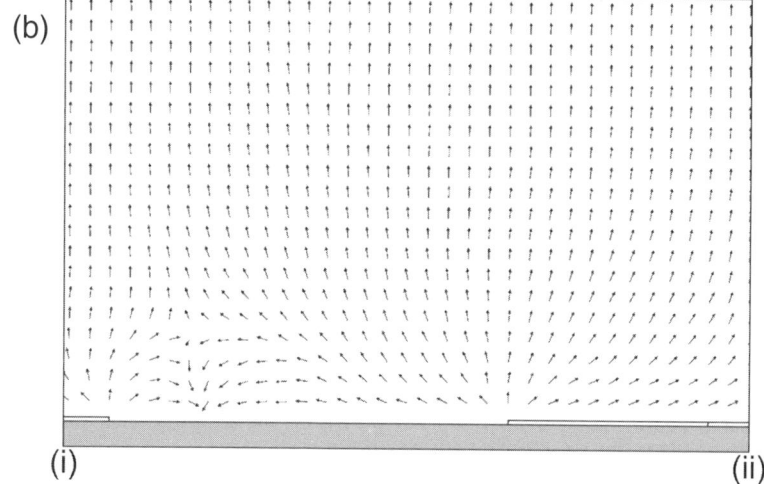

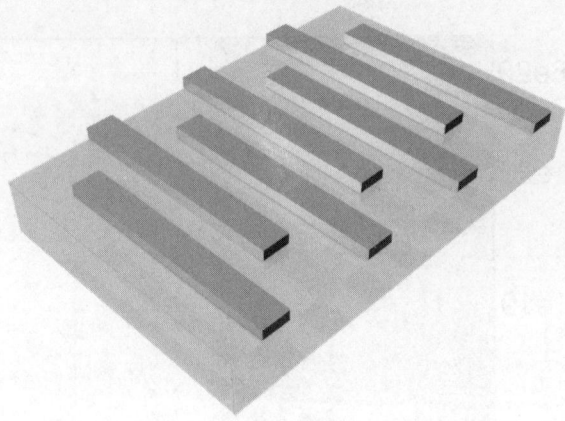

Fig. 10.9 Schematic diagram of the interdigitated electrode design, which consists of a long array of successively addressed parallel bar electrodes.

The castellated structure is simple to fabricate over large areas and this design enables the trapping of particles in controlled regular positions, both on surfaces and in solution. It has been used as an electrode array for investigation of particle characteristics, by simultaneous measurement of positive and negative DEP on the one electrode. These electrodes have also been used in the development of particle separation systems (Gascoyne *et al.* 1992; Becker *et al.* 1994; 1995; Stephens *et al.* 1996; Markx *et al.* 1996; also see Chapter Twelve).

10.3   Interdigitated finger (or bar) electrodes
Interdigitated electrode arrays have found widespread use in AC electrokinetics, primarily for DEP separation systems and for generating travelling electric fields for twDEP. In separation systems, the DEP forces are used in conjunction with hydrodynamic forces and/or gravitational forces to separate particles. In the simplest system, the DEP field is used to pull particles out of a fluid which continuously flows across the electrodes. More elaborate devices are based on field flow fractionation (FFF) combined with DEP and gravitational forces to fractionate a mixture of particles into sub-populations (Huang *et al.* 1997; Markx *et al.* 1997; Wang *et al.* 1998; 2000; Yang *et al.* 1999). In this case a negative DEP force is generated using an array of interdigitated planar electrodes that forms the bottom wall of a flow-through chamber. The electrokinetic force levitates particles of differing polarisability to different equilibrium heights within the chamber. Owing to the parabolic profile of the flowing liquid, particles experience a drag force that depends on their absolute height within the channel and consequently, particles are fractionated along the electrode array.

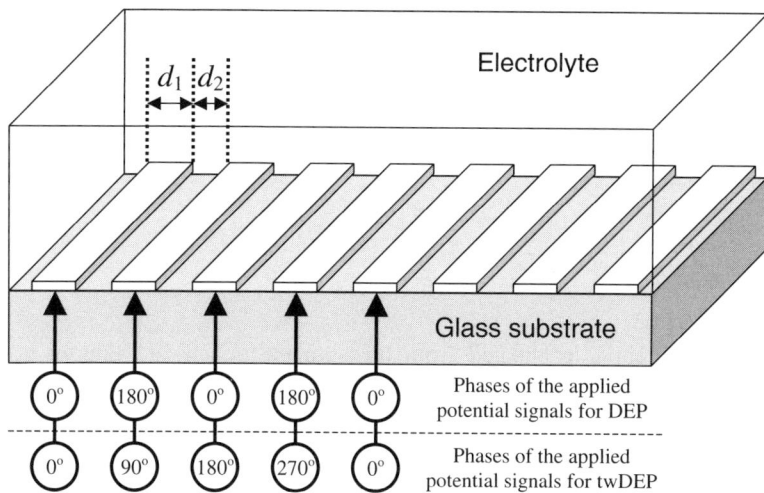

Fig. 10.10 Diagram showing how an interdigitated electrode array is configured for dielectrophoresis or travelling wave dielectrophoresis.

In travelling wave Dielectrophoresis (twDEP) particles are moved in a travelling electric field generated by interdigitated electrodes energised with a four phase signal (Hagedorn *et al.* 1992; Fuhr *et al.* 1991). There is no need to pump liquid along the device in order to produce horizontal motion. Travelling wave devices have been developed using arrays of interdigitated electrodes and have been used for particle fractionation (Talary *et al.* 1996; Morgan *et al.* 1997) and potentially as conveyor belts in Lab-on-a-chip devices.

For both DEP separation and twDEP systems, the electrode arrays are similar, consisting of a large number of thin parallel bar microelectrodes (typically 10 – 40 μm wide), as shown in figure 10.9. The potential signals applied to the electrodes in these two cases are shown in figure 10.10. The electric field for these electrode arrays has been solved analytically using Fourier series analysis (Morgan *et al.* 2001a; 2001b).

10.3.1 Analytical approximation

Assuming that the potential varies linearly between neighbouring electrodes at the level of the surface, Laplace's equation can be solved to obtain an analytical solution for the potential (Morgan *et al.* 2001a; 2001b). The potential can be written as a Fourier series

$$\phi(x, y) = \sum_{n=0}^{\infty} A_n \cos(k_n x) e^{-k_n y} \tag{10.4}$$

where $k_n = (2n+1)\pi / 2d$ and $d = (d_1 + d_2)/2$. The Fourier coefficients are given by

$$A_n = \frac{16}{\pi^2 d_2 (2n+1)^2} V_o d \cos\left(\frac{(2n+1)\pi d_1}{4d}\right) \tag{10.5}$$

and for the particular case where the interelectrode gap and electrode width are the same ($d_1 = d_2 = d$)

$$A_n = \frac{16}{\pi^2 (2n+1)^2} V_o \cos\left((2n+1)\frac{\pi}{4}\right) \tag{10.6}$$

The electric potential calculated from the sum of the infinite Fourier series is plotted in figure 10.11(a). The electric field can be calculated from the gradient of the potential using symbolic mathematics programmes such as Mathematica[®].

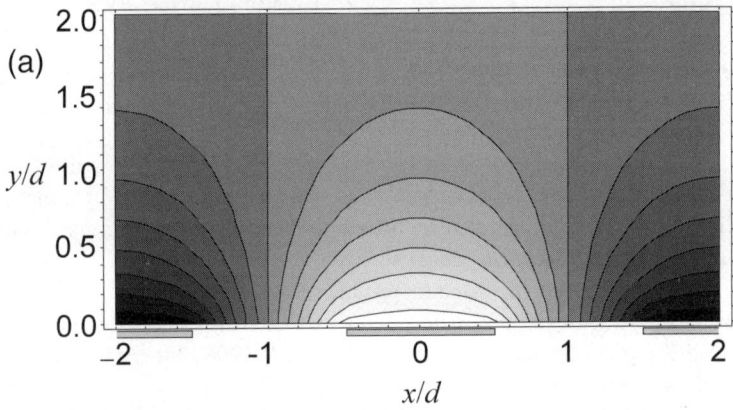

Fig. 10.11   The electrical potential (a) and electric field magnitude (b) calculated from the Fourier Series solution for the interdigitated bar electrodes.

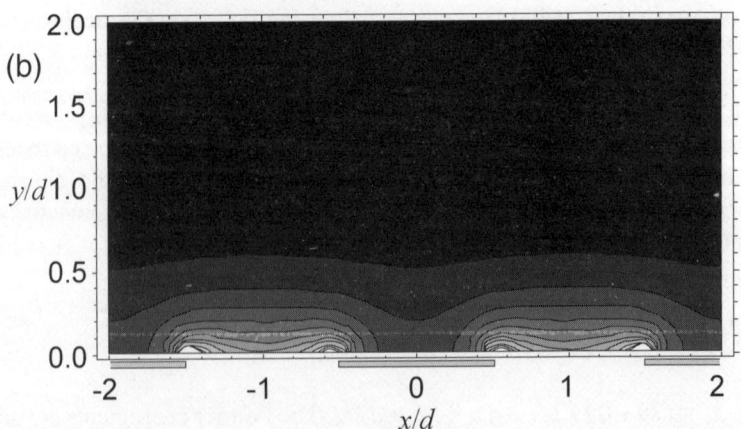

This is shown in figure 10.11(b), where it can be seen that the maximum field occurs at the electrode edges.

Figure 10.12(a) and (b) show how the term $\nabla|\mathbf{E}|^2$, which is proportional to the DEP force, varies above the electrodes. This was calculated from the infinite sum of the Fourier series, again using Mathematica® (Appendix A). Figure 10.12(a) is a vector plot showing the direction of the force whilst figure 10.12(b) is a contour plot of the magnitude of the force (on a log scale) for the complete infinite series. The figure shows that for large distances from the electrode surface, the force direction is straight downwards with a constant magnitude for any given height. When particles reach a height around d, the direction of the force is no longer vertical, vectors point towards the electrode edges and the magnitude of the force increases rapidly.

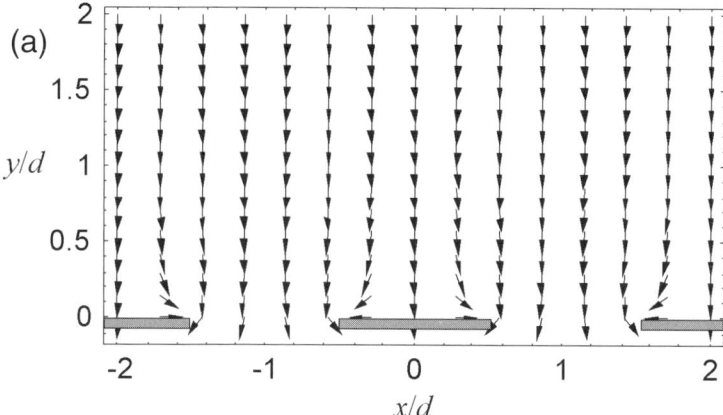

Fig. 10.12   Plot of the direction (a) and magnitude (b) of the vector part of the dielectrophoretic force calculated from the Fourier series solution to the problem of the interdigitated electrode array.

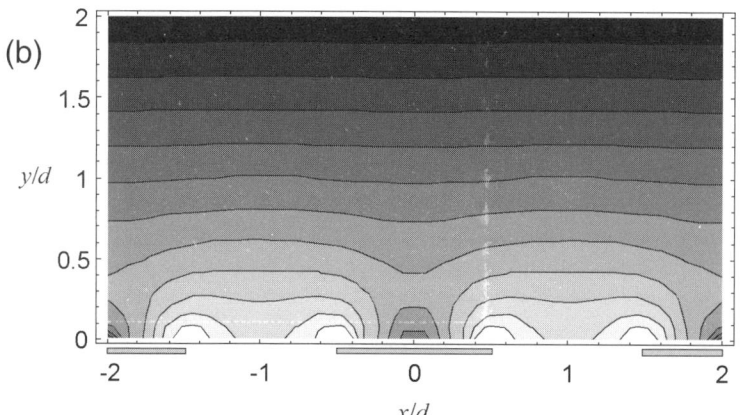

The full expression for $\nabla |\mathbf{E}|^2$ is an infinite series but can be simplified at heights greater than d above the electrodes, since all the terms except the first become very small. $\nabla |\mathbf{E}|^2$ can be approximated to the first term of the series, which for the general case $d_1 \neq d_2$ is

$$\nabla(E_x^2 + E_y^2) = -\frac{64}{\pi} \frac{V_o^2}{d_2^2 d} \cos^2\left(\frac{\pi d_1}{4d}\right) e^{-\frac{\pi y}{d}} \hat{\mathbf{y}} \tag{10.7}$$

with $d = (d_1 + d_2)/2$. For the particular case $d_1 = d_2 = d$

$$\nabla(E_x^2 + E_y^2) = \frac{32}{\pi} \frac{V_o^2}{d^3} e^{-\frac{\pi y}{d}} \tag{10.8}$$

Equations (10.7) and (10.8) are not correct for heights $y < d$, but are applicable when particles experience negative DEP, such as in DEP-FFF systems and the particles are repelled from the electrodes.

If we are interested in calculating the movement of particles close to the electrode array, then an analytical expression for $\nabla |\mathbf{E}|^2$ can be derived but only for the particular case $d_1 = d_2 = d$. This expression means that the DEP force can be calculated everywhere above the electrodes, noting that this is a two-dimensional solution.

The field gradient term for the full solution is (Morgan et al. 2001a; 2001b)

$$\nabla(E_x^2 + E_y^2) = 2\hat{\mathbf{x}}(E_x E_{x,x} + E_y E_{y,x}) + 2\hat{\mathbf{y}}(E_x E_{y,x} + E_y E_{y,y}) \tag{10.9}$$

where E_x, E_y, $E_{x,x}$, $E_{x,y}$ are given by

$$E_x(x, y) = \frac{2V_o}{\pi d}\left[\tan^{-1}\left(\frac{\sin x'}{\sinh y'}\right) \tan^{-1}\left(\frac{\cos x'}{\sinh y'}\right)\right] \tag{10.10}$$

$$E_y(x, y) = \frac{V_o}{\pi d} \ln\left(\frac{(\cosh y' + \cos x')(\cosh y' + \sin x')}{(\cosh y' - \cos x)(\cosh y' - \sin x')}\right) \tag{10.11}$$

$$E_{x,x}(x, y) = -E_{y,y}(x, y) = \frac{2V_o}{d^2}\left(\frac{\sinh y' \cos x'}{\cosh 2y' - \cos 2x'} + \frac{\sinh y' \sin x'}{\cosh 2y' + \cos 2x'}\right) \tag{10.12}$$

$$E_{x,y}(x, y) = E_{y,x}(x, y) = \frac{2V_o}{d^2}\left(\frac{\cosh y' \cos x'}{\cosh 2y' + \cos 2x'} - \frac{\cosh y' \sin x'}{\cosh 2y' - \cos 2x'}\right) \tag{10.13}$$

with $y' = \pi y / 2d$ and $x' = \pi x / 2d + \pi / 4$.

Equations (10.9) to (10.13) can be used to calculate the DEP force experienced by a particle at an arbitrary point above an interdigitated electrode array and used to determine, for example, the trajectory of a Brownian particle using the Langevin equation (see Chapter Nine).

10.3.2 Numerical solution

For the analytical solution, the potential was assumed to vary linearly between neighbouring electrodes. In reality, the potential function is defined by the surface charge density on the surface and will not be linear. Wang *et al.* (1996) approximated the potential function to a 5th order polynomial and solved for the field using Green's functions. The electric potential for the interdigitated array can be calculated numerically using more realistic boundary conditions. We have used Neumann boundary equations derived from the charge continuity equation (Green *et al.* 2002).

Figure 10.13 shows the electric field above an electrode and gap calculated using FlexPDE®. Comparison with figure 10.10(b) shows that the numerical solution is symmetrical around the electrode edge unlike the analytical solution.

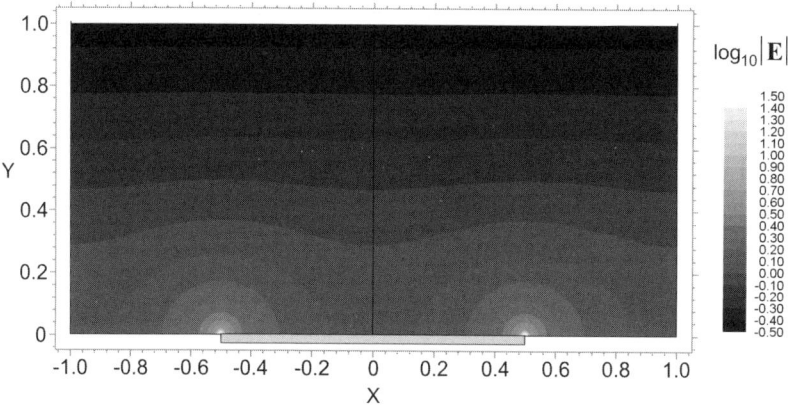

Fig. 10.13 Numerically calculated electric field for the interdigitated electrode array. In this simulation the electrode is from $x/d = -0.5$ to 0.5 and the applied potential = 1 Volt.

Figure 10.14(a) and 10.14(b) show the direction and magnitude of the vector $\nabla|\mathbf{E}|^2$, plotted on a logarithmic scale. Again the behaviour is symmetrical and above a height approximately equal to d, the vectors point straight towards the electrode plane.

Far from the electrode, the force follows an exponential function as predicted by the analytical solution. From the numerical solution, the complete expression for the DEP force in this region is

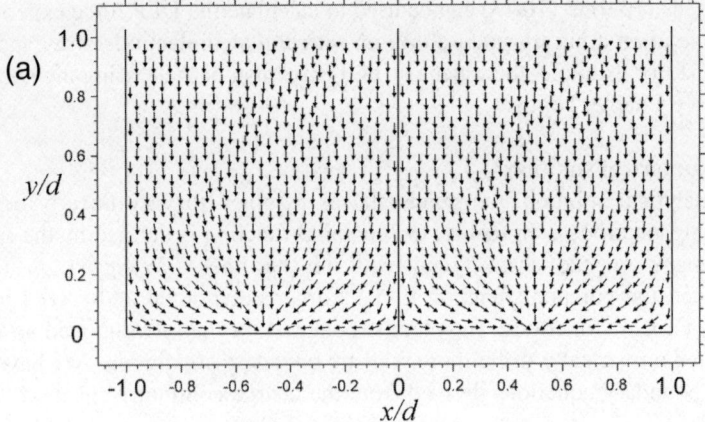

Fig. 10.14   Plot of the direction (a) and magnitude (b) of the vector part of the dielectrophoretic force, $\nabla|\mathbf{E}|^2$, calculated numerically using the finite element method. The applied voltage was 1 Volt and the electrode is from -0.5 to 0.5. The scale on the right hand side of (b) gives the $\log_{10}$ of the magnitude of the vector.

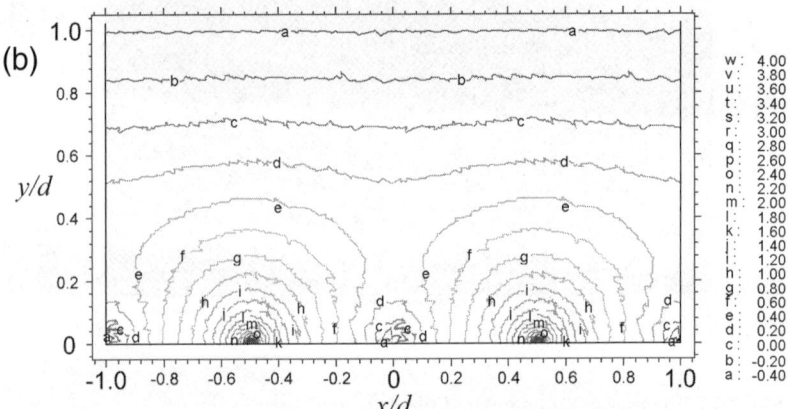

$$\nabla|\mathbf{E}|^2 = A_{DEP} \frac{V_o^2}{d^3} e^{\left(-k_{DEP}\frac{y}{d}\right)} \qquad\qquad (10.14)$$

where the value of the coefficients are $k_{DEP} = \pi$ and $A_{DEP} = 9.0088$. This equation gives the value of the force for any applied potential V_o or d as long as the characteristic electrode width to gap width ratio is 1:1 and $y > d$.

In Table 10.1 values for the exponential factor and the numerically calculated coefficient are compared, together with the analytical solutions. It can be seen from the table that the numerical and analytical values for A_{DEP} agree within 15% for

Table 10.1 Numerically calculated values of the coefficient A_{DEP} for the far field exponential approximation of the dielectrophoretic force, shown for a range of electrode/gap ratios. Also shown are the matching values calculated from the Fourier series analysis and the percentage difference between the two.

Electrode/gap $d_1 : d_2$	A_{DEP} (numerical)	A_{DEP} (Fourier series)	% difference
4 : 1	2.9954	3.039588939	1.47
2 : 1	2.7339	2.864788976	4.79
1 : 1	2.2534	2.546479089	13.0
1 : 2	1.6657	2.148591732	28.0
1 : 4	1.1462	1.79946248	56.0

large distances from the electrode. The error arises from the imposition of the linear boundary condition for the Fourier series solution. The table also shows that by increasing the electrode width relative to the gap, the DEP force can be increased by a third for the same applied voltage and that the error in the analytical expression in this case is only 1.47%.

10.3.3 Levitation under negative DEP
Interdigitated electrode arrays have been used for particle levitation and field flow fractionation. For the case when $d_1 = d_2 = d$, the levitation height of a particle can be found by balancing the dielectrophoretic force against the buoyancy force, $\mathbf{F}_{Buoyancy} = \Delta\rho_m \upsilon \mathbf{g}$. From equation (10.8) and/or (10.14) the DEP force can be written as

$$\langle \mathbf{F}_{DEP} \rangle = \frac{1}{4}\upsilon \operatorname{Re}[\tilde{\alpha}]\nabla(E_x^2 + E_y^2) = \frac{A_{DEP}}{4}\frac{V_o^2\upsilon}{d^3}\operatorname{Re}[\tilde{\alpha}]e^{-\frac{\pi y}{d}} \tag{10.15}$$

Equating this with the buoyancy force gives

$$y = \frac{d}{\pi}\ln\left[-\frac{A_{DEP}V_o^2\operatorname{Re}[\tilde{\alpha}]}{4d^3\Delta\rho_m g}\right] \tag{10.16}$$

where $A_{DEP} = 32/\pi$ when using the analytical solution for the field.

10.4   Interdigitated travelling wave electrode array

In a 4-phase travelling wave electrode array, each applied signal has an amplitude of V_o, and an angular frequency of ω. The voltage on consecutive electrodes is phase shifted by $90°$, as shown in figure 10.10, so that in this case the travelling wave moves in the negative direction. The wave has a wavelength λ equal to the distance between every fourth electrode, i.e. $\lambda = 4(d_1 + d_2) = 8d$.

Using similar principles to that used to derive the analytical solution for the DEP force, the twDEP force can be derived for distance greater than d from the electrode as (Morgan *et al.* 2001)

$$\langle F_x(t) \rangle = -v \, \mathrm{Im}[\tilde{\alpha}] \left(\frac{\pi}{4d} \right)^3 A_1^2 e^{-\frac{\pi}{2d}y} \tag{10.17}$$

$$\langle F_y(t) \rangle = -v \, \mathrm{Re}[\tilde{\alpha}] \left(\frac{\pi}{4d} \right)^3 A_1^2 e^{-\frac{\pi}{2d}y} \tag{10.18}$$

For the particular case $d_1 = d_2 = d$

$$A_1 = \frac{16V_o}{\pi^2} \left(\cos\left(\frac{\pi}{8} \right) - \cos\left(\frac{3\pi}{8} \right) \right) = 0.877354V_o \tag{10.19}$$

and substituting this into equations (10.17) and (10.18) gives

$$\boxed{\langle F_x(t) \rangle = -A_{DEP} \frac{vV_o^2}{d^3} \mathrm{Im}[\tilde{\alpha}] e^{-\frac{\pi}{2d}y}} \tag{10.20}$$

$$\boxed{\langle F_y(t) \rangle = -A_{twDEP} \frac{vV_0^2}{d^3} \mathrm{Re}[\tilde{\alpha}] e^{-\frac{\pi}{2d}y}} \tag{10.21}$$

where $A_{DEP} = A_{twDEP} = 0.372923$. It can be seen from these expressions that the horizontal, or x-component of the force is proportional to the imaginary part of the Clausius-Mossotti factor, and gives rise to the travelling wave movement of the particle. The vertical, or y-component is proportional to the real part of the Clausius-Mossotti factor and is responsible for levitation of the particle. It should be emphasised that these expressions are only valid from $y = 1.5d$ to infinity. Closer to the electrodes, higher order effects must be taken into account.

A plot showing the direction of the vector in the travelling wave force above a four-phase electrode array is shown in figure 10.15 (Green *et al.* 2002). The figure shows that for $y > d$ the force is uniform but for lower heights the particle trajectory no longer conforms to a uniform translational movement, indeed for low heights the force vectors indicate that the particle can circulate around the electrode edge, a phenomenon often seen in practice.

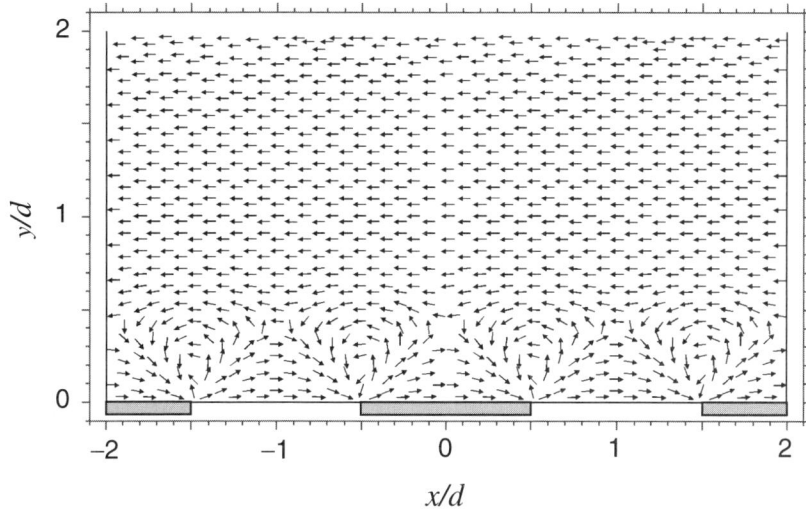

Fig. 10.15 Plot of the direction of the travelling wave force on a particle in a cross-section of the array shown in figure 10.9 and 10.10. The plot is shown for a section covering one electrode and extending to the centre of the two adjacent electrodes. The location of the unit width electrodes is shown by the grey boxes.

10.4.1   Particle levitation in a travelling wave array

As for DEP, the levitation height of a particle far from the electrodes can be determined from the balance of negative DEP force and gravity as

$$y = \frac{2d}{\pi} \ln \left[-\frac{A_{twDEP} V_o^2 \, \mathrm{Re}[\tilde{\alpha}]}{d^3 \Delta \rho_m g} \right] \tag{10.22}$$

For a spherical particle, the polarisability is given by $\tilde{\alpha} = 3\varepsilon_m \tilde{f}_{CM}$. Substituting these expressions into equation (10.20) and equating with the Stoke's drag force, gives the travelling wave induced velocity of a particle as

$$v_x = \frac{2}{9} \frac{a^2 \Delta \rho_m g}{\eta} \frac{\mathrm{Im}[\tilde{f}_{CM}]}{\mathrm{Re}[\tilde{f}_{CM}]} \tag{10.23}$$

This shows that as long as the particle is levitated to heights greater than $\sim d$, the velocity of the particle is independent of the applied voltage. Therefore, for a given particle, the velocity depends only on the frequency of the applied signal through the ratio of the imaginary and real parts of the Clausius-Mossotti factor. To increase the velocity, the real part of the Clausius-Mossotti factor can be reduced but this would also reduce the levitation height, possibly leading to anomalous behaviour.

10.5   Summary
In this chapter, we have outlined the basic principles of three different electrode designs widely used for generating non-uniform electric fields. We have outlined methods for calculating the electric field and the dielectrophoretic force for these electrode arrays and given analytical approximations to the field, where possible. In practice, any number of electrode designs can be produced in two or three dimensions and simulation of the DEP force is an important design tool for optimising the electrode structure.

10.6   References
Becker F.F., Wang X-B., Huang Y., Pethig R., Vykoukal J. and Gascoyne P.R.C. *The removal of human leukaemia cells from blood using interdigitated microelectrodes* J. Phys. D: Appl. Phys. **27** 2659-2662 (1994).

Becker F.F., Wang X-B., Huang Y., Pethig R., Vykoukal J. and Gascoyne P.R.C. *Separation of human breast cancer cells from blood by differential dielectric affinity* Proc. Natl Acad. Sci. USA **92** 860-864 (1995).

Clague D.S. and Wheeler D.K. *Dielectrophoretic manipulation of macromolecules: The electric field* Phys. Rev. E **64** art. no. 026605 (2001).

Fuhr G., Hagedorn R., Muller T., Benecke W., Wagner B. and Gimsa J. *Asynchronous travelling-wave induced linear motion of living cells* Studia Biophysica. **140** 79-102 (1991).

Gascoyne P.R.C., Huang Y., Pethig R., Vykoukal J. and Becker F.F. *Dielectrophoretic separation of mammalian cells studied by computerized image analysis* Meas. Sci. Technol. **3** 439-445 (1992).

Green N.G., Ramos A. and Morgan H. *AC electrokinetics: A survey of sub-micrometre particle dynamics* J. Phys. D: Appl. Phys. **33** 632-641 (2000).

Green N. G., Ramos A. and Morgan H. *Numerical solution of the dielectrophoretic and travelling wave forces for interdigitated electrode arrays using the Finite Element Method* J. Electrostatics (in press) (2002).

Hagedorn R., Fuhr G., Müller T. and Gimsa J. *Travelling wave dielectrophoresis of microparticles* Electrophoresis **13** 49-54 (1992).

Huang Y. and Pethig R. *Electrode design for negative dielectrophoresis applications.* Meas. Sci. Technol. **2** 1142-1146 (1991).

Huang Y., Wang X-B., Becker F.F. and Gascoyne P.R.C. *Introducing dielectrophoresis as a new force field for Field-Flow-Fractionation* Biophysical J. **73** 1118-1129 (1997).

Markx G.H., Dyda P.A. and Pethig R. *Dielectrophoretic separation of bacteria using a conductivity gradient* J. Biotechnol. **51** 175-180 (1996).

Markx G.H., Pethig R. and Rousselet J. *The dielectrophoretic levitation of latex beads, with reference to field flow fractionation* J. Phys. D: Appl. Phys. **30** 2470-2477 (1997).

Martinez G. and Sancho M. *Integral-equation methods in electrostatics* Am. J. Phys. **51** 170-174 (1983).

Masuda S., Washizu M. and Iwadare M. *Separation of small particles suspended in liquid by nonuniform traveling field* IEEE Trans. Ind. Appls. **23** 474-480 (1987).

Morgan H., Green N.G., Hughes M.P., Monaghan W. and Tan T.C. *Large-area travelling-wave dielectrophoresis particle separator* J. Micromech. Microeng. **7** 65-70 (1997).

Morgan H., Izquierdo A.G., Bakewell D.J., Green N.G. and Ramos A. *The dielectrophoretic and travelling wave forces for interdigitated electrode arrays: analytical solution using Fourier series* J. Phys. D: Appl. Phys. **34** 1553-1561 (2001a).

Morgan H., Izquierdo A.G., Bakewell D.J., Green N.G. and Ramos A. *The dielectrophoretic and travelling wave forces for interdigitated electrode arrays: analytical solution using Fourier series* Erratum J. Phys. D: Appl. Phys. **34** 2708 (2001b).

Pethig R., Huang Y., Wang X-B. and Burt J.P.H. *Positive and negative dielectrophoretic collection of colloidal particles using interdigitated castellated microelectrodes* J. Phys. D: Appl. Phys. **25** 881-888 (1992).

Pohl H.A. *Dielectrophoresis* Cambridge Uni. Press, Cambridge, UK (1978).

Schnelle T., Hagedorn R., Fuhr G., Fiedler S. and Müller T. *3-dimensional electric-field traps for manipulation of cells - calculation and experimental-verification* Biochim. Biophys. Acta **1157** 127-140 (1993).

Schnelle T., Muller T. and Fuhr G. *Trapping in AC octode field cage* J. Electrostatics **50** 17-29 (2000).

Stephens M., Talary M.S., Pethig R., Burnett A.K. and Mills K.I. *The dielectrophoretic enrichment of CD34+ cells from peripheral blood stem-cell harvests.* Bone Marrow Transplantation **18** 777-782 (1996).

Talary M. S., Burt J. P. H., Tame J. A. and Pethig R. *Electromanipulation and separation of cells using travelling electric fields* J. Phys. D: Appl. Phys. **29** 2198-2203 (1996).

Wang X-B., Huang Y., Burt J.P.H., Markx G.H. and Pethig R. *Selective dielectrophoretic confinement of bioparticles in potential-energy wells* J. Phys. D: Appl. Phys. **26** 1278-1285 (1993).

Wang X-B., Vykoukal J., Becker F.F. and Gascoyne P.R.C. *Separation of polystyrene microbeads using dielectrophoretic gravitational field-flow-fractionation* Biophys. J. **74** 2689-2701 (1998).

Wang X-B., Yang J., Huang Y., Vykoukal J., Becker F.F. and Gascoyne P.R.C. *Cell separation by dielectrophoretic field-flow-fractionation* Anal. Chem. **72** 832-839 (2000).

Wang X-J., Wang X-B., Becker F.F. and Gascoyne P.R.C. *A theoretical method of electrical field analysis for dielectrophoretic electrode arrays using Green's theorem* J. Phys. D: Appl. Phys. **29** 1649-1660 (1996).

Yang J., Huang Y., Wang X-B., Becker F.F. and Gascoyne P.R.C. *Cell separation on microfabricated electrodes using dielectrophoretic/gravitational field-flow-fractionation* Anal. Chem. **5** 911-918 (1999).

Chapter Eleven

Dielectrophoresis of sub-micrometre particles: manipulation and characterisation

Dielectrophoresis was first introduced as a term to describe the motion of particles in non-uniform electric fields in the 1950s (Pohl 1951; Pohl 1958), although observations of related phenomena, such as pearl chaining in AC fields, pre-date this by decades. The first demonstration of the use of DEP to collect living cells was made by Pohl in 1966 (Pohl and Hawk 1966). The consensus at the time was that the technique was limited to the manipulation of cells and other micro-organisms and that Brownian motion would place a lower limit on the size of particle that could be manipulated. In essence, in order to manipulate sub-micrometre particles, large field gradients are required in order to overcome the randomising effect of Brownian motion (or to provide sufficient energy to overcome the counteracting influence of diffusion). These gradients could not easily be generated with the early macroscale electrode structure that Pohl *et al.* and his contemporaries had available.

Realistically the DEP manipulation of sub-micrometre particles can only be accomplished using microelectrodes. These have two principal advantages: (i) low voltages can be used to give high electric field strengths across a wide range of frequencies, and (ii) the high surface to volume ratio of microsystems, means that any Joule heating in the device is easily and quickly dissipated. The development of photo and electron beam lithographic techniques for fabricating microdevices has led to a surge of activity in this area, so that the manipulation, characterisation and separation of sub-micrometre particles has now become almost routine.

Remarkably, in 1938 Müller (Müller 1938) considered using DEP for the manipulation of molecularly-sized particles and after careful experimentation concluded the effect would not be marked. Pioneering work was done by Debye and co-workers (Debye *et al.* 1954a; 1954b; Barber *et al.* 1954) who showed that grading of polymers and small particles by size could be achieved. Interestingly a number of attempts were made to combine DEP with capacitance measurements to determine the molecular weight and diffusion constants of polymers (Prock and McConkey 1960; Eisenstadt and Scheinberg 1972). In his book, Pohl (1978) even described a proposed design for a "molecular diffusion-dielectrophoretic separator", and after reviewing the field, concluded that the "use of dielectrophoresis for the study of molecules has a bright future."

Despite this early interest in sub-micrometre particle manipulation, there was little further attempt to work with sub-micrometre particles or molecules until Washizu and co-workers used microelectrodes to demonstrate manipulation and collection of biopolymers (Washizu and Kurosawa 1990; Washizu et al. 1994). They showed that positive DEP could be used to concentrate DNA and macromolecules onto planar microelectrodes in a flow-through device. The manipulation of DNA was aided in part by its extremely high polarisability and its relatively long length. The manipulation of other biopolymers was demonstrated in very low conductivity aqueous suspensions and with electrode gaps of 4 or 15 μm. Positive DEP was observed for all the proteins examined and the rate of collection was found to be (at least subjectively) dependent on the size of the particles. This work showed definitively that sub-micrometre particles could be manipulated and trapped by DEP forces (Washizu et al. 1994).

Subsequently, a range of particle such as viruses and sub-micrometre latex spheres have been collected and manipulated by both positive and negative DEP. Müller and co-workers (Müller et al. 1995) demonstrated that latex particles in the range 14 to 1000 nm in diameter could be manipulated by both positive and negative DEP forces using quadrupole electrode geometries fabricated with direct write electron beam lithography. The electrode gaps were in the range $0.5 - 2$ μm and field strengths of 20 MV m^{-1} were used. The role of hydrodynamic forces in manipulating and trapping sub-micrometre particles was also recognised at this time (Müller et al. 1996a). DEP has been used to characterise the dielectric properties of colloidal latex particles (Green and Morgan 1999) and to measure the binding of antibodies to colloidal particles (Hughes and Morgan 1999; Hughes et al. 1999a; 1999b).

It has also been shown that viruses can be manipulated and collected by DEP forces. Early experiments showed that the fluorescently labelled enveloped Cauliflower Yellow Mosaic Virus could be trapped by a combination of DEP and hydrodynamic forces at the centre of a quadrupole electrode array (Green et al. 1995). Influenza and Sendai viruses were collected in field funnels under negative DEP forces in quadrupole and octopole field cages (Fuhr et al. 1995; Müller et al. 1996b; Schnelle et al. 1996, Gimsa 1999). Further work showed that virus particles could be manipulated using both positive and negative DEP forces. Both the non-enveloped Tobacco Mosaic Virus (TMV), and the enveloped Herpes Simplex Virus (HSV-1) could be manipulated by positive and negative DEP forces depending on the frequency, using planar microelectrodes (Green et al. 1997; Morgan and Green 1997; Hughes et al. 1998; Hughes and Morgan 1998; Morgan et al. 1999). DEP has also been used to characterise the dielectric properties of single virus particles and the capsid (or core) of a virus (Hughes et al. 2001, Hughes et al. 2002).

The movement and trapping of DNA has attracted considerable attention. Combinations of electrophoretic and/or dielectrophoretic forces have been used to trap and manipulate DNA on planar microelectrodes (Washizu and Kurosawa 1990, Washizu et al. 1995, Asbury and van den Engh 1998, Morishima et al. 1996,

Nishioka *et al.* 1995, Bakewell *et al.* 2001). DNA can not only be manipulated in an AC field, but it can also be stretched along the field lines (*e.g.* Washizu and Kurosawa 1990, Kabata *et al.* 1993).   To this end, Washizu and co-workers have developed "fluid integrated circuit" devices to perform a range of tasks, including electrostatically driven end anchoring of DNA molecules; unidirectional orientation of DNA; determination of the size distribution of a DNA solution; and laser, mechanical or enzymatic cutting of DNA (Washizu *et al.* 1995; 1997; Yamamoto *et al.* 2000). Single DNA molecules have also been manipulated and separated (Morishima *et al.* 1997) and analysed (Oana *et al.* 1999; Yamamoto *et al.* 2000) using DEP as an enabling tool. Imaging single DNA molecules can be difficult, but it has been shown that excellent images can be obtained if Brownian motion is effectively suppressed by cooling the samples (Nishioka *et al.* 1993).

Following on from this brief review of the literature, we will now explore some of the finer details of sub-micrometre particle manipulation techniques.

11.1   Manipulation

11.1.1   Latex particles

Latex spheres are solid and homogeneous dielectric particles which, when suspended in a fluid medium, exhibit a single interfacial (Maxwell-Wagner) relaxation. In general, they are manufactured with a high density of surface carboxyl, sulphate or amino groups, which in turn gives rise to a high surface charge densiy and high values of surface conductance. As discussed in Chapter Seven, the polarisability of this type of particle at low frequencies is dominated by surface conductance effects.

Observation of the behaviour of sub-micrometre particles can be difficult since they are of the same order of magnitude in size as the wavelength of light. Fortunately, they can be manufactured containing a fluorescent dye and can therefore be seen using fluorescent imaging techniques (Lacey 1999). To increase the signal level and remove background noise, evanescent field techniques can also be used (Hughes and Morgan 1999b). In this technique, the excitation light is restricted to the surface of the electrodes, substantially reducing interference from stray light. Only those particles collecting in the region of the electrodes are illuminated and thus contribute to the observed signal.

At low frequencies, latex particles are generally more polarisable than the suspending medium (they have a high effective conductivity), and so exhibit positive DEP. At frequencies above the interfacial relaxation frequency, the particles are less polarisable and experience negative DEP. An example of the typical behaviour of these particles is shown in figure 11.1. This shows 557 nm diameter latex particles moving and collecting on a hyperbolic polynomial electrode array. At low frequencies particles collect at the high field regions on the electrode edges. Switching the field to a frequency above the Maxwell-Wagner relaxation frequency causes particles to immediately experience negative DEP so that over a relatively short period of time (~10 s) all the particles collect in the

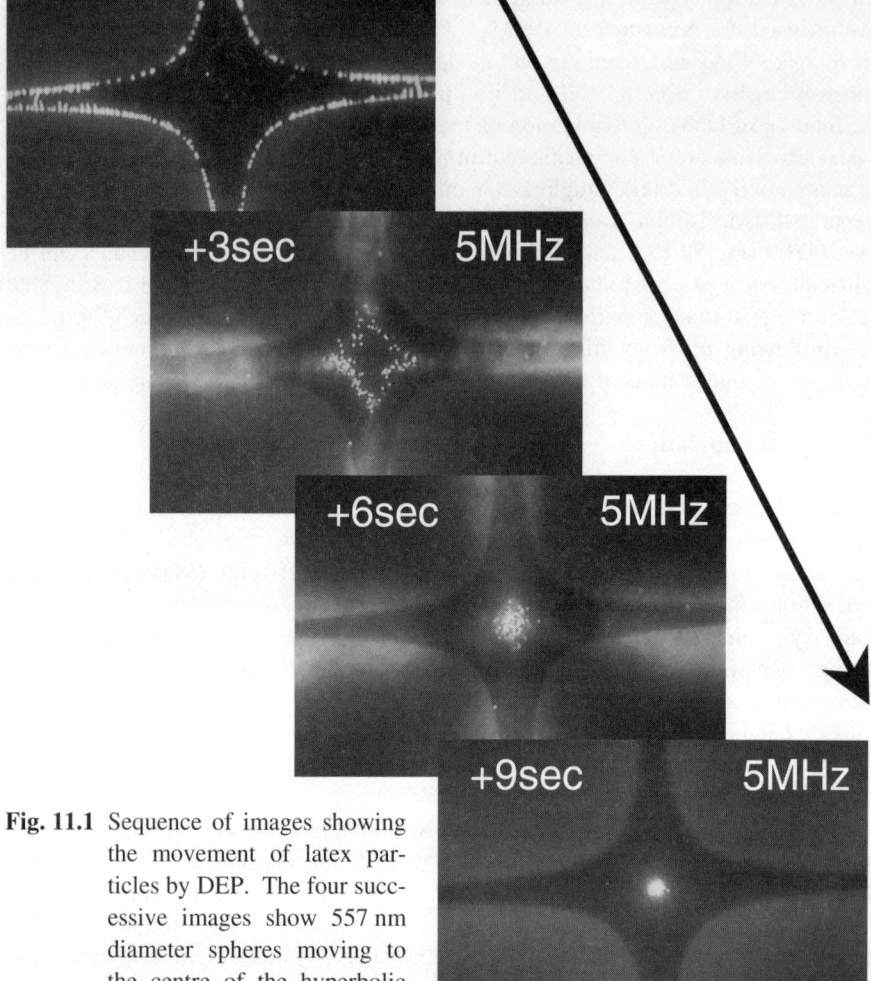

Fig. 11.1 Sequence of images showing the movement of latex particles by DEP. The four successive images show 557 nm diameter spheres moving to the centre of the hyperbolic polynomial electrode array, with central diameter 50 μm and an applied potential of 10 Volts. At 0.5 MHz the particles collect under positive dielectrophoresis. The sequence of images shows the effect of switching the frequency to 5MHz, where they experience negative dielectrophoresis. Particles are repelled from the electrode edge into the centre of the array, where they form a concentrated aggregate after approximately 10 seconds. Except for this central region, particles move away from the electrode edges, up and out of the plane of view; behaviour which matches the response predicted from the field simulations discussed in the previous chapter.

centre of the quadrupole trap. These images demonstrate that gravity has little role to play in governing the dynamics of the particles over this order of time (as discussed in Chapter Nine). Note that all the particles follow the predicted force vectors (see Chapter Ten), so that, for example, those particles sitting over the electrode edges and far from the centre, are pushed upwards and away from the surface under negative DEP.

Sub-micrometre latex spheres can also be trapped on castellated electrode arrays, and examples of this effect are shown in figure 11.2. These images clearly show both positive DEP and pearl chaining of particles at low frequencies (figure 11.2(a)). At higher frequencies, negative DEP causes particles to collect in the characteristic triangular shaped patterns observed in these devices (figure 11.2(b)).

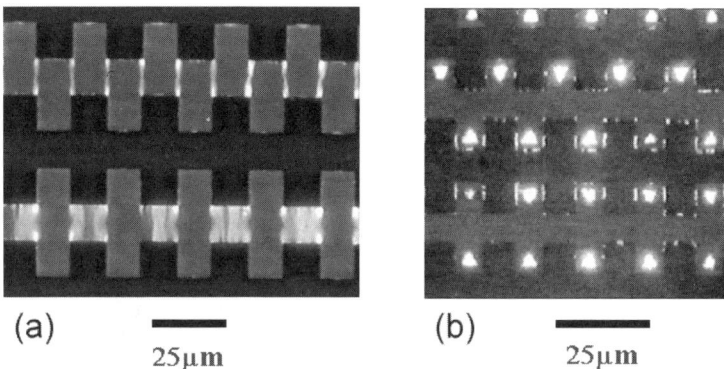

(a)   —— 25μm   (b)   —— 25μm

Fig. 11.2 Experimental photographs of 282 nm diameter spheres trapping under positive (a) and negative (b) DEP forces on a castellated electrode. In   (a) the frequency is 0.5 MHz, and in (b) it is increased to 10 MHz. Note the pearl chaining between opposite tips of the electrodes in (a).

These results are almost identical to those obtained for larger particles (> 1 μm) using castellated electrode arrays with larger characteristic dimensions (see Pethig *et al.* 1992). This indicates that, provided the DEP force is sufficiently large, Brownian motion does not necessarily become a problem, *i.e.* the deterministic force dominates over the stochastic force.

The manipulation and trapping of *single* sub-micrometre particles is also possible. It has been shown that single nanoparticles can be trapped in quadrupole traps (Hughes and Morgan 1998). The smallest single particles that have been trapped are in the 50 nm range, and it has been reported that aggregates of 14 nm diameter particles could be trapped (Müller *et al.* 1996a). It is possible that trapping can also be assisted by electric field-driven fluid flow, which pushes particles into the centre of the quadrupole field trap.

11.1.2   Viruses

From an electrical point of view, viruses can be separated into two categories; those that have an insulating membrane or envelope surrounding a conducting core (similar to a biological cell), and those that do not have a membrane, referred to as non-enveloped. The latter consist solely of a mixture of protein and DNA or RNA. An example of an enveloped virus is Herpes Simplex Virus (HSV) and a diagram of the structure of this virus is shown in figure 11.3(a) (Whitley 1996). This is one of the largest mammalian viruses and is a spherical particle approximately 260 nm in diameter. The core, called the capsid, contains the DNA. This is surrounded by a protein gel, called a tegument and the whole assembly is enveloped by the viral membrane. Non-enveloped viruses by contrast are somewhat simpler, consisting of a solid core of RNA or DNA surrounded by a protein coat. A classical non-enveloped virus is Tobacco Mosaic Virus (TMV), and a schematic cut away diagram of this virus is shown in figure 11.3(b) (Mathews 1981).

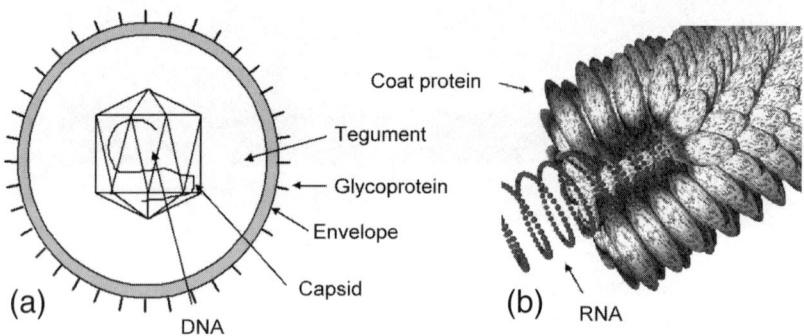

Fig. 11.3   Schematic diagram of the Herpes Simplex Virus (a) and of Tobacco Mosaic Virus (b). The HSV is an enveloped virus, 260 nm diameter, whilst the TMV consists of individual coat proteins with a single strand of helical RNA of radius 4 nm running through the protein units.

DEP has been used to manipulate and characterise both non-enveloped and enveloped viruses, *e.g.* TMV, HSV, Sendai and Influenza. Again, in common with latex spheres, the DEP properties of viruses are dominated by surface conductance effects, so that positive DEP is observed at low frequencies and negative DEP at higher frequencies.

Viruses are too small to be seen by eye and so must be fluorescently labelled in order to be observed. An important issue is whether direct attachment of fluorescent molecules to the surface proteins on the virus leads to a change in the properties of the virus, specifically the surface charge density. Since the surface conductance is a function of the surface charge density, it is to be expected that the labelling process might alter the dielectrophoretic behaviour of the particles.

In order to investigate this, TMV was labelled with a fluorescent probe using two different protocols (Green 1998). In case 1, the probe was attached to the primary lysine groups of the virus coat proteins (a positively charged amino acid). In case 2, the probe was attached to the carboxyl groups of the coat proteins (negatively charged amino acid) in a manner which was designed to minimise the change in surface charge density. Although the density of probes was small, DEP measurements showed that the surface charge density on the TMV varied depending on the labelling protocol leading to a change in the DEP behaviour and the measured dielectric properties. The measured properties of the virus following the two different labelling protocols is summarised in Table 11.1.

Table 11.1   The measured conductivity and permittivity of TMV following fluorescent labelling using two different methods (see text for details).

Labelling Method	σ_p (S m^{-1})	ε_p
Case 1	0.17 ± 0.01	40 ± 5
Case 2	0.085 ± 0.005	55 ± 10

This data shows that labelling leads to measurable differences, implying that less "invasive" methods for identifying particles are required (see later). For enveloped viruses such as HSV, a fluorescent probe can be directly incorporated into the viral membrane. These probes are generally uncharged and are unlikely to have a significant influence on the electrokinetic properties of the particle.

11.1.3 Macromolecules

(i) *Polyelectrolytes, DNA*

There is a strong interest in developing non-contact methods for positioning, holding and stretching molecules such as DNA. There are many potential uses for such a technology, not least in Lab-on-a-chip devices where different DNA fragments may need to be transported, sorted, manipulated or sequenced. DNA is a polyelectrolyte; it is highly charged, surrounded by a counterion cloud of high charge density. The molecules can be polarised in an electric field and exhibits very high effective polarisabilities. The mechanism responsible for the polarisation of DNA is still the subject of debate and dielectric spectroscopic investigations have led to the proposal of a number of mechanisms. These include Maxwell-Wagner interfacial polarisation (Takashima 1989; Saif *et al.* 1991); rotation of bound water molecules (Mashimo *et al.* 1989) and polar groups associated with the DNA (Takashima *et al.* 1984); and fluctuations of counterions along the longitudinal (Mandel 1961; 1977; Manning 1969; 1978; Oosawa 1970; Van der Touw and Mandel 1974, Bakewell *et al.* 2000) and transverse (de Xammar Oro *et al.* 1984) axes of the DNA.

Although it is difficult to draw a definitive conclusion about the polarisation mechanism(s), the evidence tends to favour counterion fluctuation along the DNA axis, particularly at low frequencies. In this mechanism, counterions are envisaged to move unhindered along a subunit of the DNA molecule in the direction of the AC electric field (which will lie parallel to the subunit longitudinal axis), until they reach a potential energy barrier. These barriers result from 'breaks' in the average conformational shape of the molecule attributable to structural features such as kinks or bends. The length L_s of each subunit is deemed to be the average macromolecular conformation between "breaks", or potential barriers, resulting from perturbations in the equipotentials (Van der Touw and Mandel 1974).

There are several variations of this model of polarisation, and in particular Manning (1969; 1978) proposed that a proportion of the counterions are so strongly attracted to the polyelectrolyte that they are said to 'condense' onto the DNA backbone. Essentially there are three distinct phases of ions (Saif *et al.* 1991, Bakewell *et al.* 2000):

- *condensed* counterions: characterised by delocalised binding to phosphate groups of the DNA and thereby neutralising a fraction of the DNA charge.
- *diffuse* counterions: responsible for neutralising the remainder of the DNA charge, with a density which decreases exponentially with distance from the axis.
- *bulk* ions or 'added salt': ordinary aqueous solution ions.

A feature of the condensed phase is that the local concentration of counterions around the DNA does not tend to zero when the bulk electrolyte concentration does. The fraction of condensed counterions ϕ can be expressed in terms of the charge density parameter ξ and ion valency z

$$\phi = 1 - \left| z^{-1} \right| \xi^{-1} \quad = 1 - \frac{4\pi\varepsilon_r\varepsilon_o k_B T b}{q^2 |z|} \tag{11.1}$$

where b is the average distance between charged sites. For a B-DNA double-helix, $b = 0.17$ nm. Using $\varepsilon_r \cong 79$ and $z = 1$, the charge density parameter at 25°C can be estimated to be $\xi = 4.17$. The polarisability α_s depends on the number of condensed counterions n giving

$$\alpha_s = \frac{z^2 q^2 L_s^2 n A}{12 k_B T} = \frac{z^2 q^2 L_s^2 A}{12 k_B T} \left(\frac{\phi L_s}{zb} \right) \tag{11.2}$$

A is the stability factor of the ionic phase and includes mutual repulsion between fixed charges on the backbone and the effect of Debye screening,

$$A = \left(1 - 2(|z|\xi - 1)\ln(\kappa b)\right)^{-1} \tag{11.3}$$

where κ is the inverse Debye length.

Combining equations (11.1), (11.2) and (11.3) leads to a prediction for the subunit length L_s in terms of a measured dielectric decrement $\Delta\varepsilon'$

$$L_s = \sqrt{\frac{9\Delta\varepsilon'}{\pi\varepsilon_r(|z|\xi - 1)AN_A c_p b}} \tag{11.4}$$

where c_p is the concentration of phosphate groups on the DNA (Bakewell *et al.* 2000). Typical subunit lengths are in the range 50 nm to 200 nm so that the polarisability of DNA molecules can be calculated. Alternatively, the subunit length can be estimated from dielectric spectroscopy measurements (Bakewell *et al.* 2000). Typical values for the polarisability of DNA lie between $1 - 100\times10^{-31}$ F m^2 per DNA macromolecule (Bakewell *et al.* 2000; Saif *et al.* 1991; Suzuki *et al.* 1998). This can be compared with the effective polarisability of a 100 nm diameter latex sphere, which is 10×10^{-31} F m^2.

Owing to its high polarisability, the dielectrophoretic manipulation of DNA can be accomplished at relatively low field strengths. Consequently DEP-based methods have been proposed for DNA sorting. For example, Ajdari and Prost (1991) proposed a DEP-enhanced free-flow electrophoresis device which would give one or two orders of magnitude improvement in selectivity over gel electrophoresis. Despite the possible technological potential little was done in this area until Washizu and co-workers developed DEP-based DNA manipulation devices. They showed that DNA could be anchored to an electrode and, through the application of an AC electric field, that the DNA could be stretched. They used this technique to produce "belts of DNA" suspended between two electrodes which allowed molecular "surgery" to be performed on single molecules. In one experiment, single molecules of RNA polymerase were visualised (using fluorescence microscopy) sliding along stretched and oriented DNA molecules (Kabata *et al.* 1993).

In an elegant series of experiments, DNA molecules were stretched between two electrodes and accurate measurements of the spatial position of sequence specific probes determined using restriction enzymes (Oana *et al.* 1999). The restriction enzyme *Eco*RI was used which recognises the sequence GAATTC. This enzyme was fluorescently labelled and complexed with a dielectrophoretically stretched DNA molecule. It was found that the enzyme could be located and that its binding site corresponded to the known molecular sequence of the DNA. Washizu and co-workers have also shown how a stretched DNA molecule can be cut at a specific site using enzymes (Yamamoto *et al.* 2000). The authors used the

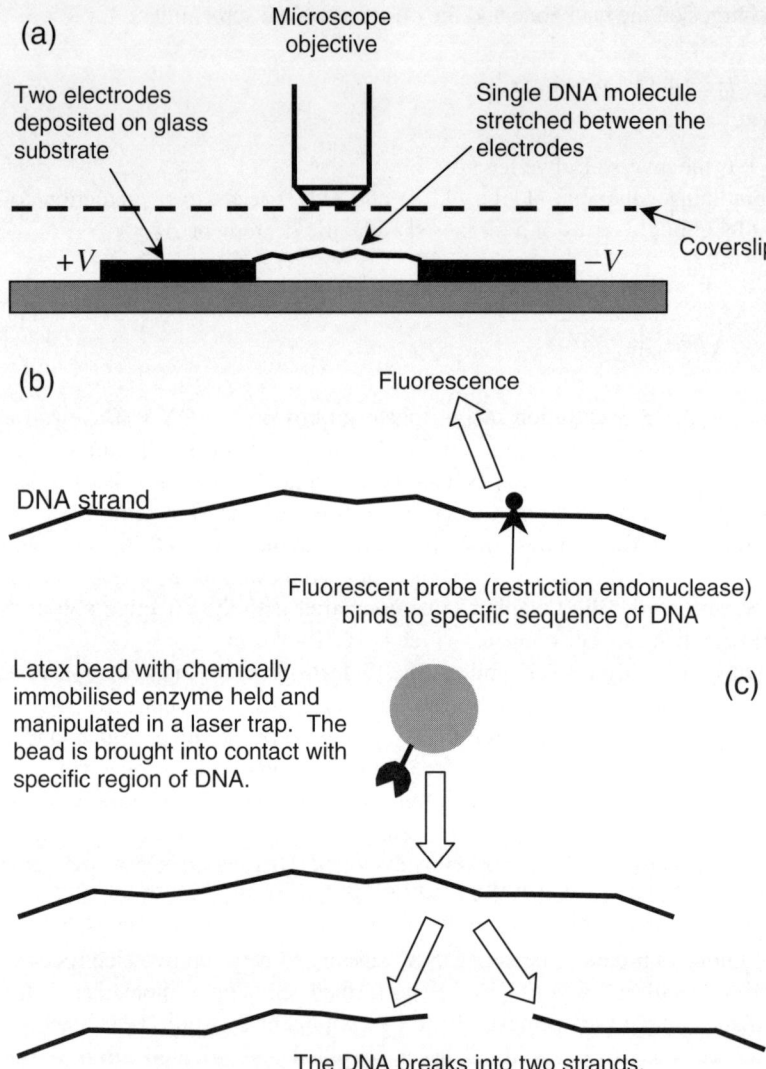

(a)

Microscope
objective

Two electrodes
deposited on glass
substrate

Single DNA molecule
stretched between the
electrodes

$+V$ $-V$

Coverslip

(b)

Fluorescence

DNA strand

Fluorescent probe (restriction endonuclease)
binds to specific sequence of DNA

Latex bead with chemically
immobilised enzyme held and
manipulated in a laser trap. The
bead is brought into contact with
specific region of DNA.

(c)

The DNA breaks into two strands.

Fig. 11.4 Schematic diagram showing how "molecular surgery" is performed on
DNA. A single molecule of DNA is stretched between two electrodes
by dielectrophoresis above the substrate (a). Fluorescent probes can
be used to spatially localise sequences on the DNA strand (b).
(c) shows how a latex bead can be manipulated using laser tweezers to
chemically cleave the DNA molecule into two pieces at a specific
sequence. Redrawn from Yamamoto *et al.* (2000).

enzymes DNaseI and HindIII to cut the DNA. DNaseI cuts DNA at any point while the restriction enzyme HindIII cuts at a specific sequence. These enzymes were chemically attached to a small (1 μm diameter) latex bead, which was manipulated using optical traps against a stretched DNA molecule. The non-specific enzyme DNase1 could cut the DNA as soon as the bead contacted the molecule. However, the restriction enzyme would only cut the DNA at particular points corresponding to the correct nucleotide sequence (restriction sites). The principle of these stretch and manipulate experiments of DNA are illustrated in figure 11.4. This shows how the DNA molecule is first stretched between two electrodes using an AC electric field, keeping it off the substrate allowing "space-resolved molecular surgery" to be performed. Figure 11.4(c) shows how the DNA molecule is cut using an enzyme which is immobilised on a latex bead. The bead is manipulated using a laser tweezers and when the bead is brought into contact with the DNA molecule, it breaks at a particular position along its length.

Such elegant experiments serve to highlight the use of electrokinetic techniques for single molecule experiments and the potential for incorporating this technology into Lab-on-a-chip devices.

(ii) *Proteins*
Understanding and exploring the dielectrophoretic behaviour of protein molecules warrants further attention. Following on from the early work of Debye (1954a; 1964b), Washizu and co-workers (Washizu *et al.* 1994) demonstrated that different protein molecules could be collected at an electrode using DEP forces. The collection of large molecular weight porin trimers by DEP has also been reported. (Fuhr *et al.* 1996). The frequency dependence of the dielectrophoretic properties of proteins, however, has not been published. Morgan and Hughes (unpublished work) showed that the protein Avidin exhibited frequency-dependent behaviour, whereas a very similar protein, Streptavidin, did not. Using hyperbolic polynomial electrodes, fluorescently labelled Avidin could be collected by either positive DEP

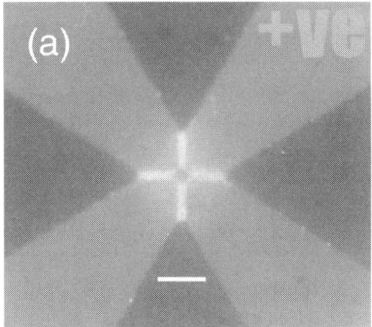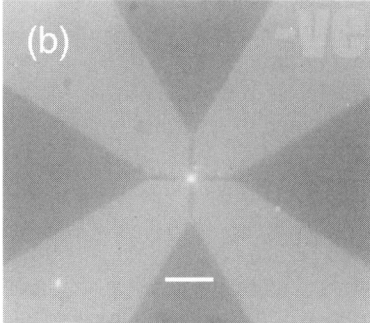

Fig. 11.5 Positive (a) and negative (b) dielectrophoresis of avidin at 2 and 20 MHz respectively on polynomial electrodes. The white scale bars indicate 20 μm.

at the electrode edges or negative DEP in the centre of the electrode, depending on the frequency. Figure 11.5 shows this effect. Avidin is a highly stable protein and has a high affinity for the small molecular weight ligand biotin. Its molecular weight is 63,000 and its size is approximately 5 nm on all faces. During the experiments, no clumping of the protein could be observed, although the possibility that aggregates of protein were formed, could not be entirely ruled out. The observed differences in the electrokinetic behaviour between the two proteins could be due to the differences in charge density, since Avidin has an iso-electric point of 10.5 and Streptavidin has an iso-electric point of approximately 5.5 to 6.0.

The dielectrophoretic manipulation and collection of proteins could, in part, be aided by the fact that many protein molecules possess permanent dipole moments. For example, measurements of the dielectric properties of proteins have indicated that the dipole moments of proteins range from 100 Debye (Myoglobin = 170 Debye) to over 1000 Debye (Horse serum γ-pseudo-globulin = 1100 Debye), see Onkley (1943).

The dielectric properties of macromolecules have been described in detail by a number of authors (e.g. Pethig 1979; Takashima 1989). Essentially, protein molecules consist of large numbers of amino acids (the basic building blocks) linked together into long chains, called polypeptide chains. The three-dimensional structure of the protein is governed to a large degree by the way in which the amino acid residues in the chains cross-link with other residues in the same or different polypeptide chain. Depending on this interaction, the structure may form either an α-helix or β-sheet three-dimensional configuration. Figure 11.6(a), shows the repeat structure of the backbone of a polypeptide chain. The different atoms exhibit different degrees of electronic delocalisation, which gives rise to a net dipole moment as shown in the figure. A typical value for the dipole moment is around 3.4 to 3.6 Debye. If we now consider that the polypeptide chain consists of a large number of such groups connected together, this clearly results in the formation of a large permanent dipole moment. Pethig (1979) illustrated that the dipole moments in an α-helical confirmation are additive, as shown in figure 11.6(b). (For comparison note how the peptide moments in the **double** helix of DNA cancel to give zero net dipole moment.) The α-helix contains 3.6 peptide residues per turn, so that just ten turns gives rise to a dipole moment of 120 Debye.

In addition to the main helix, the polar side chains will also give rise to dipole moments, some of which may add to, or detract from the α-helix dipole. When a protein molecule is solvated (in water) many of the amino acids on its surface ionise, to a greater or lesser degree, depending on their iso-electric points. For example, the iso-electric point of the basic amino acids such as lysine or arginine are 9.5 and 10.76 respectively, so that at pH 7.0 these amino acids will be (almost) fully ionised and carry a net positive charge. In the presence of an AC field such molecules will be highly polarisable and a solution of these molecules will have a high dielectric constant. As the pH increases and approaches the iso-electric point, the molecules will become less ionised and the permittivity of the solution will drop. To a first approximation a protein molecule can be thought of as a globule

(a)

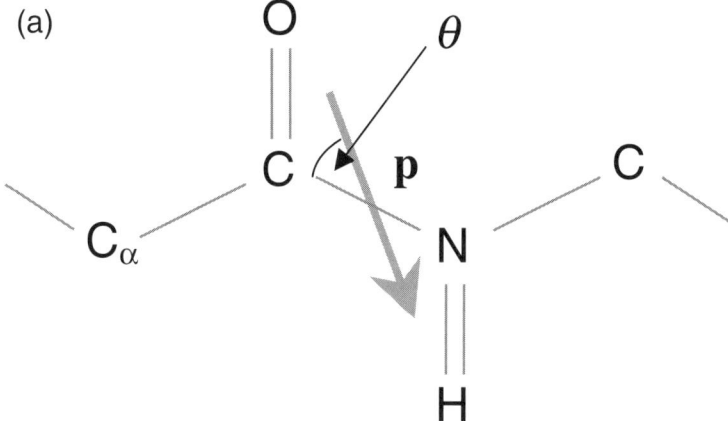

Fig. 11.6   Schematic diagram of how the dipole moments of the monomer in a polypeptide change add. (a) shows the dipole moment of the repeat structure in the backbone of the polypetide and (b) shows how the individual dipole moments are additive.

(b)

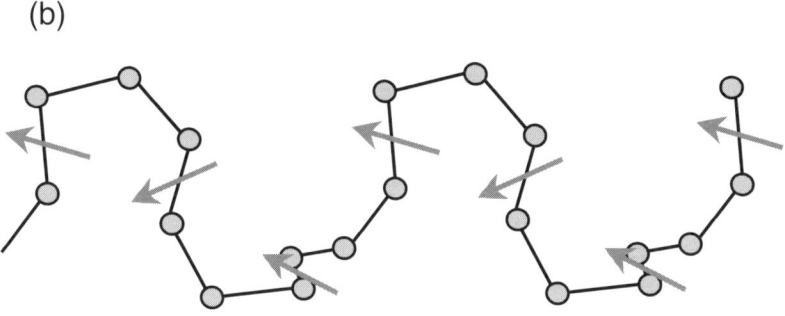

covered in amino acids of various polarisabilities. The ionisable amino acids are nearly always on the surface, so that depending on the pH the protein molecule will carry a net positive or negative charge (or no charge at all at its iso-electric point). The net dipole moment of the protein is then the vectorial sum of the dipole moments due to these charges and the dipole moments due to the various α-helices.

Although this model of a fixed dipole orienting in an electric field may be correct, other interpretations of the dielectric properties of proteins have also been suggested. It is also important to remember that the protein is a colloidal particle and so is surrounded by a double layer. It will therefore polarise in an electric field, leading to the creation of an induced dipole such as for a latex sphere. This explanation for the protein dipole was first proposed by Debye and Falkenhagen

(1928). Any model which is based on double layer polarisation must also consider surface conductance, which is likely to dominate over other effects. Schwan (1957) developed a model to explain the dielectric properties of protein molecules based on Maxwell-Wagner interfacial polarisation mechanisms and this required a large protein conductivity to explain the experimental results (Hasted 1973).

Kirkwood and Shumaker (1952) postulated that the fluctuation of surface-confined protons associated with amino acids was responsible for the large dielectric constant of protein solutions, a model that predicts a pH dependent dielectric dispersion. Grant *et al.* (1978) concluded that the most likely mechanisms were orientation and rotation of the permanent dipole or the proton fluctuation model. The current consensus is that the measured dielectric properties of proteins are due to orientation and rotation of the permanent dipole, which arises from the fixed charge on the protein (Feldman 2002). However, whichever mechanism is ultimately responsible for the dipole moment of the protein, the dielectrophoretic behaviour of proteins has hardly been explored and this is clearly an area that warrants further study.

11.2   Characterising particle dielectric properties with AC electrokinetics

A major application of dielectrophoresis is for characterising the intrinsic dielectric properties of single particles. As we discussed in Chapters Three and Four, in the mid to high frequency range the dielectric and dielectrophoretic properties of particles are governed by Maxwell-Wagner interfacial polarisation. At low to intermediate frequencies long-range polarisation of the double layer occurs. Measurements of the dielectrophoretic properties of particles as a function of frequency can give information about the different polarisation mechanisms.

A number of techniques for measuring the dielectrophoretic properties or spectra of particles have been developed over the years. From the results, values of the dielectric properties of the particles can be obtained. Most of the techniques have been developed for, and applied to the characterisation of particles with sizes greater than 1 μm. Many of the techniques are equally applicable to the characterisation of sub-micrometre particles, although their use in practice has yet to be proven in many cases.

Absolute measurements of the dielectrophoretic force are difficult to achieve for colloidal particles owing to the effects of Brownian motion and electric field-induced fluid flow. Measurements of the dielectrophoretic properties of particles can be done on an ensemble, such as measuring the rate of collection of a suspension of particles at an electrode, or through observation of the behaviour of *single* particles. One obvious problem with the latter method is that the particles are often too small to be seen with the naked eye. In this case particles are labelled with a fluorescent probe.

The techniques used for characterising the AC electrokinetic properties of particles can broadly be divided into the following categories:

Single particle measurements
   (i)   *Zero force measurements*
   (ii)   *Particle levitation*
   (iii) *Particle velocity measurements*

Ensemble average measurements
   (i)   *Collection rates*
          - Optical absorption and fluorescence
          - Impedance measurements
   (ii)   *Light scattering techniques*
   (iii) *Orientation techniques*

In addition to these DEP-based methods, the technique of electrorotation is now beginning to be applied to characterise sub-micrometre particles.

11.2.1   Single particle measurements

(i) *Zero force measurements*

This technique involves observing the motion of single particles in a non-uniform electric field. The field is generated using a microelectrode device and particle motion observed using a fluorescence microscope. Depending on the frequency, particles move either towards an electrode under positive DEP or away from an electrode under negative DEP. At a particular frequency (or frequencies) the DEP force will be zero and the particle will remain stationary. This occurs when the real part of the effective polarisability of the particle is exactly equal to that of the suspending medium (the point at which $\mathrm{Re}[\tilde{f}_{CM}] = 0$). If the latter is known, then the effective complex permittivity of the particle can be calculated. The technique is relatively simple and was developed by Marszalek and co-workers to measure the dielectric properties of sedimenting cells (Fikus *et al.* 1987; Marszalek *et al.* 1989a; 1989b; Marszalek *et al.* 1991).

When this technique is used to characterise homogeneous solid particles (such as latex spheres) it gives unambiguous values for both the permittivity and conductivity of a particle (*e.g.* Green and Morgan 1999). Providing certain assumptions are made, the technique can also be used for shelled particles such as cells (Marszalek *et al.* 1991; Gascoyne *et al.* 1993, Gimsa *et al.* 1991). Measurement of the frequency at which the dielectrophoretic force is zero are relatively simple to make, even for particles down to 50 nm diameter, provided of course they can be seen. The data can provide information about the internal and surface dielectric properties of the particle.

In a typical experiment, the zero force or cross-over frequency is measured for different suspending medium conductivities spanning two to three decades. For a solid homogeneous spherical particle, a frequency *vs.* conductivity map can be plotted, as shown in figure 11.7. Only positive DEP is observed for frequencies

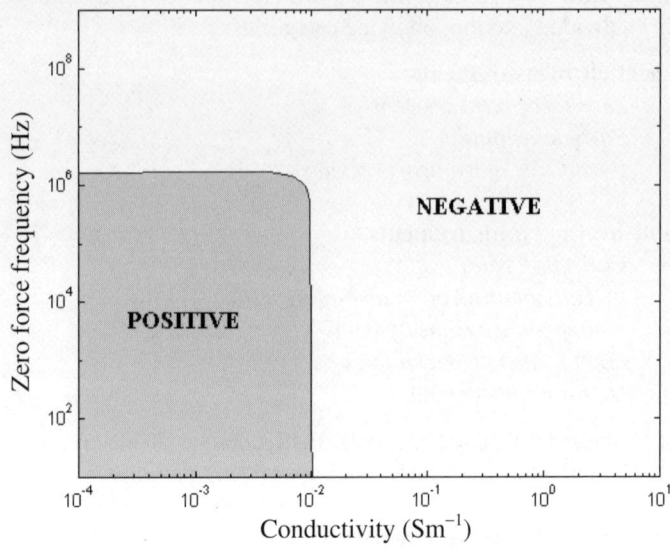

Fig. 11.7 A theoretical plot showing how the zero force point (cross-over frequency) varies with electrolyte conductivity for a solid homogeneous particle. The plot was calculated for a 557 nm diameter particle, with relative permittivity 2.55 and conductivity 10 mS m^{-1}. Note that at a threshold conductivity (10 mS m^{-1} in this figure) the particle polarisability is always less than the electrolyte and only negative DEP is observed.

and conductivities corresponding to the shaded area. In all other regions, for example at high conductivities when the particle's effective polarisability is always less than the suspending medium, only negative DEP is observed. The variation in the cross-over frequency as a function of electrolyte conductivity for different particle conductivities and permittivities, is illustrated in figure 11.8. In figure 11.8(a), the particle permittivity is varied, keeping the conductivity constant, whilst in figure 11.8(b), the particle conductivity has been altered. Note that in practice, the spectrum obtained with latex spheres is more sensitive to variations in particle conductivity than particle permittivity.

For a single homogeneous particle the cross-over frequency can be defined as

$$f_o = \frac{1}{2\pi} \sqrt{-\frac{(\sigma_p - \sigma_m)(\sigma_p + 2\sigma_m)}{(\varepsilon_p - \varepsilon_m)(\varepsilon_p + 2\varepsilon_m)}}$$

(11.5)

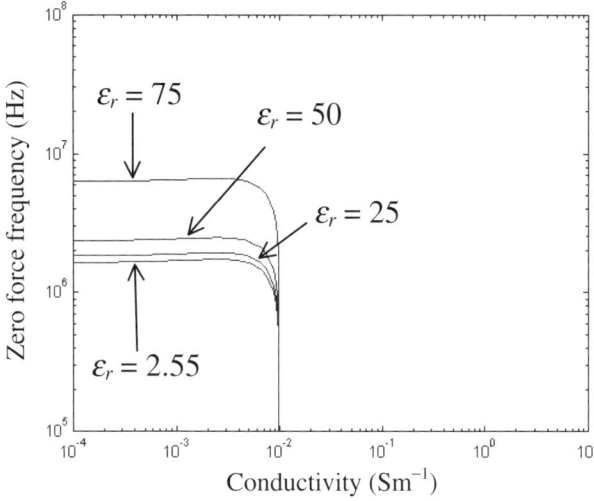

Fig. 11.8 The variation of the zero force line with particle dielectric properties for a 557 nm sphere suspended in a suspending medium of relative permittivity 78. (a) shows how changing the particle relative permittivity alters the zero force frequency at a particular suspending medium conductivity. (b) shows how increasing or decreasing the particle conductivity, increases and decreases the range of both frequency and suspending medium conductivity over which positive dielectrophoresis is observed. (Reprinted with permission from Green and Morgan (1999), copyright (1999) American Chemical Society.)

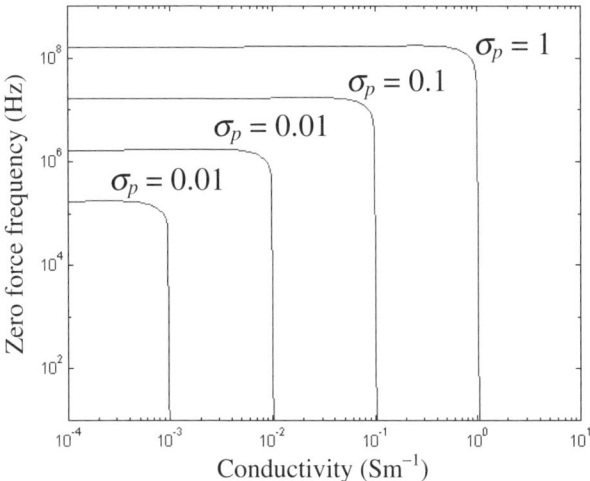

Recasting this equation enables the surface conductance K_s, to be calculated directly from these plots

$$K_s = \frac{a}{4}\left(-\sigma_m + \sqrt{9\sigma_m^2 - 4(2\pi f_0)^2(\varepsilon_p - \varepsilon_m)(\varepsilon_p + 2\varepsilon_m)}\right) \qquad (11.6)$$

where f_o is the cross-over frequency.

The dielectric properties of sub-micrometre latex particles have been measured using this technique (Green and Morgan 1999). Variations in the surface conductance of latex beads could be measured following surface modification such as coupling of antibodies (Hughes and Morgan 1999a; Hughes et al. 1999a; 1999b). The dielectric properties of homogeneous non-enveloped viruses such as TMV have been also been determined (Morgan and Green 1997) using cross-over methods. The measurement of shelled particles is more difficult, since an unambiguous fit cannot be obtained in this case. However, subject to certain assumptions, the dielectric properties of enveloped viruses (HSV) and hollow protein particles (HSV-1 capsids) have also been measured (Hughes et al. 2001; Hughes et al. 2002).

A graph of the zero-force or cross-over data for a 557 nm diameter latex particle is shown in figure 11.9. It can be seen that the data matches the response predicted from figures 11.7 and 11.8 (as shown by the solid line) except at high suspending medium conductivities. It has been shown that the surface conductance of sub-micrometre latex particles is in the range 0.5 – 2 nS (Green and Morgan 1999), consistent with data obtained for larger particles (Arnold et al. 1987). For small particles (< 250 nm diameter) the experimental data cannot be completely explained using the Maxwell-Wagner interfacial relaxation model with constant surface conductance. The behaviour of the particles shows that the surface conductance increases with increasing suspending medium conductivity.

The best fit to the data is found using the model described in Chapter Seven (section 7.1.1) which divides the surface conductance into two independent components, one due to conduction in the Stern layer and one due to conduction in the diffuse layer. Using this model, values for both the conductance of the Stern layer and the zeta potential of the particles can be obtained (Hughes et al. 1999a; Hughes and Morgan 1999).

Owing to the fact that the DEP spectrum is extremely sensitive to changes in the surface characteristics of the particles, it is to be expected that the binding of molecules, such as proteins to the particle, will lead to measurable changes in the AC electrokinetic properties. The first report concerning the effect of binding proteins to sub-micrometre particles was by Pohl (1977). He measured the DEP collection spectrum of silica gel particles in the range of 0.6 μm to 1 μm in diameter as a function of the amount of adsorbed trypsin. The collection rate of particles was found to decrease with increasing trypsin concentration, which was interpreted as a proportional reduction in particle surface charge density. Simple analysis of the collection rate data showed that the amount of trypsin adsorbed on

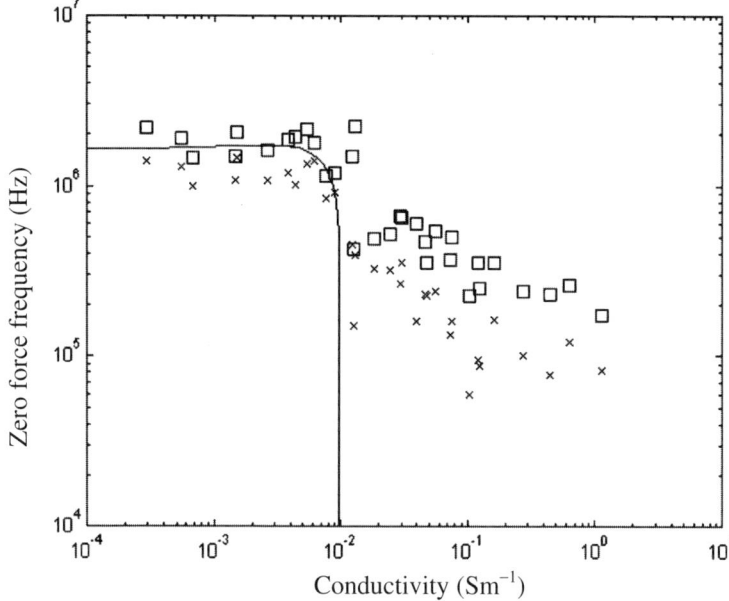

Fig. 11.9 Plot of the mean experimental cross-over frequency against suspending medium conductivity for 557nm diameter spheres in KCl. (×: point up to which only positive DEP was observed. : point above which only negative DEP was observed). Also shown is the best fit to the data calculated with the following parameters for the particle: $\sigma_p = 10$ mS m^{-1}, $\varepsilon_p = 2.55$ and $\varepsilon_m = 78.4$. (Reprinted with permission from Green and Morgan (1999), copyright (1999) American Chemical Society.)

the particles could be directly related to the concentration of the protein in solution through a Langmuir adsorption isotherm.

Subsequent work has shown that both DNA and antibodies can be covalently coupled onto the surface of latex particles (Burt *et al.* 1996; Hughes and Morgan 1999a; Hughes *et al.* 1999a; 1999b). DEP measurements indicate that the effect of binding proteins to the surface is to reduce the surface charge density of the particle and therefore the surface conductance and cross-over frequency (Hughes *et al.* 1999a; 1999b). Owing to the fact that the polarisability of sub-micrometre particles is highly sensitive to surface conductance (through the $2K_s/a$ term in the Clausius-Mossotti factor), binding of as little as a few hundred molecules gives rise to significant changes in the DEP response.

The dielectrophoretic properties of viruses have been measured using cross-over techniques. As described earlier, these are inhomogeneous particles, consisting of a number of different shells or layers, each with a different dielectric property. Consequently they are more difficult to model, but if some

Table 11.2 Values of the dielectric parameters for Herpes Simplex Virus type 1 (HSV-1) as determined from DEP cross-over data. The best fit values are shown in bold; limits are shown in normal type. "*ins.*" means that the model was insensitive to the parameter below a certain threshold. (Reproduced from Hughes *et al.* 2002)

	K_s (nS)	σ_{int} (mS m^{-1})	$\varepsilon_{r,int}$	$\varepsilon_{r,mem}$	ζ (mV)
Fresh	0.3 ± 0.1	100 ± 5	75 ± 25	7.5 ± 1.5	70 ± 5
1 day @ 4°C	0.2 ± 0.02	85 ± 1	75 ± 15	7 ± 0.5	67 ± 2
Trypsin	< 0.05	83 ± 3	75 ± 20	7.5 ± 0.5	40 ± 5
Saponin	0.3 ± 0.05	$40 - 60$	65 ± 10	10 ± 5	75 ± 20
Valinomycin	< 1 (*ins.*)	$\sigma_{med} + 40$	78 ± 2	26 ± 2	74 ± 4

approximations are made, then the dielectric properties of the particles can be obtained. For example, Tobacco Mosaic Virus (TMV) can be approximated to a long thin solid homogenous particle. Modelling it as a prolate ellipsoid, the zero-force *vs* conductivity data can be analysed to give the particle's dielectric properties.

Zero-force measurements have also been used to analyse the properties of enveloped particles such as the Herpes Simplex Virus and the capsid of this virus (Hughes *et al.* 2001; 2002). The shell model (Irimajiri *et al.* 1979, Huang *et al.* 1992) was used to describe the properties of the virus particle. Freshly harvested virus can be modelled as a particle with an internal conductivity of 0.1 S m^{-1}, internal permittivity = $75\varepsilon_o$, membrane permittivity = $7.5\varepsilon_o$, Stern layer surface conductance (K_s) = 0.3 nS and ζ = 70 mV. Because surface effects dominate the conductivity of sub-micrometre particles, *in general* the membrane conductivity cannot be determined for shelled particles (Hughes *et al.* 1998). These dielectric parameters are consistent with values previously reported for cells; the surface conductance is comparable to a value of 0.54 nS determined for erythrocytes (Gascoyne *et al.* 1997); also the membrane permittivity is similar to the value determined for erythrocytes (Gascoyne *et al.* 1993), but the internal conductivity is lower than that measured for a typical cell.

Measurements were also made on virus particles following storage at 4°C, or after exposure to various chemical agents, such as valinomycin (a trans-membrane potassium ionophore), trypsin (an enzyme that removes surface glycoproteins) and saponin (a chemical which increases the membrane permeability by changing the internal conductivity). The measured changes in the dielectric properties of the virus are in line with expectations based on the physiological effects of these agents. Table 11.2 summarises the dielectric parameters for HSV as estimated by fitting the dielectrophoretic cross-over spectra using the single shell model.

Particle behaviour at high suspending medium conductivities
At high electrolyte conductivities, the behaviour of the particles is no longer dominated by Maxwell-Wagner effects. A low frequency cross-over can be measured at high electrolyte conductivities (Green and Morgan 1999), which could be due to a macroscopic polarisation of the entire diffuse double layer tangential to the particle, as discussed in Chapter Seven, section 7.3 and 7.4.

There is no comprehensive theory that explains the low-frequency AC electrokinetic behaviour of colloidal particles. However, to a first approximation, an empirical fit to the data can be obtained by adding a second independent Debye

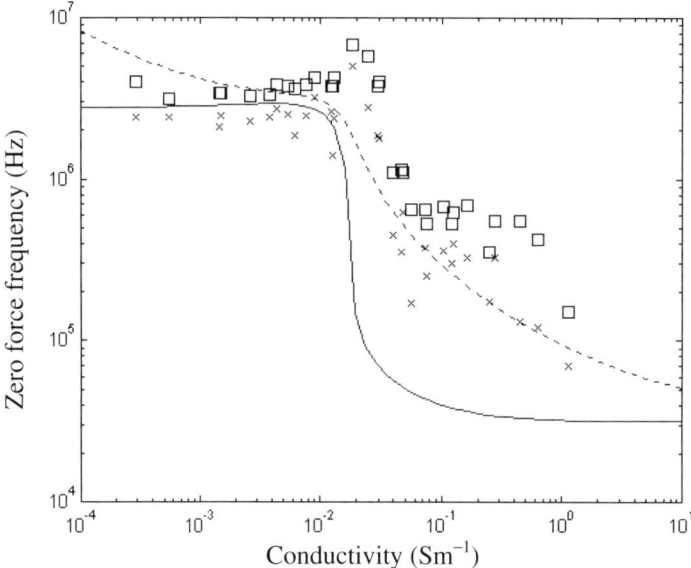

Fig. 11.10   Plot of the cross-over frequency against electrolyte conductivity for 557 nm diameter spheres in KCl. (✕: point up to which only positive DEP was observed.   : point above which only negative DEP was observed.) Also shown is the theoretical cross-over frequencies calculated by adding a second Debye dispersion to account for the α-relaxation seen at higher medium conductivities. Curve (a) is calculated using the relaxation time given by equation (7.17) and curve (b) using equation (7.26). (Reprinted with permission from Green and Morgan (1999), copyright (1999) American Chemical Society.)

relaxation of the form

$$\frac{1}{1+i\omega\tau_\alpha} \tag{11.7}$$

where the time constant of the relaxation is given by either equation (7.17) or (7.26). Using bulk values for ion mobilities and measured values of zeta potentials, the low-frequency zero-force data was calculated, and for the 557 nm diameter beads a best-fit to the low frequency data is shown in figure 11.10. Two cases are shown, one with the electroosmotic contribution to the ion flux around the particle (equation (7.26)) and the other without (equation (7.17)). The predicted response does not match the experimental data fully, pointing to the need for a comprehensive theory which takes into account the entire frequency dependent polarisation mechanism of the particle.

(ii) Particle levitation
The cross-over frequency technique is limited in its application to multi-shelled particles. In order to extend the frequency range of measurements, particle levitation techniques were developed. In principle, these can give a direct measurement of the force on a particle as a function of frequency. The simplest arrangement is an electrode array designed to produce a stable balance between a repulsive negative DEP force and a sedimenting gravitational force (Veas and Schaffer 1969, Jones and Bliss 1977, Bahaj and Bailey 1979, Jones and Kallio 1979, Jones 1995). When a particle experiences negative dielectrophoresis it is repelled from an electrode or an array of electrodes. Under these conditions, the particle moves to a stable potential energy minimum where the gravitational force is exactly balanced by the repulsive DEP force. (Stable trapping of particles can be accomplished using negative DEP forces, provided a local minimum in the electric field magnitude can be arranged, which is allowed in curl-free fields.) Several electrode designs have been used successfully to levitate particles and some typical geometries are shown in figure 11.11. In the design shown in figure 11.11(a), the steady-state levitation height is governed by the balance of a negative DEP force against gravity, whilst in the design shown in figure 11.11(b) an active feedback control system is used to balance a positive DEP force against gravity.

Levitation experiments have been used successfully to measure the dielectric properties of a range of large particles (*e.g.* Kallio and Jones 1980; Huang and Pethig 1991; Fuhr *et al.* 1992; Hartley *et al.* 1999; Cui *et al.* 2000). However, in order to measure the dielectric properties of colloids using levitation methods, the density of the particle must be sufficient for steady-state levitation to be reached within a reasonable time scale. Latex particles have a density slightly greater than water (typically 1050 kg m^{-3}), which means that the buoyancy force is not high. In addition, for sub-micrometre particles Brownian motion can make measurements rather difficult, particles can "escape" from the trap if the impulse from Brownian motion is sufficiently large. However, some sub-micrometre biological particles

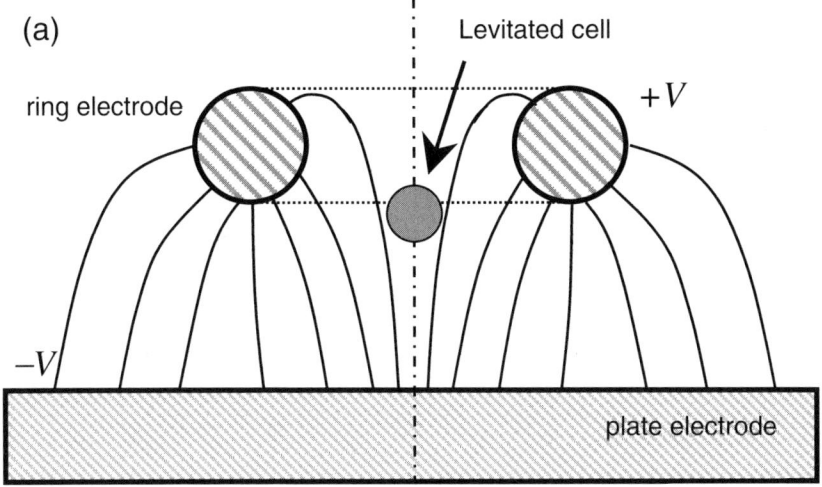

Fig. 11.11   Diagrams of particle levitators, redrawn from Jones 1995 (with permission).   (a) shows a design for a passive levitator, and (b) shows the arrangement of a feedback controlled levitator.

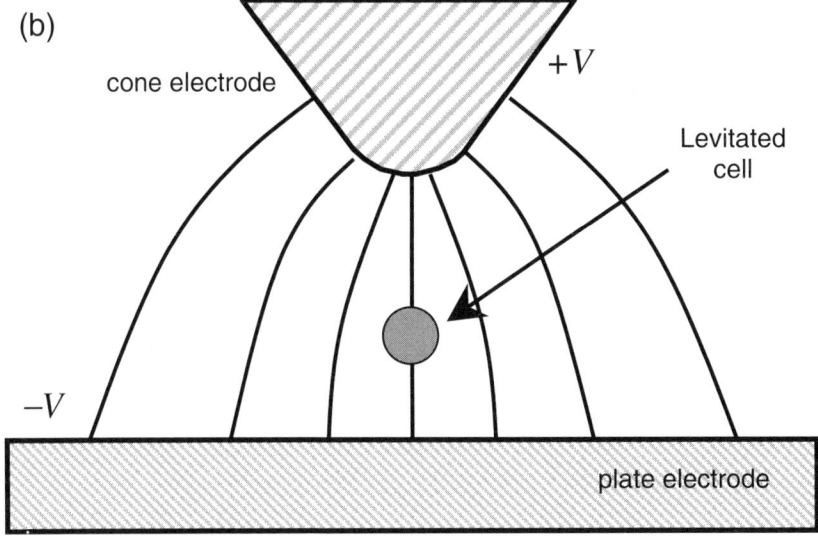

are considerably more dense than water. For example, the Herpes Simplex Virus has a density of 1400 kg m^{-3}; the buoyancy force is much higher than a latex particle and it is therefore possible to estimate the polarisability of the particle using levitation techniques with a planar quadrupole electrode (Hughes and Morgan 1998).

The DEP behaviour of particles undergoing levitation in the electrode arrays, shown in figure 11.11, can be modelled using the dipolar approximation. However, in electrode designs with highly non-uniform electric fields (such as the planar quadrupole), multipolar theory will give a more accurate representation of the effects (Washizu and Jones 1981; Jones and Washizu 1996; Washizu and Jones 1994). In this case, a trapped particle will experience different forces in the radial and axial directions. The axial force term is determined by the quadrupole, whilst the centring force in the radial direction consists of dipolar and quadrupolar moments. Hartley *et al.* (1999) have compared theoretical and experimental data on the levitation of particles in quadrupole traps. They numerically modelled the quadrupolar field and calculated the body forces acting on a particle and found excellent correlation with an analytical solution to the point-charge quadrupole. Their experimental results on the negative DEP levitation of particles in the size 15 μm to 40 μm were found to be in agreement with the model. They used levitation height data to correlate particle radius with measured values and found close correlation (for fixed particle polarisability). Voldman *et al.* (2000) also modelled the quadrupole holding forces on particles in similar traps and determined the maximum fluid flow that could be applied to a particle before it left the trap. Such methods could also be used to measure the frequency-dependent levitation height and thus the polarisability of particles. However, for particles smaller than 1 μm in sizes the effects of Brownian motion would have to be considered in any analysis.

Stable levitation of particles using positive DEP force is impossible without active feedback control, since such a system would require a field intensity maximum at a point in free space, which is not allowed in curl-free electrostatic fields. Therefore, in order to measure the positive DEP response of cells, a dynamic feedback-controlled levitation technique was developed (Kaler and Pohl 1983; Jones and Kraybill 1986; Kaler and Jones 1990). The conditions required for stable levitation of a particle experiencing positive DEP are that the particle must be stable with respect to radial and axial motions. Radial stabilisation is achieved using a focussed electric field with a radially decreasing intensity, whilst axial stabilisation is maintained using a feedback controlled levitation system by controlling the voltage applied to a pair of electrodes, usually of a cone-plane configuration, as shown in figure 11.11(b) (Jones and Kraybill 1986, Jones 1995). This technique was used to measure the dielectric properties of biological particles such as protoplasts and fibroblasts, but it has not been used to analyse the properties of colloidal particles. The technique could also induce fluid motion, which has to be taken into account, particularly at low frequencies or at high suspending medium conductivities.

In order to extend the frequency range over which measurements could be made using this system, a dual frequency excitation technique, which stabilises the particle under negative DEP conditions, was developed (Kaler *et al.* 1992). These authors observed major discrepancies between their data for the DEP force at low frequencies (< 1 kHz) and that predicted by the shell model, which they attributed to electrophoresis. It is likely that low frequency AC electroosmosis of the fluid may also have given anomalous force spectra.

The technique of feedback controlled levitation has been successfully applied to measure the dielectric properties of single cells, and has potential for analysing single sub-micrometre particles. Clearly there are a number of problems in extending the technique. The main issues that must be addressed are the requirements for much higher field strengths to trap particles against Brownian motion and the resulting electric field-induced hydrodynamic effects. In addition, the feedback system requires continuous imaging of the particle in the trap with a camera. This could be adapted for colloidal particles using fluorescence or dark field imaging techniques.

In a variation to the levitation technique, Pohl developed a method for single particle characterisation by measuring the voltage required to release a cell from an electrode, where it was held by positive DEP against the force of gravity (Chen and Pohl 1974). A pair of parallel wires was used, one above the other, and cells were trapped on the uppermost wire where they were held against the force of gravity. The voltage was reduced until the cells were released and fell away from the electrode. The voltage at release was taken to be a measure of the polarisability of the cells. Such a method appears simple, but unfortunately any absolute measure of the particle polarisability will be extremely difficult since the data is influenced by cell-substrate interactions.

In a recent development, laser traps have been combined with DEP traps to measure forces on particles (Fuhr *et al.* 1998). A particle was trapped in the central part of an octopole field cage by negative DEP. Laser tweezers were then used to displace the particle a given distance. The DEP trap voltage was increased until the particle jumped out of the trap. The DEP restoring force acting on the particle was calculated, thus calibrating the laser trapping force. Forces in the tens of pN range were measured for trapped 5 μm diameter latex beads. Although the authors used the technique for calibrating the laser trap, clearly it could be extended to measure the frequency-dependent forces on single particles. In this case an independent calibration of the optical trap strength is required.

(iii) *Particle velocity measurements*

One of the simplest techniques for characterising the dielectrophoretic and dielectric properties of single particles, is measuring their velocity in a well-defined electric field geometry. The DEP force acting on a particle is always balanced by the Stokes force, so that in a region of constant electric field gradient, particles will move with a constant velocity.

Early work using this method was performed by Dimitrov and Zhelev (1987) who measured the velocity of cells moving within an electric field gradient generated using a coaxial electrode system. In the calculations of the DEP forces, the authors also made allowances for particle-particle interaction at short distances. They observed frequency-dependent dielectrophoretic mobilities for a range of single particles but they were unable to relate the data to the permittivity and conductivity of the particles. The velocity of single cells moving between two cylindrical wires was measured by Mahaworasilpa *et al.* (1994). Mid way between the wires, the electric field was calculated analytically. They found good agreement between experiment and theory for cells of diameters between 15 to 20 µm. In similar experiments, Huang *et al.* (1992) measured the velocity of yeast cells moving within the central region of a hyperbolic polynomial electrode array. Within the central region of this type of electrode array, the field gradient and thus the DEP force varies at a constant rate. Observations and measurements of the translational 2D velocity of single cells were made as a function of applied voltage and found to obey the predicted V^2 dependence, although with a non-zero intercept. This was attributed to retardation effects arising from boundary layer effects and convection. Full DEP spectra of cells were measured over four decades of frequency for both live and dead cells and the data was found to fit the predictions of the two-shell model very accurately. Using similar methods, Watarai *et al.* (1997) measured the velocity of single 3 µm diameter latex particles moving laterally on the surface of a hyperbolic polynomial electrode array. These authors were able to measure a DEP force spectrum over 2.5 decades of frequency for two different suspending medium conductivities.

The technique of single particle velocity measurements has potential for characterising the dielectric properties of sub-micrometre particles, providing the effects of Brownian motion are considered. However, the major problem is distinguishing between the direct effect of the electric field on the particle and the secondary effects of electric field-induced fluid motion, particularly AC electro-osmosis. This can be very pronounced in low conductivity electrolytes and at low frequencies, leading to anomalous results.

The DEP properties of sub-micrometre colloidal particles has been characterised by measurement of particle mobility in a well defined field (unpublished data). The velocity of single particles was measured as they moved under positive DEP towards the edge of a parallel strip electrode (figure 11.12(a)). The instantaneous velocity of the particles was obtained by analysing video recordings of particle trajectories, and a plot of the velocity as a function of distance from the electrode edge is shown in figure 11.12(b). The electric field gradient for this electrode geometry was calculated numerically, and the data analysed to obtain average values of the dielectrophoretic mobility of the particles. The fit is shown by the solid line in the figure, giving for these particles an average dielectrophoretic mobility of 2.5×10^{-23} m^4 V^{-2} s^{-1}. Measurements of the mobility as a function of the applied frequency showed that the DEP force decreased over a certain frequency window, which was interpreted as a relaxation process.

The DEP force acting on a particle has also been measured through observation of the critical velocity above which particles can enter a region of electrodes where they would otherwise be repelled by negative DEP (Schnelle *et al.* 1999). The authors were able to show correlation between numerically calculated critical velocities and the measured velocities for latex particles of between 3 µm and 33 µm diameter. Further analysis of the data could be made to

(a)

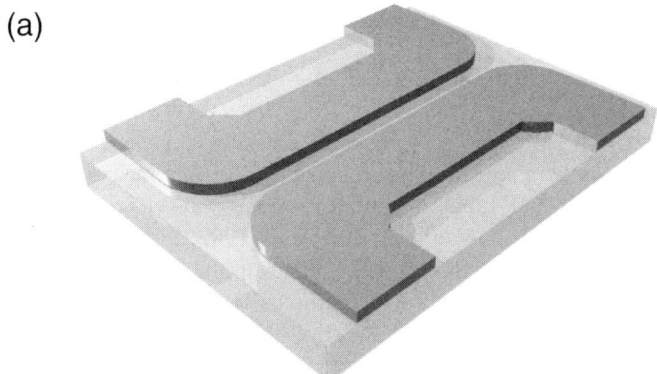

Fig. 11.12   Measurements of the dielectrophoretic velocity of colloidal
particles in a non-aqueous medium. (a) shows a schematic diagram
of the electrode array. (b) shows the instantaneous velocity of
particles experiencing positive dielectrophoresis and moving
towards the electrode edges, plotted against distance from the
edge. The solid line is the theoretical velocity calculated from the
field simulation for this electrode array. (Data reproduced with
permission from The Technology Partnership Ltd.)

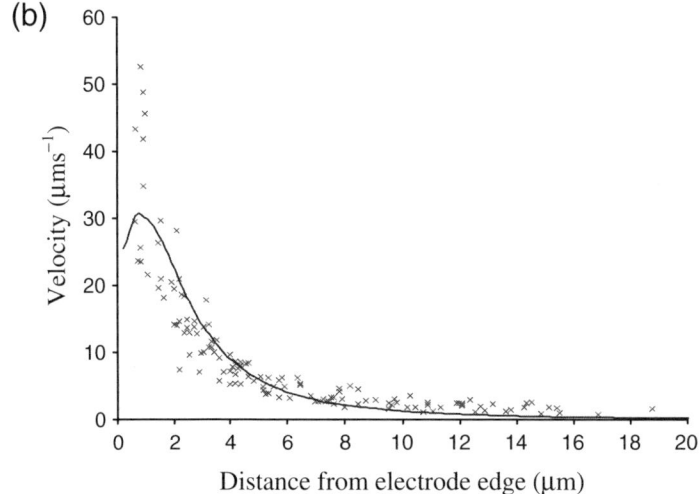

determine the maximum value of DEP force acting on the particles. Measurements
of the dielectrophoretic properties of colloidal particles is possible using this
method, but owing to the high electric fields used, again care must be taken to
distinguish between electric field-driven fluid flow and dielectrophoresis of the
particle.

11.2.2   Ensemble average measurements

(i) *Collection rates*

Observation and measurement of the rate of collection of cells on electrodes under positive DEP forces was first reported by Pohl (for a review see Pohl 1978). It is a relatively simple method and only requires a means of producing a non-uniform electric field (such as a coaxial wire) together with some means of recording cell numbers. A simple and rapid method of measuring cell collection rates was developed by Pethig and co-workers (Burt *et al.* 1989; Price *et al.* 1988; Talary and Pethig 1994). In this method, small changes in the intensity of the transmitted light passing through a suspension of particles held between two interdigitated castellated electrode arrays was measured.   In one design (Talary and Pethig 1994), a small chamber was constructed using two planar electrode arrays separated by a 300 μm spacer. A laser beam is focussed into a 100 μm diameter spot and then onto the particle suspension between the electrode array. The output light is monitored (together with that from a reference chamber) and used as a direct measurement of the DEP force. A schematic diagram of the system is shown in figure 11.13.

When particles collect at the walls of the chamber by positive DEP, the light output increases, and *vice versa* for negative DEP, when particles are pushed into a dense band in the centre of the chamber. The DEP collection rate of a range of organisms was measured across seven decades of frequency. The technique was used to develop practical conditions for cell separation (*e.g.* Markx *et al.* 1994), but was not used to directly determine the dielectric properties of particles. Using similar principles, Brown *et al.* (1997) designed a microscope and camera system to record the rate of retention of cells on a DEP filter by optically measuring cell counts downstream of the device.

Image analysis methods have been applied to the measurement of the DEP collection of cells on electrode arrays (Gascoyne *et al.* 1992; 1994). This technique is particularly suitable for analysis of cell sub-populations (200 to 300 cells) on an electrode array. Using a camera, microscope and computer for analysis, images of cells were recorded before and during the application of an electric field. Image processing enabled identification of cells and characterisation of movement into and out of regions of positive and negative DEP on a castellated electrode array. The DEP collection of cells was recorded as a function of frequency, and values of the cross-over frequency obtained from this data were used to calculate the dielectric properties of the cells (Gascoyne *et al.* 1994).

Collection rate measurements have also been used to characterise the properties of sub-micrometre particles (Hughes and Morgan 1999b, Bakewell and Morgan 2001). In one design, an evanescent field method was developed to ensure that only particles collecting on the electrodes were analysed. The total fluorescence signal emitted by particles in the region of the electrodes, was used as a measure of the particle number. The Maxwell-Wagner relaxation of the particles could be clearly identified as a frequency-dependent change in the collection rate.

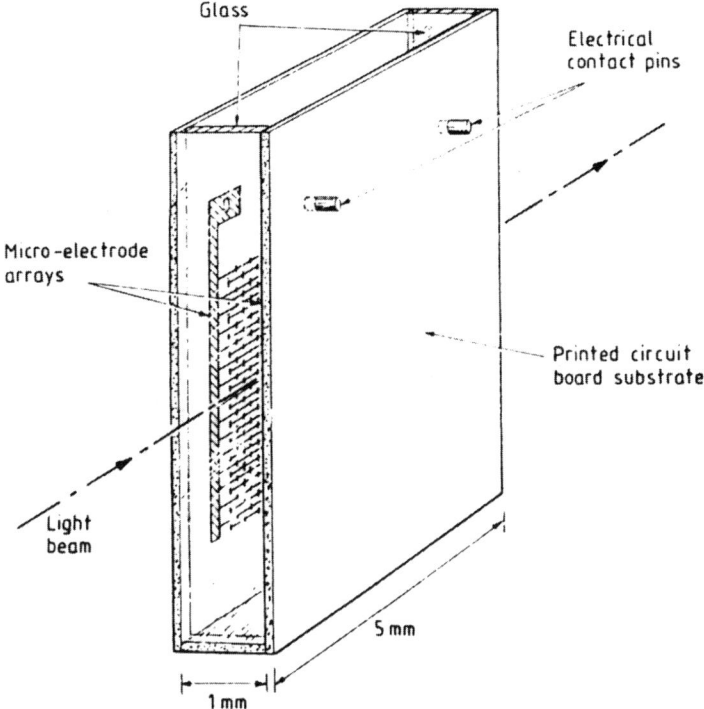

Fig. 11.13 Schematic diagram of the optical chamber used for the detection and measurement of dielectrophoretic particle collection. (Figure reprinted with permission from Talary and Pethig (1994), © 1994 IEEE.)

The collection of sub-micrometre particles onto electrodes can also be measured using optical methods, for example fluorescence. Washizu *et al.* (1994) used microelectrodes to demonstrate the trapping of a range of biopolymers (proteins and DNA) which had been fluorescently labelled. These authors were able to qualitatively estimate the fluorescent intensity over the electrode as a function of time for different concentrations of protein collecting by DEP.

In other experiments, the collection of fluorescently-labelled TMV between interdigitated electrodes was measured in order to estimate the minimum deterministic force required for DEP manipulation (Morgan and Green 1997). Asbury and van der Engh (1998) demonstrated that the time course of DNA trapping onto the edges of interdigitated electrodes could be measured using fluorescence. The collection rate was related to the total light emitted from the area of observation. A major experimental problem in this case was photobleaching of the fluoroprobe. To avoid this, the authors took a series of short duration images using an electronic shutter. Gold electrodes and very low frequencies (30 Hz) were used so that electrohydrodynamic effects are likely to strongly influence the collection rate data. However, they were able to quantify movement of particles on and off an electrode. Fluorescent techniques have also been used to quantify the DEP collection of latex particles onto an interdigitated electrode array (Bakewell and Morgan 2001). In this work, image processing was applied to analyse sequential images of fluorescent latex particles collecting on an electrode array. The total number of particle collecting (in a given area of electrode) as a function of time was plotted. Results showed that the on and off rates could be fitted to a first order exponential.

Measuring particle polarisabilities from collection rate data
Although collection rate techniques are able to demonstrate differences between cell types, and also frequency dependent effects, it is difficult to correlate measurements of optical absorbence directly with the polarisability of particles. However, given the ease with which particle collection rates can be measured, there is clearly the possibility of developing a method for analysing the data to obtain values of particle polarisability. To this end, a theoretical framework capable of predicting the spatial and temporal evolution of particle collection at electrodes has recently been developed (Bakewell and Morgan 2001). The model is based on a numerical solution to the Fokker-Planck equation (FPE). As outlined in Chapter Nine, the FPE describes the behaviour in space and time of the probability distribution of particles in solution when subjected to an arbitrary external force.

The FPE can be solved (numerically or analytically) to give a particle concentration profile provided that all the constants are known. The diffusion constant (and friction factor) is determined by the size of the particle, temperature, viscosity, etc. The DEP force depends on the particle volume and its effective polarisability, together with the electric field gradient, which can be calculated analytically or numerically. The collection rate of particles on a surface is related to the probability density function and can be determined from collection rate data. This means that within the constraints of the model, the solution to the Fokker Planck equation depends on a single unknown, the particle polarisability.

The FPE has been solved using numerical techniques and estimates of the time constant of the particle collection rate, the steady-state concentration, etc. obtained. Comparison of simulation with experimental collection rates shows a degree of correlation, particularly with regard to the time constants to steady state (Bakewell and Morgan 2001). Further refinement of the model should lead to a comprehensive method for analysing collection rate data allowing the frequency-dependent effective polarisability of an ensemble of particles to be determined.

Impedance measurements
Considering the early success of optical methods for quantifying the collection of particles at electrodes, it is surprising that it took a number of years before impedance measuring techniques were developed. Milner *et al.* (1998) and Allsopp *et al.* (1999) demonstrated that by measuring the impedance of the DEP

collection electrodes, they could monitor particles collecting under positive DEP. Subsequently, Suehiro *et al.* (1999) developed a simple but elegant model that was able to quantify a measured change in impedance with the number of particles collecting on an electrode array. The model was based on an equivalent circuit analysis. Single particles were collected as pearl chains and modelled as consecutive series impedance elements; the total impedance of the system was modelled as a parallel combination of the suspending electrolyte and the parallel pearl chains. The principle of the measurement and analysis is shown in figure 11.14. The equivalent circuit is made of two components, the impedance of the pearl chains (Z_{Ch}) and the impedance of the electrode-suspending medium (Z_E). Modelling a biological particle (cell) as a capacitor C_P and resistor R_P in parallel with total impedance Z_P, the impedance of a pearl chain is the series sum of these impedances. The total impedance of the chains bridging the electrodes is the parallel sum of Z_P, as shown in the figure. In the analysis it was assumed that if the total number of pearl chains is small, then the impedance of the electrode plus electrolyte does not change. An expression for the total number of particles collected N, can be derived from simple circuit analysis to give (Suehiro *et al.* 1999)

$$N = m^2 \frac{R_P}{R_T} = m^2 \frac{C_P}{C_T}$$

where the m is the number of particles forming a pearl chain (a constant which is set by the electrode gap) and the total capacitance C_T and resistance R_T of the system, which are measured.

The collection of cells on an electrode consisting of interdigitated fingers was measured in a flow-through system. A suspension of cells was passed over an electrode array at a constant rate and an AC voltage applied to the electrodes to induce DEP collection; this voltage was also used to simultaneously measure the impedance. Excellent correlation was obtained between the measured collection rate and the model. A quantitative estimate of the cell concentration in suspending liquid could be made down to 10^5 cells ml^{-1} (Suehiro *et al.* 1999).

Such a technique could be extended to measure the effective impedance of the cells in suspension, since the ratio of the measured capacitance to conductance is given by the product of the cell resistance and capacitance, which is a characteristic of a particular cell type. It has been shown that impedance techniques can, in principle, be used to monitor the collection of sub-micrometre particles onto electrode arrays (Allsopp 1999), but it remains to be seen if the method can measure the effective polarisability of collecting sub-micrometre particles.

Impedance-measuring techniques are in their infancy at present, but it is expected that rapid developments will be made in this area over the next decade, particularly with the advent of devices incorporating microelectrodes capable of rapid measurement of the impedance of single particles (Ayliffe *et al.* 1999, Fuller *et al.* 2000, Gawad *et al.* 2001).

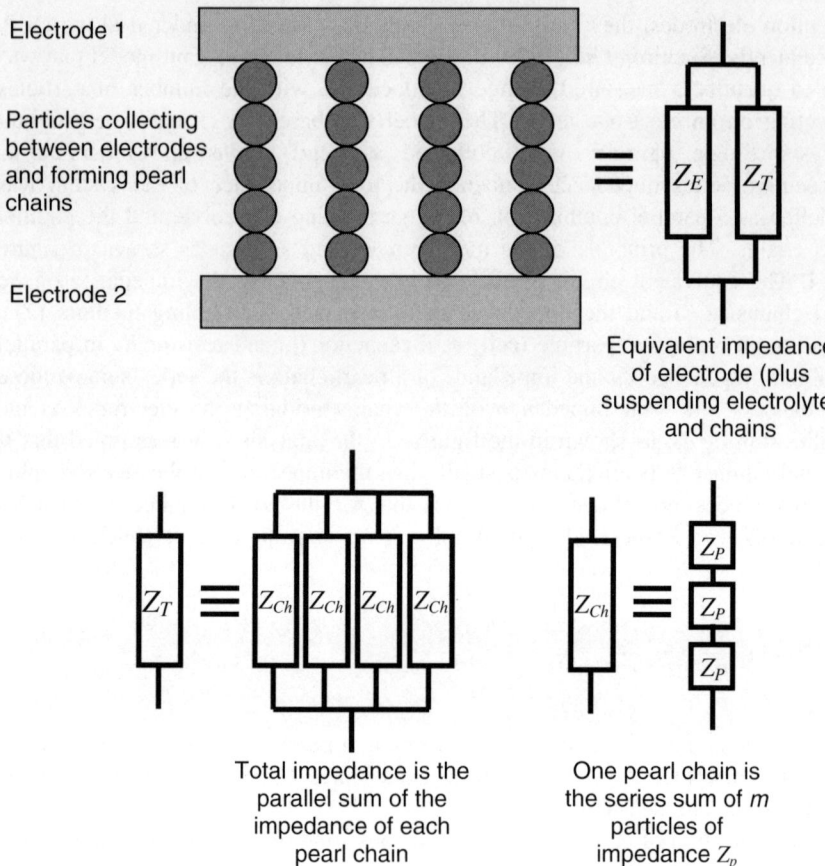

Fig. 11.14 Diagram showing the principle of impedance detection of particle
collection on an electrode array. The total impedance is the parallel
sum of the electrode-electrolyte impedance plus the impedance due to
all the pearl chains. (Figure redrawn from Suehiro *et al.* 1999).

(ii) *Light scattering techniques*
The first demonstration of particle velocity measurements using light scattering
techniques was reported by Kaler *et al.* (1988). The technique measures the
velocity of a particle by recording the Doppler frequency shift of laser light
scattered from a suspension of particles. The technique of light scattering is well
known and is explained in a number of texts (Chu 1976). The experimental data is
obtained in the form of an autocorrelation function τ, and for small scattering
angles the velocity of particles is proportional to the inverse auto correlation
function. In the work of Kaler *et al.* (1988), a photomultiplier was used to measure

the autocorrelation function for yeast cells in suspension moving in an isomotive electrode geometry over a range of frequencies. The resulting velocity spectra (related to the DEP force) showed several peaks, which the authors assigned to various relaxation phenomena.

This early work was improved upon (Eppmann *et al.* 1996; 1999; Gimsa *et al.* 1997) with the development of phase analysis light scattering to measure the dielectrophoretic mobility of particles. In this technique, two lasers with slightly different wavelengths are used to create an interference region with a moving fringe pattern. When particles cross the fringe pattern the scattered light intensity varies, generating a Doppler signal. Phase demodulation of the Doppler signal can be performed if the frequency difference between the two lasers is used as a reference. This technique is called dielectrophoretic phase analysis light scattering (DPALS).

Using this technique, the movement of 15 μm diameter latex particles between two pin electrodes was measured to obtain the particle velocity and dielectrophoretic force spectrum (Eppmann *et al.* 1999). The results matched the theory, and the spectrum was analysed to obtain the surface conductance of the particles. DPALS is a promising method for characterising colloidal particles, providing a number of experimental problems can be overcome, including sedimentation of the particles. Measuring particle velocities in isomotive electrode geometries would also improve the system and, in principle, the technique can be used to analyse particle movement down to 10 nm diameter (Eppmann *et al.* 1999).

(iii) *Orientation techniques*
It has been known for some time that non-spherical particles orient in an AC electric field. The orientation depends on the effective polarisability of the particle (see Chapter Four) and has been observed for a wide range of biological particles (Teixeira-Pinto *et al.* 1960; Schwarz *et al.* 1965; Saito *et al.* 1966; Griffin and Stowell 1966; Griffin 1970; Fomchemkov and Gavrilyuk 1977; 1978; Ascoli *et al.* 1978; Iglesias *et al.* 1985; Miller and Jones 1993). Early work even showed that virus particles could be oriented in a low frequency AC field (Lauffer 1939; O'Konski and Zimm 1950).

The orientation of a particle is caused by a torque acting on the particle. For the general case of the prolate ellipsoid (see Chapter Four), the orienting torque can be calculated (Jones 1995) as a function of frequency. The ellipsoid can orient with one of its three axis aligned with the field. Calculation of the stable orientation position requires the electric field-induced torque to be determined from the dipole moment of the particles. The resulting expression is given in Miller and Jones (1993) and Jones (1995); the expected orientation of particles can be calculated for different frequencies and suspending medium conductivities.

Good agreement between theory and experiment has been obtained. For example, Miller and Jones (1993) showed that for *Euglena gracilis* and llama erythrocytes, values for the internal permittivity and conductivity of the cells could be obtained through a subjective best fit between data and theory. Recently Ignatov *et al.* (1997; 1998) used electro-orientation to determine the biological activity of bacteria in suspension. They illuminated a suspension of bacteria with non-polarised light and measured the optical density of the suspension along and across the orientation direction. It was shown that the orientation spectrum could be used to determine whether bacteria were metabolising a substrate, and that the spectrum changed depending on the concentration of substrate in the suspending medium. Cells that lacked suitable metabolising enzymes showed no such behaviour.

Perhaps the most widely studied non-spherical sub-micrometre particle is DNA. The orientation of DNA in electric fields has been observed by a number of workers. In an electric field the DNA aligns along the field lines. Since it exhibits optical anisotropy, this orientation has been studied using optical birefringence (Hogan *et al.* 1978) and linear dichroism (Stellwagen 1981). Using fluorescently labelled DNA, Washizu and co-workers (Suzuki *et al.* 1998) studied the electric field-induced fluorescence anisotropy as a function of field strength and frequency. As well as alignment, they noticed dielectrophoretic attraction of the DNA to the electrodes. From the anisotropy they calculated a value of the apparent polarisability of the DNA (1×10^{-32} F m^2 for pUC18 plasmid DNA, 2.7 kb, circular, 900 nm diameter).

Such a technique could, in principle, be applied to characterise the dielectric properties of a range of non-spherical particles provided their orientation can be reliably observed. Since the optical path length in microsystems is short, optical birefringence and linear dichroism techniques are difficult to use, so that fluorescence microscopy would be the preferred method.

11.3   Electrorotation

The technique of electrorotation has been developed as a method for characterising the dielectric properties of single cells and other particles and several reviews have been written on the subject (Gimsa *et al.* 1991, Fuhr *et al.* 1991). As described in Chapter Four, the method measures the imaginary part of the effective polarisability of the particle. The technique involves observation (by eye usually) of the rotation of a particle in a circularly-polarised electric field. Particles rotate at speeds of up to a few revolutions per second, which makes data collection and analysis fairly straight forward. However, in order to observe motion by eye the particles must be optically anisotropic. For example, they need to be partly non-spherical and/or exhibit regions of different optical contrast (refractive index). This means that observation of cells is possible, but the rotation rate of latex spheres is much more difficult to measure unless they are irregularly shaped. The electrorotation rate of elongated latex spheres (3×5 µm) has been measured (Burt *et al.* 1996; Hughes *et al.* 1999c). It is possible to measure the rotation rate of sub-micron latex spheres by "decorating" the particles with small marker beads. For example, 0.83 µm carboxylate modified latex particles were decorated with 0.2 µm amino fluorescent latex particles using standard coupling protocols (unpublished data). The stoicheiometry was such that, on average, each large bead had three

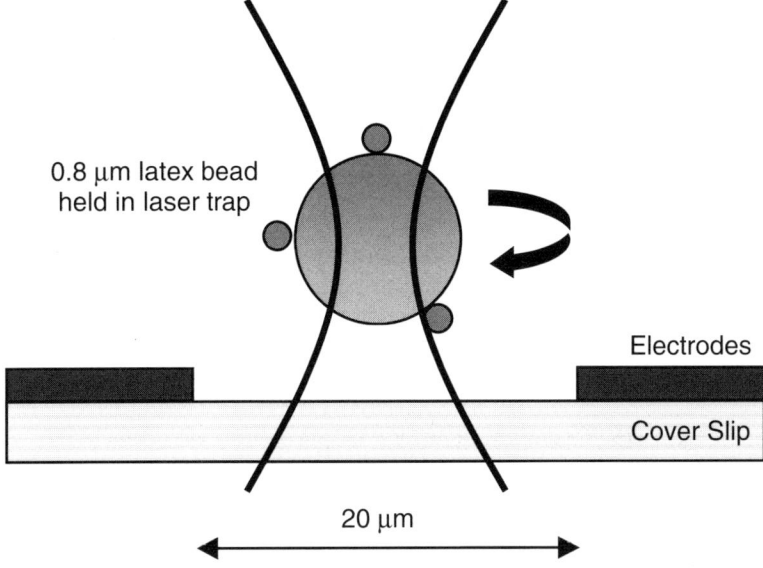

Fig. 11.15   Schematic diagram of a latex bead held in an optical trap at the focal point of a laser beam and rotated by electrorotation.

small beads on its surface, as shown in figure 11.15. Electrorotation electrodes were fabricated on thin glass cover slips so that they could be used in an optical trap. Single beads could be selected and held in the centre of a polynomial electrode array using the trap as shown in the figure. The trap stabilises the particle against DEP forces and EHD forces so that small electrode gaps can be used. The rotation rate can be measured from the power spectrum of the light scattered by the particle onto a quadrant photodiode (Berry and Berg 1997). Using this arrangement the rotation rate of sub-micrometre particles can be characterised with relative ease and extremely high velocities can be achieved (up to 1000 Hz) at low voltages. An example of a rotation spectrum for a latex particle (0.83 μm diameter) is shown in figure 11.16, together with a best fit obtained using the standard theory for modelling a solid particle (see Chapter Four).

Recently dynamic light scattering has been coupled with electrorotation in the development of a technique called electrorotation light scattering (ERLS) (Gimsa 1995; Prüger 1997; Eppmann 1996; Prüger 1998). This technique can be used to measure the rotation of sub-micrometre particles, except that the stochastic behaviour of small particles causes a decorrelation of the intensity of the signal. As for any electrorotation, application of ERLS requires that the particles be optically anisotropic. The technique offers the possibility of rapid automated measurements of the properties of dielectric particles and also enables time-resolved characterisation of events. However, the ultimate sensitivity of the

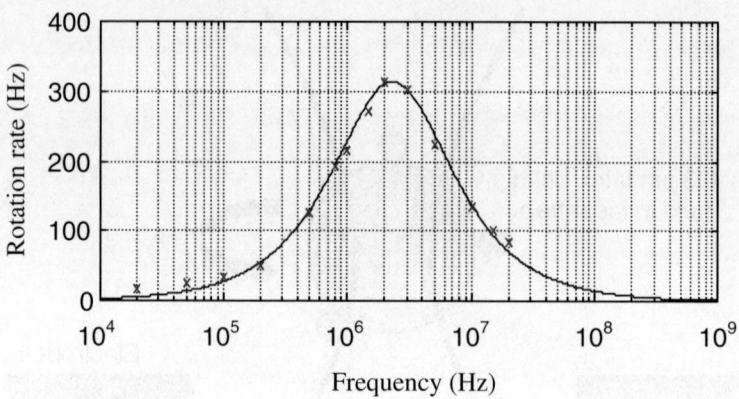

Fig. 11.16 Rotation spectrum for a 830 nm diameter latex particle.

technique places lower limits on the size of particles that can be practically measured. The deterministic force could, in theory, be increased by maximising the field strength, but as with DEP this causes unwanted fluid movement. For small particles, the rotational term cannot easily be distinguished from the background translational diffusion so that a theoretical lower limit of 800 nm diameter was postulated (Eppmann 1996). Despite this the ERLS spectra of the influenza virus has been obtained (Gimsa 1999), although it was concluded that the signal probably originated from aggregates of a few viruses.

11.4 Summary
It is now an established fact that dielectrophoresis can be used to manipulate a range of micro- and nano-scale particles. Single particles can be trapped and their dielectric properties measured. Collections of particles can also be manipulated and various physical characterisation methods used to determine their properties. In the following chapter, we will show how small changes in particle properties can be exploited to produce selective separation.

11.5 References
Allsopp D.W.E., Milner K.R. Brown A.P and Betts W.B. *Impedance technique for measuring dielectrophoretic collection of microbiological particles* J. Phys. D: Appl. Phys. **32** 1066-1074 (1999).

Ajdari A. and Prost J. *Free-flow electrophoresis with trapping by transverse inhomogeneous field* Proc. Natl. Acad. Sci. USA **88** 4468-4471 (1991).

Arnold W.M., Schwan H.P and Zimmermann U. *Surface conductance and other properties of latex particles measured by electrorotation* J. Phys Chem. **91** 5093-5098 (1987).

Asbury C.L. and van der Engh G. *Trapping of DNA in non-uniform oscillating electric fields* Biophys. J. **74** 1024-1030 (1998).

Ascoli C., Barbi M., Frediani C. and Petracchi D. *Effects of electromagnetic field on the motion of Euglena gracilils* Biophys. J. **24** 610-612 (1978).

Ayliffe H.E., Frazier A.B. and Rabbit R.D. *Electric impedance spectroscopy using microchannels with integrated metal electrodes* IEEE J. Microelectromech. Systems. **8** 50-56 (1999).

Bahaj A.S. and Bailey A.G. *Dielectrophoresis of small particles* Proc. Ind. Appl. Soc. (IEEE) Annual Meeting, Cleveland, OH, USA 154-157 (1979).

Bakewell D.J., Ermolina I., Morgan H., Milner J.J. and Feldman Y. *Dielectric relaxation measurements on 12 Kbp plasmid DNA* Biochim. Biophys. Acta. **1493** 151-158 (2000).

Bakewell D.J. and Morgan H. *Measuring the frequency dependent polarisability of colloidal particles from dielectrophoretic collection data* IEEE Trans. Dielectrics & Electrical Insulation **8** 566-571 (2001).

Barber W.A., Debye P. and Eckstein B.H. *Field-induced diffraction method for molecular weight determination* Phys. Rev. **94** 1412 (1954).

Berne B.J. and Pecora R. *Dynamic Light Scattering* Wiley, New York (1976).

Berry R.M. and Berg H.C. *Absence of a barrier to backwards rotation of the bacterial flagellar motor demonstrated with optical tweezers* Proc. Natl. Acad. Sci. **94** 14433-14437 (1997).

Brown A.P., Harrison A.B., Betts W.B. and O'Neill J.G. *Measurement of the dielectrophoretic enrichment on grid electrode using image analysis.* Microbios. **91** 55-65 (1997).

Burt J.P.H., Al-ameen T.A.K. and Pethig R. *An optical dielectrophoresis spectrometer for low-frequency measurements on colloidal suspensions.* J. Phys. E: Sci. Instrum. **22** 952-957 (1989).

Burt J.P.H., Chan K.L., Dawson D., Parton A. and Pethig R. *Assays for microbial contamination and DNA analysis based on electrorotation* Ann. de Biologie Clinique **54** 253-257 (1996).

Chu B. *Laser Light Scattering* Academic Press, New York (1974).

Chen C.S. and Pohl H.A. *Biological dielectrophoresis: the behaviour of lone cells in non-uniform electric fields* Ann. N.Y. Acad. Sci. **238** 176-185 (1974).

Cui L., Holmes D. and Morgan H. *The dielectrophoretic levitation and separation of particles in travelling wave electric fields* Electrophoresis **22** 3893-3901 (2001).

Debye P. and Falkenhagen H. Dispersion von Leitfähigkeit und Dielektrizitätskonstante bei starken Electrolyten Phys Z. **29** 401-416 (1928).

Debye P., Debye P.P., Eckstein B.H., Barber W.A. and Arquette G.J. *Experiments on polymer solutions in inhomogeneous electrical fields* J. Chem. Phys. **22** 152-153 (1954a).

Debye P., Debye P.P. and Eckstein B.H. *Dielectric high-frequency method for molecular weight determinations* Phys. Rev. **94** 1412 (1954b).

Dimitrov D.S. and Zhelev D.V. *Dielectrophoresis of individual cells – experimental methods and results* Bioelectrochem. Bioenerg. **17** 549-557 (1987).

Duhkin S.S. *Dielectric phenomena and the double layer in disperse systems and polyelectrolytes* John Wiley and Sons, New York (1974).

Eisenstadt M. and Scheinberg I.H. *Dielectrophoresis of macromolecules: determination of the diffusion constant of poly-γ-benzyl-L-glutamate* Science **176** 1335-1337 (1972).

Eppmann P., Prüger B. and Gimsa J. *Particle characterisation by AC electrokinetic phenomena 2. Dielectrophoresis of latex particles measured by dielectrophoretic phase analysis light scattering (DPALS).* Colloids and Surface A: Physicochemical and Engineering Aspects **149** 443-449 (1999a).

Eppmann P., Gimsa J., Prüger B. and Donath E. *Dynamic light scattering from oriented, rotating particles: A theoretical study and comparison to electrorotation data* J. Phys. III, France **6** 421-432 (1996b).

Feldman Y. (*Personal Communication*)

Fikus M., Marszalek P., Rozycki S. and Zielinsky J.J. *Dielectrophoresis and electrofusion of Neurospora crassa slime* Stud. Biophys. **119** 73-79 (1987).

Fomchemkov V.M. and Gavrilyuk B.K. *Dielectrophoresis of cell suspensions* (in Russian) Stud. Biophys. **65** 35-46 (1977).

Fomchemkov V.M. and Gavrilyuk B.K. *The study of dielectrophoresis of cells using the optical technique of measuring* J. Biol. Phys. **6** 29-68 (1978).

Fuhr G., Arnold W.M., Hagedorn R., Müller T., Benecke W., Wagner B. and Zimmermann U. *Levitation, holding and rotation of cells within traps made by high-frequency fields* Biochim. Biophys. Acta **1108** 215-223 (1992).

Fuhr G., Schnell Th., Müller T., Hitzler H., Monajembashi S. and Greulich K-O. *Force measurements of optical tweezers in electro-optical cages* Appl. Phys. A **7** 385-390 (1998).

Fuhr G., Schnelle Th., Hagedorn R. and Shirley S.G. *Dielectrophoretic field cages: techniques for cell, virus and macromolecule handling* Cellular Engineering **1** 47-57 (1995).

Fuhr G., Zimmermann U. and Shirley S.G. *Cell motion in time-varying fields: principles and potential* in Zimmermann U. and Neil G.A. (Eds) *Electromanipulation of cells* CRC Press Inc, Boca Raton (1996).

Fuller C.J., Hamilton J., Ackler H., Krulevitch P., Boser B., Eldredge A., Becker F.F., Yang J. and Gascoyne P.R.C. *Microfabricated multi-frequency particle impedance characterisation system* in van den Berg A., Olthuis W. and Bergveld P. (Eds) *Total Analysis Systems 2000*, Kluwer Academic Publisher 265-268 (2000).

Gascoyne P.R.C., Wang X-B., Huang Y. and Becker F.F. *Dielectrophoretic separation of cancer cells from blood* IEEE Trans. Ind. Appls. 33 670-678 (1997).

Gascoyne P.R.C., Pethig R., Satayavivad J., Becker F.F. and Ruchirawat M. *Dielectrophoretic detection of changes in erythrocyte membranes following malarial infection* Biochim. Biophys. Acta **1323** 240-252 (1997).

Gascoyne P.R.C., Huang Y., Pethig R., Vykoukal J. and Becker F.F., *Dielectrophoretic separation of mammalian cells studied by computerized image analysis* Meas. Sci. Technol. **3** 439-445 (1992).

Gascoyne P.R.C., Pethig R., Burt J.P.H., and Becker F.F. *Membrane changes accompanying the induced differentiation of Friend murine erythroleukemia cells studied by dielectrophoresis* Biochim. Biophys. Acta **1149** 119-126 (1993).

Gascoyne P.R.C., Noshari J., Becker F.F. and Pethig R. *Use of dielectrophoretic collection spectra for characterising difference between normal and cancerous cells* IEEE Trans. Ind. Appl. **30** 829-834 (1994).

Gawad S., Schild L. and Renaud Ph. *Micromachined impedance spectroscopy flow cytometer for cell analysis and particle sizing* Lab on a Chip **1** 76-82 (2001).

Gimsa J., Prüger P., Eppmann P. and Donath E. *Electrorotation of particles measured by dynamic light scattering - a new dielectric spectroscopy technique* Colloids Surf. A **98** 243-249 (1995).

Gimsa J. *New Light scattering and field-trapping methods access the internal electric structure of submicron particles, like influenza viruses* Ann. N.Y Acad. Sci. 287-298 (1999).

Gimsa J., Eppmann P. and Prüger B. *Introducing phase analysis light scattering for dielectric characterization: Measurement of traveling-wave pumping* Biophys J. **73** 3309-3316 (1997).

Gimsa J., Glaser R. and Fuhr G. in Schütt W., Klinkmann H., Lamprecht I. and Wilson T. (Eds) *Physical Characterisation of Biological Cells* Verlag Gesundheit, Berlin 295-323 (1991).

Gimsa J., Marszalek P., Löwe U. and Tsong T.Y. *Dielectrophoresis and electrorotation of neurospora slime and murine myeloma cells* Biophys. J. **60** 749-760 (1991).

Grant E.H., Sheppard R.J and South G. *Dielectric Behaviour of Molecules in Solution* Clarendon Press, Oxford (1978).

Green N.G., Morgan H. and Wilkinson C.D.W. *Dielectrophoresis of virus particles.* Proc. St Andrews meeting of the SEB 77 (1995).

Green N.G., Milner J.J. and Morgan H. *Manipulation and trapping of nanoscale bioparticle using dielectrophoresis* J. Biochem. Biophys. Methods **35** 89-102 (1997).

Green, N.G. and Morgan H. *Dielectrophoretic Investigation of Sub-micrometre Latex Spheres* J. Phys D: Appl. Phys. **30** 2626-2633 (1997).

Green N.G. and Morgan H. *Dielectrophoresis of sub-micrometre latex spheres.* J. Phys. Chem. **103** 41-50 (1999).

Griffin J.L. *Orientation of human and avian erythrocyte in radio-frequency fields* Exp. Cell Res. **61** 113-120 (1970).

Griffin J.L. and Stowell R.E. *Orientation of Euglena by radio frequency field* Exp. Cell Res. **44** 684-668 (1966).

Hanai T., Koizumi N. and Irimajiri A. *A method for determining the dielectric constant and the conductivity of membrane-bounded particles of biological relevance* Biophys. Struct. Mechanisms **1** 285-294 (1975).

Hartley L.F., Kaler K.V.I.S. and Paul R. *Quadrupole levitation of microscopic dielectric particles* J. Electrostatics **46** 233-246 (1999).

Hinch, E.J., Sherwood, J., Chen, W.C. and Sen P.N. *Dielectric response of a dilute suspension of spheres with thin double layers in an asymmetric electrolyte* J. Chem. Soc: Faraday Trans. **80** 535-551 (1984).

Hogan M., Dattagupta N. and Crothers D.M. *Transient electric dichroism of rod like DNA molecule* Proc. Nat. Acad. Sci., USA **75** 195-199 (1978).

Huang Y., Hölzel R., Pethig R. and Wang X-B.   *Differences in the AC electrodynamics of viable and non-viable yeast cells determined through combined dielectrophoresis and electrorotation studies*   Phys. Med. Biol.   **37**   1499-1517   (1992).

Huang Y. and Pethig R.   *Electrode design for negative dielectrophoresis*   Meas. Sci. Technol.   2   1142-1146   (1991).

Hughes M.P. and Morgan H.   *Dielectrophoretic trapping of single sub-micron scale bioparticles*   J. Phys. D: Appl. Phys.   **31**   2205-2210   (1998).

Hughes M.P., Morgan H., Rixon F.J, Burt J.P.H. and Pethig R.   *Manipulation of herpes simplex virus type 1 by dielectrophoresis*   Biochim. Biophys. Acta   **425**   119-126   (1998).

Hughes M.P. and Morgan H.   *Dielectrophoretic characterisation and separation of antibody coated sub-micron latex spheres*   Anal. Chem.   **71**   3441-3445   (1999a).

Hughes M.P. and Morgan H.   *An evanescent field technique for colloidal particle dielectrophoresis studies*   Meas. Sci. Technol.   **10**   759-762   (1999b).

Hughes M.P., Flynn M.F. and Morgan H.   *Dielectrophoretic measurements of sub-micrometre latex particles following surface modification*   Institute of Physics Conference Series   **163**   81-84   (1999a).

Hughes M.P., Morgan H. and Flynn M.F.   *The dielectrophoretic behaviour of sub-micron latex spheres: influence of surface conductance*   J. Colloid Int. Sci.   **220**   454-457   (1999b).

Hughes M.P., Archer S. and Morgan H.   *Mapping the electrorotational torque in planar microelectrodes*   J. Phys. D: Appl. Phys.   **32**   1548-1552   (1999c).

Hughes M.P., Morgan H. and Rixon F.J.   *Dielectrophoretic Manipulation and Characterisation of Herpes SimplexVirus-1 Capsids*   European Biophys. J.   **30**   268-272   (2001).

Hughes M. P., Morgan H. and Rixon F. J.   *Measurements of the properties of Herpes Simplex Virus Type 1 virions with dielectrophoresis*   Biochim. Biophys. Acta. (in the press)   (2002).

Iglesias F.J., Lopex M.C., Santamaria C. and Dominguez A.   *Orientation of Schizosaccharomyces pombe non-living cells under alternating uniform and non-uniform electric fields*   Biophys J.   **48**   721-726   (1985).

Ignatov O.V., Khorkina N.A., Shchyogolev S.Yu., Khlebtsov N.G., Rogacheva S.M. and Bunin V.D.   *Electro-optical properties of microbial cells as affected by acrylamide metabolism*   Anal. Chim. Acta   **347**   241-247   (1997).

Ignatov O.V., Khorkina N.A., Singirtsev I.N., Bunin V.D. and Ignatov V.V.   *Exploitation of the electro-optical characteristics of microbial suspension for determining p-nitophenol and microbial degradative activity*   FEMS Microbiology Lett.   **165**   301-304   (1998).

Irimajiri A., Hanai T. and Inouye A.   *A dielectric theory of "multi-stratified shell" model with its application to a lymphoma cell*   J. Theor. Biol.   **78**   251-269   (1979).

Inoue T., Pethig R., Al-ameen T.A.K., Burt J.P.H. and Price J.A.R.   *Dielectrophoretic behaviour of Micrococcus lysodeikticus and its protoplasts*   J. Electrostatics   **21**   215-223   (1988).

Jones T.B.   *Electromechanics of particles*   Cambridge University Press, Cambridge (1995).

Jones T.B. and Kallio G.A.   *Dielectrophoretic levitation of spheres and shells*   J. Electrostatics **6** 207-224 (1979).

Jones T.B. and Bliss G.W.   *Bubble dielectrophoresis*   J. Appl. Phys. **48** 1412-1417 (1977).

Kaler K.V.I.S., Xie J-P., Jones T.B. and Paul R.   *Dual frequency dielectrophoretic levitation of Canola protoplasts*   Biophys. J. **63** 58-69 (1992).

Kabata H., Kurosawa O., Ara I., Washizu M,. Margarson S.A., Glass R.E. and Shimamoto N.   *Visualisation of single molecules of RNA polymerase sliding along DNA*   Science **262** 1561-1563 (1993).

Kaler K.V. and Pohl H.A.   *Dynamic dielectrophoretic levitation of living individual cells*   IEEE Trans. Ind. Appl. **19** 1089-1093 (1983).

Kaler K.V.I.S., Fritz O.G. and Adamson R.J.   *Dielectrophoretic velocity measurements using quasi-elastic light scattering*   J. Electrostatics **21** 193-204 (1988).

Kaler K.V. and Jones T.B.   *Dielectrophoretic spectra of single cells determined by feedback-controlled levitation*   Biophys. J **57** 173-182 (1990).

Kallio G.A. and Jones T.B.   *Dielectric constant measurements using dielectrophoretic levitation*   IEEE Trans. Ind. Appl. **16** 69-75 (1980).

Kirkwood J.G. and Shumaker J.B.   *The influence of dipole moment fluctuations on the dielectric increment of proteins in solution*   Proc. Natl. Acad. Sci., USA **38** 855-862 (1952).

Jones T.B. and Kraybill J.P.   *Active feedback controlled dielectrophoretic levitator*   J. Appl. Phys. **60** 1247-1252 (1986).

Lacey A.J. (Ed)   *Light Microscopy in Biology. A practical Approach*   Oxford Uni. Press, Oxford   (1999).

Lauffer M.A.   *Electro optical effects in certain viruses*   J. Am. Chem. Soc. **61** 2412-2416   (1939).

Lyklema J.   *Fundamentals of Interface and Colloid Science*   Academic Press, London (1991).

Mahaworasilpa T.L., Coster H.G.L. and George E.P.   *Forces on biological cells due to applied alternating (AC) electric fields*   Biochim. Biophys. Acta **1193** 118-126 (1994).

Mandel M.   *The electric polarisation of rod-like, charged macromolecules*   Mol. Phys. **4** 489-496 (1961).

Mandel M.   *Dielectric properties of charged linear macromolecules with particular reference to DNA*   Ann. N.Y. Acad. Sci., **303** 74-87 (1977).

Mandel M. and Odjik T.   *Dielectric properties of polyelectrolyte solutions*   Ann. Rev. Phys. Chem. **35** 75-107 (1984).

Manning G.S.   *Limiting laws and counterion condensation in polyelectrolyte solutions I. Colligative properties*   J. Chem. Phys. **51** 924-933 (1969).

Manning G.S.   *The molecular theory of polyelectrolyte solutions with applications to the electrostatic properties of polynucleotides*   Q. Rev. Biophys. **11** 179-246 (1978a).

Manning G.S. *Limiting laws and counterion condensation in polyelectrolyte solutions V. Further development of the chemical model* Biophys. Chem. **9** 65-70 (1978b).

Markx G.H., Talary M.S. and Pethig R. *Separation of viable and non-viable yeast using dielectrophoresis* J. Biotechnology **32** 29-37 (1994).

Marszalek P., Zielinsky J.J. and Fikus M. *Experimental verification of a theoretical treatment of the mechanism of dielectrophoresis* Bioelectrochem. Bioenerg. **22** 289-298 (1989a).

Marszalek P., Zielinsky J.J. and Fikus M. *A new method for the investigation of cellular dielectrophoresis* Z. Naturforsch. Sec. C. Biosci. **44c** 92-95 (1989b).

Marszalek P., Zielinsky J.J., Fikus M. and Tsong T.Y. *Determination of electric parameters of cell membrane by a dielectrophoresis methods* Biophys. J. **59** 982-987 (1991).

Mashimo S., Umehara T. and Kuwabura S. *Dielectric study on dynamics and structure of water bound to DNA using a frequency range 10 MHz - 10 GHz* J. Phys. Chem. **93** 4963-4967 (1989).

Mathews R.E.F. *Plant Virology* Academic Press, London (1981).

Milner K.R., Brown A.P., Allsopp D.W.E. and Betts W.B. *Dielectrophoretic classification of bacteria using differential impedance measurements* Electron Lett. **34** 66-67 (1998).

Miller R.D. and Jones T.B. *Electro-orientation of ellipsoidal erythrocytes Theory and experiment* Biophys. J. **64** 1588-1595 (1993).

Morgan H. and Green N.G. *Dielectrophoretic manipulation of rod-shaped viral particles* J. Electrostatics **42** 279-293 (1997).

Morgan H., Hughes M.P. and Green N.G. *Separation of sub-micron bio-particles by dielectrophoresis* Biophys J. **77** 516-525 (1999).

Morishima K., Fukuda T., Arai F., Matsuura H. and Yoshikawa K. *Non-contact transportation of DNA molecule by dielectrophoretic force for micro DNA flow system* Proc. IEEE Int. Conf. Robotics and Automation 2214-2219 (1996).

Morishima K., Fukuda T., Arai F. and Yoshikawa K. *Manipulation of DNA molecule utilizing the conformational transition in the higher order structure of DNA* Proc. IEEE Int. Conf. Robotics and Automation 1454-1459 (1997).

Müller F.H. *Dielektrische Polarisation von Flüssigkeiten in ungleichförmigem Felde* Wiss Veröffentl. Siemens-Werken **17** 20-37 (1938).

Müller T., Gerardino A.M., Schnelle Th., Shirley S.G., Fuhr G., De Gasperis G., Leoni R. and Bordoni F. *High-frequency electric-field trap for micron and sub-micron particles* Il Nuovo Cimento. **17** 425-432 (1995).

Müller T., Gerardino A.M., Schnelle Th., Shirley S.G., Bordoni F., De Gasperis G., Leoni R. and Fuhr G. *Trapping of micrometre and sub-micrometre particles by high-frequency electric fields and hydrodynamic forces* J. Phys. D: Appl. Phys. **29** 340-349 (1996a).

Müller T., Fiedler S., Schnelle Th., Ludwig K., Jung A. and Fuhr G. *High frequency electric fields for trapping viruses* Biotechnology Tech. **10** 221-226 (1996b).

Nishioka M., Sakaguchi Y. and Mizuno A. *Manipulation of DNA molecules at low temperature* Proc. Ann. Meeting of Inst. Electrostatics Japan 217-220 (1993).

Nishioka M., Tanizoe T., Katsura S. and Mizuno A. *Micro manipulation of cells and DNA molecules* J. Electrostatics **35** 83-91 (1995).

O'Brien R.W. *The high frequency dielectric dispersion of a colloid* J. Coll. Int. Sci. **113** 81-93 (1986).

O'Konski C.T. and Zimm B.H. *New methods for studying electrical orientation and relaxation effects in aqueous colloids: Preliminary results with tobacco mosaic virus* Science **111** 113-116 (1950).

Oana H., Ueda M. and Washizu M. *Visualisation of a specific sequence on a single large DNA molecule using fluorescence microscopy based on a new DNA-stretching method* Biochim. Biophys. Res. Comm **265** 140-142 (1999).

Onkley J.L. in Cohn E.J. and Edsall J.T. (Eds) *Proteins, Amino Acids and Peptides, Chapter 22* Reinhold, New York (1943).

Oosawa F. *Counterion fluctuation and dielectric dispersion in linear polyelectrolytes* Biopolymers **9** 677-688 (1970).

Pethig R. *Dielectric and electronic properties of biological materials* John Wiley & Sons, Chichester (1979).

Pohl H.A. *Dielectrophoresis* Cambridge Uni. Press, Cambridge (1978).

Pohl H.A. *The motion and precipitation of suspensoids in divergent electric fields* J. Appl. Phys. **22** 869-871 (1982).

Pohl H.A. and Hawk I. *Separation of living and dead cells by dielectrophoresis* Science **162** 647-649 (1966).

Price J.A.R., Burt J.P.H. and Pethig. R. *Application of a new optical technique for measuring the dielectrophoretic behaviour of micro-organisms* Biochim. Biophys. Acta **964** 221-230 (1988).

Prock A. and McConkey G. *Inhomogeneous field method for the study of large polarizable particles* J. Chem. Phys. **32** 224-236 (1960).

Prüger B., Eppmann P., Donath E. and Gimsa J. *Measurement of inherent particle properties by dynamic-light scattering: Introducing electrorotation light scattering* Biophys. J. **72** 1414-1424 (1997).

Prüger B., Eppmann P. and Gimsa J. *Particle characterisation by AC-electrokinetic phenomena 3. New developments in electrorotational light scattering (ERLS)* Colloids and Surfaces A **136** 199-207 (1998).

Rosen L.A. and Saville D.A. *Dielectric spectroscopy of colloidal dispersions: comparison between experiment and theory* Langmuir **7** 36-42 (1991).

Saif B., Mohr R.K., Montrose C.J. and Litovitz T.A. *On the mechanism of dielectric relaxation in aqueous DNA solutions* Biopolymers **31** 1171-1180 (1991).

Saito M., Schwan H.P. and Schwarz G. *Response of non-spherical biological particles in an alternating electric field* Biophys. J. **6** 313-327 (1966).

Schnelle T., Müller T., Fiedler S., Shirley S.G., Ludwig K., Hermann A. and Fuhr G. *Trapping of viruses in high-frequency electric field cages* Naturwissenschaften **83** 172-176 (1996).

Schnelle Th., Müller T., Gradl G., Shirley S.G. and Fuhr G. *Paired electrode system: dielectrophoretic particle sorting and force calibration* J. Electrostatics **47** 121-132 (1999).

Schwan H.P. *Electrical properties of tissues and cells* in Adv. Biol. Med. Phys. (Eds.) Lawrence J.H. and Tobias C.A.   Academic Press, N. York   **5**   147-209 (1957).

Schwarz G. *A theory of low-frequency dielectric dispersion of colloidal particles in electrolyte solutions* J. Phys. Chem. **66** 2636-2642 (1962).

Schwarz G., Saito M and Schwan H.P. *On the orientation of non-spherical particles in an alternating electric field* J. Chem. Phys. **43** 3562-3569 (1965).

Stellwagen N.C. *Electric birefringence of restricted enzyme fragments of DNA: optical factor and electric polarisability as a function of molecular weight* Biopolymers **20** 399-434 (1981).

Suehiro J., Yatsunami R., Hamada R. and Hara M. *Quantitative estimation of biological cell concentration suspended in aqueous medium by using dielectrophoretic impedance measurement method* J. Phys. D: Appl. Phys **32** 2814-2820 (1999).

Suzuki S., Yamanashi T., Tazawa S., Kurosawa O. and Washizu M. *Quantitative analysis of DNA orientation in stationary AC electric fields using fluorescence anisotropy* IEEE Trans. Ind. Appl. **34** 75-82 (1998).

Takashima S. *Electrical properties of biopolymers and membranes* Adam Hilger, Philadelphia (1989).

Takashima S. *Dielectric dispersion of DNA*   J. Mol. Biol. **7** 455-467 (1963).

Takashima S., Gabriel C., Sheppard R.J. and Grant E.H. *Dielectric behaviour of DNA solution at radio and microwave frequencies (at 20°C)* Biophys. J. **46** 29-34 (1984).

Talary M.S. and Pethig R. *Optical technique for measuring the positive and negative dielectrophoretic behaviour of cells and colloidal suspension* IEEE Proc.-Sci. Meas. Technol. **141** 395-399 (1994).

Teixeira-Pinto A.A., Nejelski Jr L.L., Cutler J.L. and Heller J.H. *The behaviour of unicellular organisms in an electromagnetic field* Exp. Cell Res. **20** 548-564 (1960).

van der Touw F. and Mandel M. *Dielectric increment and dielectric dispersion of solutions containing simple charged linear macromolecules* Biophys. Chem. **2** 218-241 (1974).

van der Wal A., Minor M., Norde W., Zehnder A.J.B. and Lyklema J. *Conductivity and dielectric dispersion of gram-positive bacterial cells* J. Coll. Int. Sci. **186** 71-79 (1997).

Veas F. and Schaffer M.J. *Stable levitation of a dielectric liquid in a multi-frequency electric field.* In Proc. of the Int. Symp. on Electrohydrodynamics Cambridge MA 113-115 (1969).

Voldman J., Braff R.A., Toner M., Gray M. and Schmidt M.A. *Holding forces of single-particle dielectrophoretic traps* Biophys. J. **80** 531-541 (2000).

Washizu M., Suzuki S., Kurosawa O., Nishizaka T. and Shinohara T. *Molecular dielectrophoresis of biopolymers* IEEE Trans Ind. Appl. **30** 835-843 (1994).

Washizu M. and Kurosawa O. *Electrostatic manipulation of DNA in microfabricated structures* IEEE Trans. Ind. Appl. **26** 1165-1172 (1990).

Washizu M., Suzuki S., Kurosawa O., Nishizaka T. and Shinohara T. *Molecular dielectrophoresis of biopolymers* IEEE Trans. Ind. Appl. **30** 835-843 (1994).

Washizu M., Kurosawa O., Arai I., Suzuki S. and Shimamoto N. *Applications of electrostatic stretch-and-positioning of DNA* IEEE Trans. Ind. Appl. **31** 447-456 (1995).

Washizu M., Yamamoto T., Kurosawa O. and Shimamoto N. *Molecular surgery based on microsystems* Transducers **97** 1-4 (1997).

Whitley R.J. in Fields B.N., Knipe D.M., Howley P.M., Chanock R.M., Melnick J.L., Monath T.P., Roizman B. and Straus S.E. (eds) *Field's Virology* Lippincott-Raven, Philadelphia 2297-2342 (1996).

Yamamoto T., Kurosawa O., Kabat H., Shimamoto N. and Washizu M. *Molecular surgery of DNA based on electrostatic manipulation* IEEE Trans. Ind. Appl. **36** 1010-1016 (2000).

Ying H., Hölzel R., Pethig R. and Wang X-B. *Differences in the AC electrodynamics of viable and nonviable yeast-cells determined through combined dielectrophoresis and electrorotation studies* Phys. Med. Biol. **37** 1499-1517 (1992).

de Xammar Oro J.R. and Grigera J.R. *Dielectric properties of aqueous solutions of sonicated DNA above 40 MHz* Biopolymers **23** 1457-1463 (1984).

Chapter Twelve

Dielectrophoretic separation of sub-micrometre particles

One of the most important applications of AC electrokinetics is particle separation. It was first shown over 30 years ago that differences between the dielectrophoretic force exerted on different biological particles suspended in non-uniform electric fields could be used as the basis for particle sorting (Pohl and Hawk 1966; Crane and Pohl 1968; Pohl and Crane 1971; Mason and Hammond 1971; Pohl 1977). In this early work, both batch and continuous separation schemes were proposed and evaluated, and the separation of live and dead yeast cells was clearly demonstrated. The potential for using DEP to concentrate or precipitate material from a carrier solute was demonstrated even earlier (Pohl and Schwar 1959). A cylindrical three-dimensional device for separating biological material (micelles) was developed by Black and Hammond (1965a; 1965b). They used low frequency 60Hz AC fields to separate material in a batch system, but significantly, achieved efficiencies approaching 100% for low particulate concentrations. Continuous separation of particles by dielectrophoresis was shown by Pohl and Plymale (1960) using an isomotive electric field geometry.

The separation of particles by dielectrophoresis relies on the fact that one particular sub-population of particles has unique frequency-dependent dielectric properties, which is different from any other population. The relative magnitude and direction of the dielectrophoretic force exerted on a given population of particles depends on the conductivity and permittivity of the suspending medium, together with the frequency and magnitude of the applied field. Therefore, differences in the dielectric properties of particles manifest themselves as variations in the DEP force magnitude and/or direction, producing separation.

In his later work, Pohl (1977; 1978) laid many of the foundations of DEP technology and described many simple electrode geometries suitable for the separation of particles. These electrodes were macroscopic 3-dimensional structures, typically a point and plane electrode or a thin wire suspended coaxially within a conducting tube through which particles flowed. Such designs are fine in principle but have limited practical application, primarily because high voltages must be used to generate the high DEP forces required to move particles over relatively large distances. A significant advance in the development of practical and efficient separation devices was made when it became possible to build planar micro-electrode systems using micro-electronic fabrication methods. These small-

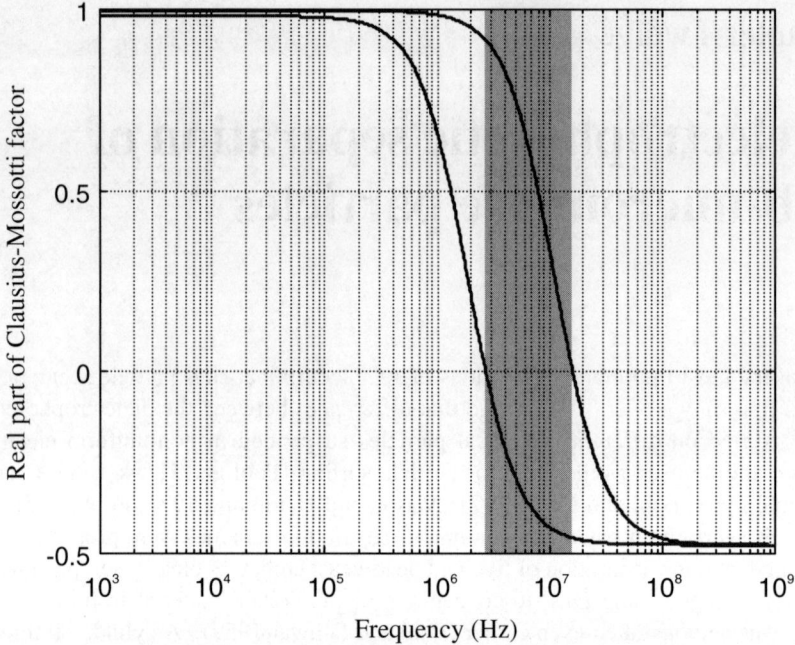

Fig. 12.1 Plot of the Clausius-Mossotti factors for two different spherical particles as a function of frequency. The particles exhibit different dielectrophoretic properties and over a certain range of frequencies, indicated by the shaded region, they can be effectively separated; one type experiences positive DEP and the other negative DEP.

scale devices form the basis of a range of novel and sophisticated dielectrophoretic/microfluidic systems. In this chapter we review separation systems and discuss their potential for the separation of colloidal and nano-particles.

12.1   Principle of spatial separation

Consider two homogeneous spherical particles, each with a different effective polarisability and exhibiting a different DEP behaviour, as shown by the Clausius-Mossotti plot of figure 12.1. The consequence of this, is that within a particular frequency band, as shown in the figure, the DEP force experienced by the two particle populations may be in diametrically opposite directions. Therefore, on an electrode array that has clearly defined regions for positive and negative DEP collection, spatial separation of the particles will occur within this frequency band. This is schematically demonstrated in figure 12.2, which shows the commonly observed pattern of particle behaviour seen on a castellated electrode array. At a particular frequency, particles move in opposite directions into one of two regions of minimum potential energy, depending on the sign of the DEP force. This gives

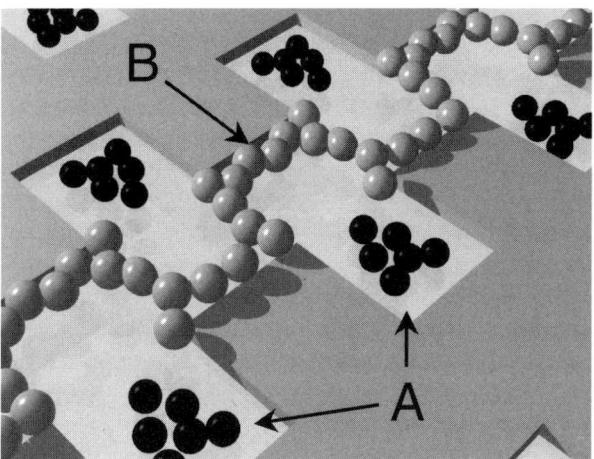

Fig. 12.2 Schematic diagram showing two different particle types separating
by dielectrophoresis on a castellated electrode. The black particles
experience negative DEP and move to the low field regions in the
bays between the castellations (point A). The grey particles
experience positive DEP and collect at the high field regions
(point B) and form pearl chains between opposing castellation tips.

rise to spatial separation of particles on the electrode. The fact that separation of
particles happens within such short distances simultaneously, all over the one
electrode array, is the reason that this type of electrode has been so widely used to
observe and characterise the dielectrophoretic behaviour of particles (*e.g.* see
Pethig 1996).

In addition to spatial separation, particles can also be separated, or more
correctly "fractionated" according to their relative velocities in a system which
combines DEP forces with other forces, in particular hydrodynamic forces. For
small particles, we saw in Chapter Four that the instantaneous velocity is
proportional to the instantaneous dielectrophoretic force, *i.e.* $\mathbf{v} = \mathbf{F}_{DEP} / f$. For a
spherical particle and substituting for $\mathbf{F}_{DEP}$ gives equation (5.36) for the particle
velocity

$$\mathbf{v} = \frac{\pi a^3 \varepsilon_m \, \mathrm{Re}[\tilde{f}_{CM}] \nabla |\mathbf{E}|^2}{6 \pi \eta a} \tag{12.1}$$

The dielectrophoretic mobility is

$$\mu_{DEP} = \frac{a^2 \varepsilon_m \, \mathrm{Re}[\tilde{f}_{CM}]}{6 \eta} \tag{12.2}$$

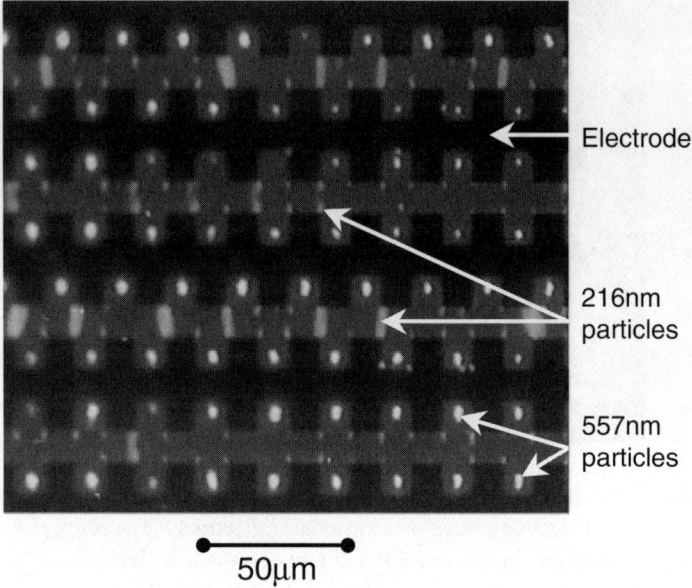

Fig. 12.3 Photograph of 557 nm diameter particles experiencing negative
DEP and 216 nm particles experiencing positive DEP on a 10 µm
castellated electrodes behaving as shown schematically in figure
12.2. (Reprinted from Morgan *et al.* (1999) with permission from
the Biophysical Society – original in colour.)

in the direction of $\nabla|\mathbf{E}|^{2}$. It can be seen from this equation that for a spherical
particle the dielectrophoretic mobility depends both on the real component of the
polarisability and the square of the particle radius.

12.1.1   Spatial separation of latex spheres

The simultaneous movement of particles into regions of both positive and negative
DEP on castellated electrodes was first demonstrated by Pethig *et al.* (1992).
Spatial separation of blood cells (Gascoyne *et al.* 1992) and blood from bacteria
was also performed on such an electrode array (Wang *et al.* 1993). The spatial
separation of sub-micrometre particles on a castellated electrode array has also
been demonstrated (Morgan *et al.* 1999). In this case the electric field strength was
higher than that required to separate cells, and both the suspending medium
conductivity and the frequency of the applied field must be carefully chosen to
avoid adverse effects from electrohydrodynamics. Figure 12.3 shows images of
216 nm and 557 nm diameter latex particles separating on a 10 µm feature size
castellated electrode. At a particular frequency, the 216 nm spheres experience
positive DEP and form pearl chains between opposing electrode tips, whilst
simultaneously the 557 nm particles experience negative DEP and become trapped

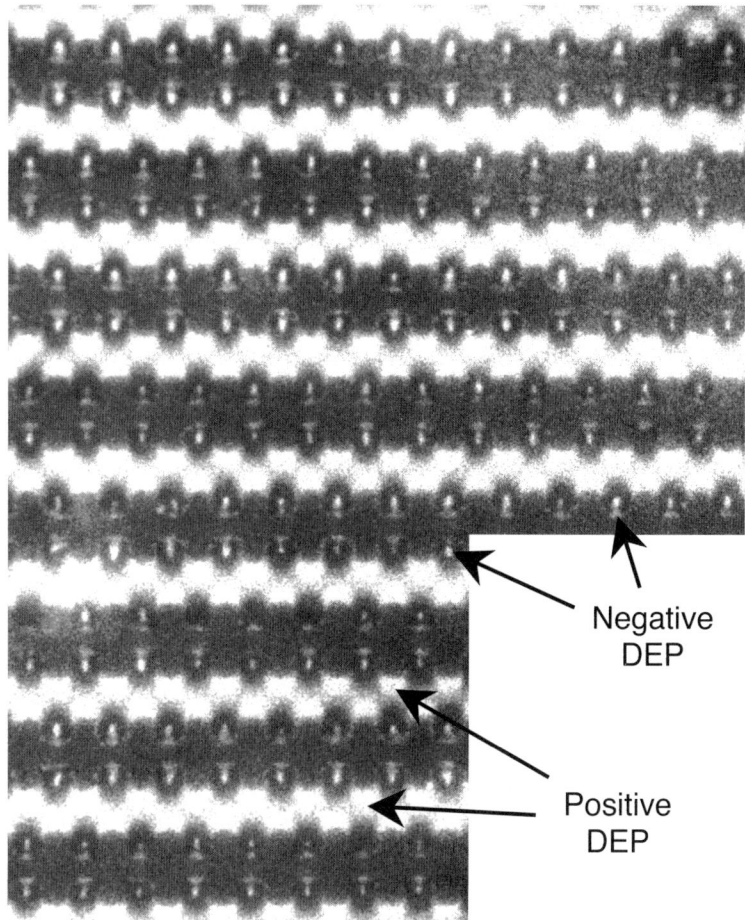

Fig. 12.4(a) A photograph showing the separation of a heterogeneous mixture of 93nm diameter latex beads into subpopulations on castellated electrodes. Beads with lower surface charge move to the low field regions between the electrode castellations. Beads with higher surface charge are trapped at the electrode tips. (Reprinted with permission from Green and Morgan (1997), copyright IOP Publishing Ltd.)

in triangle-shaped patterns in the inter-electrode bays, as detailed in the figure (*c.f.* figure 12.2 and figure 10.7(a)).

The spatial separation of a heterogeneous population of sub-micrometre particles (all of nominally identical size) can also be accomplished using castellated electrode arrays (Green and Morgan 1997). Figure 12.4(a) shows a photograph of 93nm diameter latex spheres separating under DEP and dividing

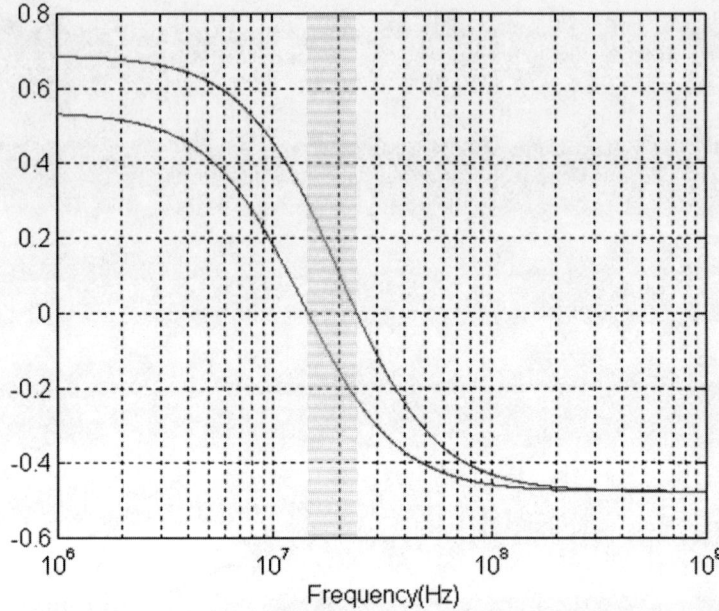

Fig. 12.4(b) A plot of the Clausius-Mossotti factor for a mixture of 93 nm
diameter latex spheres. The two curves demonstrate a spread in the
particle surface charge density that gives rise to a distribution in
the dielectrophoretic behaviour of particles in the mixture. For the
frequency shown by the shaded region, the mixture can be
separated into two sub-populations. (Reprinted with permission
from Green and Morgan (1997), copyright IOP Publishing Ltd.)

into two sub-populations. The nominally homogeneous population of particles in
fact has a distribution in surface charge density. This manifests itself as a
distribution in the effective polarisabilities of the particles, as shown by the
Clausius-Mossotti plots of figure 12.4(b). This figure shows that at a frequency of
20 MHz, a percentage of the particles experience positive DEP, whilst the
remainder experience negative DEP, leading to separation.

12.1.2   Separation of surface modified particles

Proteins and other molecules can be attached chemically to functionalised latex
spheres using well-characterised coupling methods (Hermanson 1996). Chemical
attachment often leads to a change in the surface charge density of the particles
(Hughes and Morgan 1999, Hughes et al. 1999a, Hughes et al. 1999b). This leads
to a change in the surface conductance and a consequential change in the
dielectrophoretic mobility. Since the particle conductivity is an inverse function of
radius, smaller particles exhibit greater changes in effective polarisability with
changing surface charge density. This means that particles of identical size, but

with differing amounts of protein molecules immobilised onto the surface, experience different magnitudes and/or direction of DEP force (*i.e.* they can be separated; the degree of separation increasing as the particle size is reduced).

Figure 12.5 shows an example of the separation of a sample consisting of a mixture of two identically sized latex spheres (216 nm diameter) both experiencing positive DEP (at low frequencies). One particle type was modified by immobilising a monolayer of protein molecules onto the surface, leading to a reduction in the effective polarisability of the particles. The other population was unmodified. The figure shows how unmodified particles are held strongly onto the electrodes (double arrow), whilst the modified particles are attracted less strongly and form a

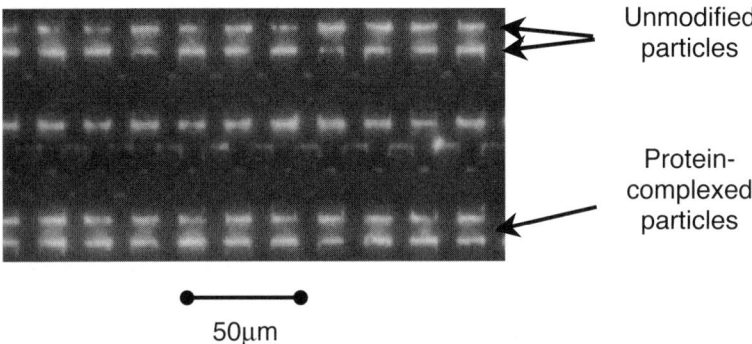

Unmodified particles

Protein-complexed particles

50μm

Fig. 12.5 Photograph showing separation of unmodified and protein conjugated 216 nm diameter latex particles due to positive DEP forces of different magnitude. The protein-conjugated beads form pearl chains between the tips of the castellated electrode array. The unmodified particles (brighter in the picture) experience a significantly stronger positive DEP force and are attracted to the electrode tips. (Reprinted from Morgan *et al.* (1999) with permission from the Biophysical Society – original in colour.)

diffuse cloud of particles located between alternate electrode (single arrow). For the modified particles, the DEP force is low and the magnitude of the diffusion flux (pushing the particles apart from one another) is of the same order as the DEP flux pulling them onto the end of the electrode.

The polarisability of sub-micrometre particles is a strong function of the surface charge characteristics so that there is the potential for developing highly selective AC electrokinetic-based separation systems, which could be used for antigen binding recognition or DNA hybridisation assays.

12.1.3 Biological particles

Small biological particles such as viruses, DNA and macromolecules exhibit frequency-dependent DEP behaviour, although the origin of the polarisation mechanisms is not always clear. Again, this frequency-dependent behaviour can be exploited to achieve separation of these particles by dielectrophoresis. For example, the spatial separation of two different viruses (Tobacco Mosaic Virus and Herpes Simplex Virus) using a polynomial electrode is shown in figure 12.6 (Morgan *et al.* 1999). The photograph shows TMV held on the electrode edges under positive DEP, whilst at the same time HSV experiences negative DEP and collects in the centre of the quadrupole trap.

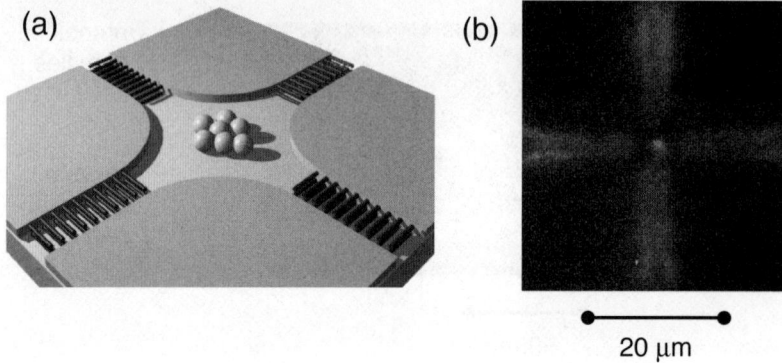

20 μm

Fig. 12.6 Schematic diagram (a) and experimental photograph (b) of Tobacco Mosaic Virus (TMV) and Herpes Simplex Virus (HSV) separating on polynomial electrodes. The rod-shaped TMV experience strong positive DEP and collect between adjacent electrodes, separating from the HSV which experiences negative DEP and collects in the centre of the electrode array. (Reprinted from Morgan *et al.* (1999) with permission from the Biophysical Society – original in colour.)

In Chapter Eleven it was shown that the protein Avidin exhibits a frequency-dependent DEP behaviour, implying that there is the possibility of separating proteins and other molecules by DEP. Although the spatial separation of macromolecules on an electrode array has not been shown, nevertheless DEP has been used to separate different types of macromolecules, as discussed in the following section.

12.2 Flow-through separation systems

It is relatively easy to demonstrate the spatial separation of particles on an electrode array; however, the implementation of a practical separator requires rather more ingenuity. In particular, some means of processing particles on and off a device is required. There has been a concerted effort over the last decade and a half to produce efficient DEP separators and a number of continuous or bulk separation strategies have been devised. The simplest form of DEP separator is one that simply pulls particular particles out of a suspending medium flowing over an electrode array.

The principle of such a system is shown in figure 12.7. Particles are carried into the device as a suspension in a carrier medium that must have an appropriate conductivity, permittivity, density and viscosity. The former pair of parameters controls the DEP force; the latter pair controls the rate of sedimentation of particles. When the electrodes are energised, the required particles are pulled out of solution by positive DEP forces, and provided the flow velocity is not too high, remain trapped on the electrodes. Particles which either experience negative DEP or no force at all, are eluted in the carrier stream.

In the simplest implementation, the separation chamber is constructed from two opposing interdigitated micro-electrode arrays. These electrodes generate a DEP force that decreases exponentially from the surface (see Chapter Eight for details of electric field calculations). Particles are pulled out of the carrier medium onto the electrode array at a rate that depends on the frequency and flow rate. In this way particles experiencing negative DEP will be separated from those experiencing positive DEP. Ignoring Brownian motion, the forces acting on a particle can be divided into the horizontal and vertical components. For large particles, we can see that in the vertical axis (y), the DEP force acts either with or

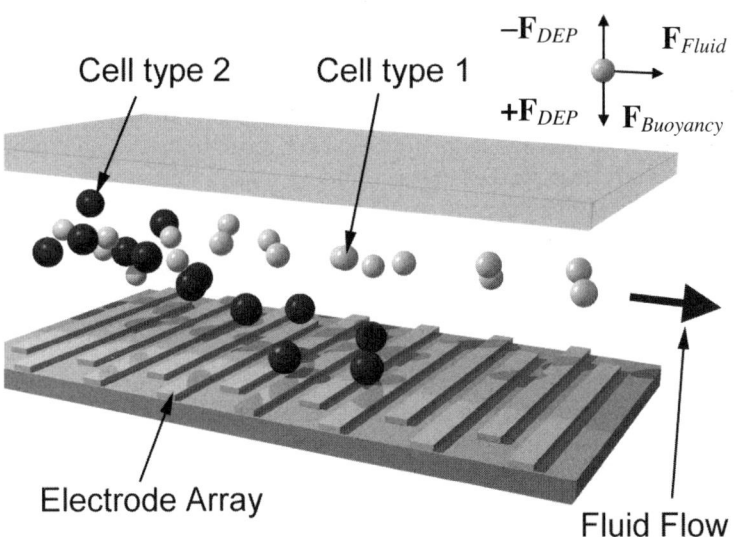

Fig. 12.7   Schematic diagram showing the principle of operation of a simple dielectrophoretic separator. As the mixture flows across the electrode array, cell type 2 experiences positive DEP and are attracted to the electrodes. They separate from cell type 1 which experience negative DEP (or no force) and are repelled from the electrodes, passing through the chamber.

against the buoyancy force, as shown in figure 12.7. Ignoring near-wall forces and particle-particle interactions, the equation of motion in this axis is

$$m\frac{dv_y}{dt} = F_{DEP,y} + F_{Buoyancy} \tag{12.3}$$

In the horizontal plane (x direction) and at a reasonable distance from the electrode so that the horizontal component of the DEP force can be ignored, the only force acting on the particle is the Stokes force due to the movement of the fluid. To a first approximation the velocity of the particle can be considered identical to the fluid. We have seen that in a narrow channel the fluid adopts the parabolic or Pouiseille flow profile with a velocity given by equation (5.11)

Using these equations, typical particle trajectories can be calculated. For the sake of comparison with experiment it is convenient to calculate the volume flow rate through the device given by

$$Q = \frac{wh^3}{12\eta}\frac{p_o}{l_0} \tag{12.4}$$

Here p_o is the pressure drop, l_o the length, w the depth and h the height of the device (n.b. $h = 2d$ in equation (5.14)). A simulation of a particle trajectory in such a device is shown in figure 12.8. The calculation was performed assuming that the particle and carrier medium are of the same density so that sedimentation did not occur and using the analytical 2D solution for the electric field (see Chapter Ten). It can be seen from the simulation that the DEP force brings particles down onto the edge of an electrode, as seen experimentally.

Theoretically, all particles of a given size and polarisability will collect at the same point in the device. However, a distribution in polarisabilities, either through variations in internal properties or surface characteristics, leads to a distribution of collection points across several electrodes. For sub-micrometre particles, the effect of Brownian motion acts to broaden the band of collection points. For high concentrations of particles, the system is complicated and any analysis should be made on the basis of fluxes rather than the trajectories of single particles.

12.2.1 Designs of particle separators

A device such as that illustrated in figure 12.7, clearly has limited throughput. It is 2-dimensional and the electric field only extends a few tens of microns into the carrier medium. In order to scale up processing for high volume throughput, grid shaped electrode arrays have been developed. A method utilising screens to provide non-uniform electric fields over a large area was proposed by Verschure and Ijlst (1966). In a later design (Archer *et al.* 1993) an electric field was established between consecutive conducting plates, each separated by an insulating material. Such a device has been used to separate micro-organisms (Archer *et al.*

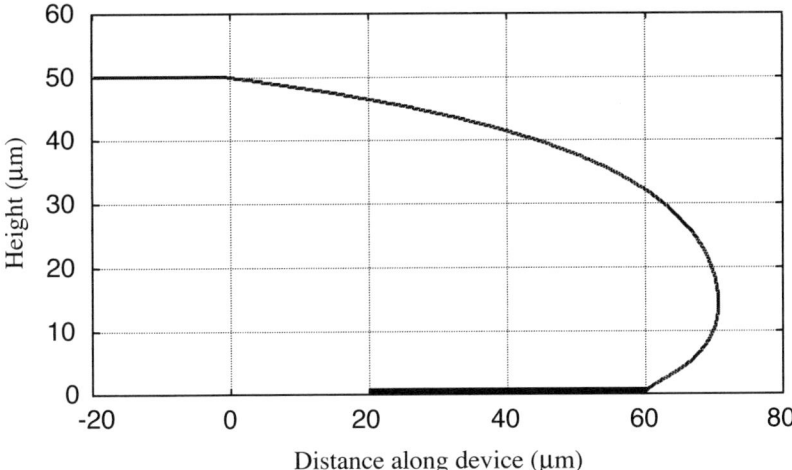

Fig. 12.8 Plot of the trajectory of a particle passing through a 100 μm high chamber with parallel bar electrodes of width and separation of 20 μm on the base. The particle is experiencing positive DEP and is pulled downwards and towards the electrode edge. The trajectory was calculated for a device of length $l_o = 10$ cm and a flow rate of 1 ml h^{-1}.

1993). More recently, a microfabricated filter has been demonstrated to be capable of selective retention of viable cells in a high conductivity medium (Docoslis *et al.* 1997). The filter consists of long strip micro-electrodes deposited on a Si wafer, as shown in figure 12.9. Holes were made between the electrodes using Si wet etching techniques. The filter was able to discriminate between viable and non-viable cells on the basis of their polarisability and an efficiency of nearly 100% at flow rates of 10 ml h^{-1} was achieved.

These simple DEP separation devices can be used to extract or filter particles experiencing positive DEP from a carrier medium. For sub-micrometre particles the DEP force is much lower, so that providing either the device length is increased and/or the applied potential is increased, colloidal particles can also be concentrated and retained on the electrodes.

Batch DEP separation devices have been developed, notably by Markx *et al.* (1994a; 1994b). The basic separation device takes the form of that shown in figure 12.7, except that in the designs of Markx *et al.*, interdigitated castellated electrode arrays were used. A suspension of particles is injected into the device, and under zero-flow conditions the electrodes are energised with a suitable frequency and voltage so that **all** the particles are attracted onto the electrodes. Clean suspending medium is then continuously pumped through the device and the frequency altered until one of the particle types within the mixture is selectively released and eluted from the device. Either changing the frequency and/or lowering the voltage allows a second particle type to be released, and so on. Using this technique it was

Fig. 12.9 A dielectrophoretic filter consisting of long strip electrodes fabricated on a silicon substrate with wet-etched holes in between. (Figure courtesy of K.V.I.S. Kaler, University of Calgary; from Docoslis *et al.* 1997.)

demonstrated that a range of micro-organisms could be selectively separated (Pethig 1996). Variations on the technique include flowing a suspending medium whose conductivity or permittivity is slowly altered whilst maintaining a constant frequency and voltage (Markx *et al.* 1996). Continuous rather than batch separation methods have also been developed through the selective switching of fields and frequencies.

The first report of a device capable of working in continuous mode was by Pohl *et al.* (1981), using isomotive electrodes. In his book Pohl (1978) also outlined the working principle of a macroscale molecular DEP separator which was based on a wire coaxial electrode system, although he does not report any results for the device. Generally the yields obtained with the early devices were very low compared with the near 100% efficiency achieved in the latest devices (Markx and Pethig 1995; Becker *et al.* 1994; 1995).

Separation devices are now beginning to be combined with other sensing systems to produce Lab-on-a-chip devices. In this case several different functional units are combined into the one device, for example a sample processing device followed by a detector of some kind. For example, Cui *et al.* (2000) showed how a fibre optic illumination and detection system was integrated into a twDEP separation system. Cheng *et al.* (1998) describe how a DEP-based particle separator can be combined with electronically addressable DNA/RNA hybridisation arrays to produce a laboratory on a chip capable of isolating bacteria from blood samples and subsequently identifying the bacteria through analysis of the DNA or RNA content of the organism.

12.3 DEP Field Flow Fractionation

The principles outlined in section 12.2 for the separation of particles by DEP utilises the fluid as a carrier to move particles on and off the electrodes or to remove particles which are held less weakly than others on an electrode array. However, DEP forces can be combined with hydrodynamic forces in a separation method known as Field Flow Fractionation (FFF).

In the general technique of FFF, particles flow through a thin chamber within which the fluid adopts the classic parabolic flow profile as shown in figure 12.10. In the absence of any fluid flow or other perturbation field, the steady-state particle concentration profile is uniform. Application of a force perpendicular to the flow direction disturbs this equilibrium position, and a new concentration profile is

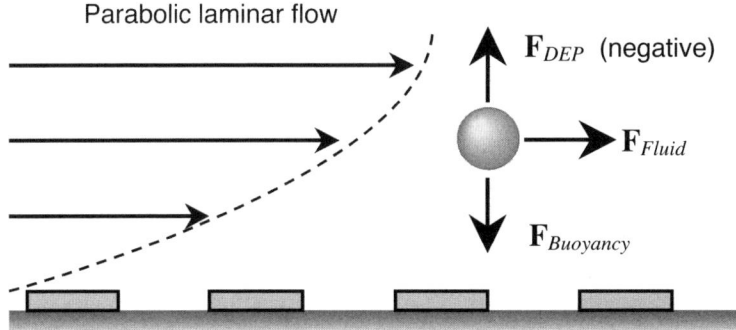

Fig. 12.10   Schematic diagram of the forces a particle experiences in a gravitational DEP-FFF separation system. The particle height above the electrodes is determined by a balance of the negative DEP force it experiences and the buoyancy force. This leads to the vertical separation of different particle types. The parabolic flow profile causes different particles to move at different velocities, depending on their height, leading to spatial separation along the array.

established which is dictated by some inherent physical property of the particle, such as density or charge. Because of the parabolic flow profile, the liquid velocity is a function of channel height and the superposition of the fluid velocity gradient on the non-uniform particle distribution leads to separation. FFF systems have been developed for analysis and fractionation of colloids and macromolecules using forces such as sedimentation, cross-flow, thermal diffusion, and electrophoretic forces (Berg and Purcell 1967a; 1967b; Berg *et al.* 1967; Caldwell *et al.* 1972; Giddings 1989; Giddings 1993). Micromachined electric-field flow fractionation systems have also been developed (Gale *et al.* 1998).

12.3.1 Dielectrophoretic Field Flow Fractionation (DEP-FFF)

In DEP-FFF particles are separated according to a combination of their effective polarisability and density (Huang *et al.* 1997; Markx *et al.* 1997). Particles are repelled from the electrodes under a dielectrophoretic levitation force, which acts on a suspension of particles. This force is combined with a parabolic fluid flow to achieve separation. Owing to the fact that particles experience negative DEP, this system has two major advantages over previous separation devices. In the first case, higher suspending medium conductivities can be used, secondly particle-substrate interactions, which inevitably lead to sticking of particles, are avoided.

A suspension of particles flows through a long separation chamber with a small separation between top and bottom, as shown in figure 12.7. The chamber bottom consists of a substrate onto which is fabricated an array of interdigitated electrodes. Under conditions of negative DEP, particles with different polarisabilities and/or densities are levitated to different heights, the equilibrium position being given by the point at which the (negative) buoyancy force exactly balances the repulsive (negative) DEP force, as shown in figure 12.10. Ignoring hydrodynamic effects the governing equation is simply

$$\frac{4}{3}\pi a^3 (\rho_p - \rho_m)\mathbf{g} + \mathbf{F}_{DEP} = 0 \tag{12.5}$$

where ρ_p and ρ_m is the density of the particle and suspending medium respectively. For an interdigitated electrode, and in the far field, we have seen from Chapter Ten that the electric field gradient can be calculated analytically as

$$\nabla |\mathbf{E}|^2 = \frac{32}{\pi} \frac{V_o^3}{d^3} e^{(-\pi y/d)} \tag{12.6}$$

where d is the electrode gap and width.

Combining equations (12.5) and (12.6) gives the equilibrium levitation height of a particle as

$$y = \frac{d}{\pi} \ln \left[-\frac{24 V_o^2 \varepsilon_m \operatorname{Re}[\tilde{f}_{CM}]}{\pi d^3 \Delta \rho g} \right] \tag{12.7}$$

Owing to the parabolic flow profile of the fluid, particles exit the chamber at different times, *i.e.* they are fractionated. In contrast to other DEP separation methods, where particles remain on the same plane and are either eluted or remain trapped, DEP-FFF exploits the velocity gradient in the flow profile to achieve highly selective separation. Since separation of particles is based on a balance of DEP force and gravitational force, the term dielectrophoretic/gravitational FFF has been adopted for this technique. Recent examples of applications include the separation of latex particles (Markx *et al.* 1997; Wang *et al.* 1998) and blood cells

(Yang *et al.* 1999; Wang *et al.* 2000).

12.3.2   Particle-wall interaction forces in FFF

A detailed analysis of DEP-FFF for cells and larger particles must include second order effects such as the correction to the Stokes drag when the particle is near the channel wall and also the effects of hydrodynamic "lift forces". In classical gravitational FFF, particle differentiation is achieved by balancing a force (*e.g.* electrophoresis or dielectrophoresis) against gravity. Particle-wall interaction forces repel particles away from the wall at the bottom of the chamber, *i.e.* upwards to give what is termed a lift force. However, these forces will occur between particles and any of the surfaces in the device.

When a particle approaches a wall, the velocity of the particle $\mathbf{v}$ is less than that of the fluid by a factor called the retardation coefficient (Happel and Brenner 1983; Williams *et al.* 1992). As the particle approaches the wall, the restriction of fluid flow between the particle and the wall creates a pressure gradient between the two. The fluid flow between the particle and the surface is less than the undisturbed fluid $\mathbf{u}$ at the same point. The resulting effect on a particle at distance δ (from particle centre to the wall) is that its velocity is modified by a factor $\varUpsilon$ (Williams *et al.* 1992)

$$\mathbf{v} = \varUpsilon(\delta/a)\mathbf{u} \tag{12.8}$$

This factor approaches zero as the distance to the wall decreases, *i.e.* as $(\delta/a) \to 0$, and rapidly approaches unity as the particle moves away from the wall. For the sizes of channels used in electrokinetic separation systems, the correction factor is extremely small unless the particle is almost touching the wall.

In addition to retardation effects, experiments show that particle-wall interaction forces can significantly influence particle transit times through separation systems. The combined effect of the interaction forces is not completely understood, but Williams *et al.* (1996) completed a comprehensive study of lift forces for polystyrene micro-spheres in field flow fractionation systems. He showed that for sedimentation FFF, the observed force could be accounted for by the sum of three independent components. These forces are the classical lift force due to fluid inertia, an as yet unidentified lift force due to near wall interaction and an electrokinetic force. For typical micro DEP-FFF systems, the fluid has no inertia so that the hydrodynamic lift force can be ignored. A short-range electrokinetic lift force can occur due to the combined action of double layer interaction and the van der Waals force between the particle and the wall, *i.e.* the DLVO interaction as discussed in detail in Chapter Six.

The electrokinetic lift force is (Bike and Prieve 1992; van de Ven *et al.* 1993)

$$F_{Ek} = \frac{27\pi}{16}\varepsilon_m^3 \left(\frac{v_p}{\sigma_m}\right)^2 \frac{a^2}{(a+\delta_s)^4}(\zeta_p + \zeta_w)\zeta_p \tag{12.9}$$

where δ_S is the distance from the particle surface to the wall, ζ is the zeta potential, (subscripts p and w denote particle and wall respectively), and v_p is the velocity of the particle at the point closest to the wall. Except for when the particle is very close to the wall, this force is negligible so that the near wall lift force remains the dominant effect. This force was empirically obtained by Williams *et al.* (1992). However, Huang *et al.* (1997) found considerable discrepancy between the experimentally determined levitation heights for rougher biological cells than predicted by this empirical relationship.

Observation of the levitation of weakly adhering vesicles in shear flow have been made by Lorz *et al.* (2000). These authors showed that vesicles can be repelled from the surface by forces which are up to two orders of magnitude greater than predicted by simple theories. These forces have been assigned to small variations in the contact disk between the vesicle and substrate, *i.e.* the vesicle tilts relative to the flow direction. The repulsion force causes unbinding of the vesicle, which then hovers above the surface before flowing with the fluid. Sukumaran and Seifert (2001) numerically simulated the repulsion force for vesicles subject to shear flow and showed that their results confirmed experiment and that the force could easily be sufficient to overcome gravity at moderate flow rates. Owing to the complex nature of particle-wall interaction forces it is nearly impossible to obtain an exact analytical expression for cell velocity in a DEP-FFF chamber at low voltages (low levitation heights). In addition, other forces such as electric field induced fluid flow have to be taken into account to model the system fully. Despite these problems the technique holds out promise as a new method for separating particles, and as recently demonstrated by Yang *et al.* (1999) and Wang *et al.* (2000), the technique is ideally suited to cell separation where several thousand cells per minute can be processed.

12.3.3 Sub-micrometre particles

Whilst the gravitational DEP-FFF method works well for large particles such as cells which have relatively rapid sedimentation times, it is impractical for smaller particles such as viruses or macromolecules. To achieve separation of colloidal particles, DEP forces can be used to generate a **flux** of particles towards or away from the walls of a separation chamber (*i.e.* the electrode array). This DEP flux is counteracted by a diffusion flux moving in the opposite direction leading to a steady state particle distribution, as shown schematically in figure 12.11. When the particle concentration profile is perturbed through the application of an AC electric field perpendicular to the fluid flow direction, separation occurs. As with gravitational DEP-FFF, the parabolic velocity profile of the fluid causes particles to fractionate and elute as sub-populations.

DEP-FFF of colloidal particles bears many similarities to electrical field flow fractionation (E-FFF), where a DC field perturbs a particle distribution due to electrophoretic forces. However, in the case of E-FFF the perturbation field is linear everywhere, whereas in a DEP-FFF device the field falls exponentially (in the first order approximation) from the walls of the channel. The main

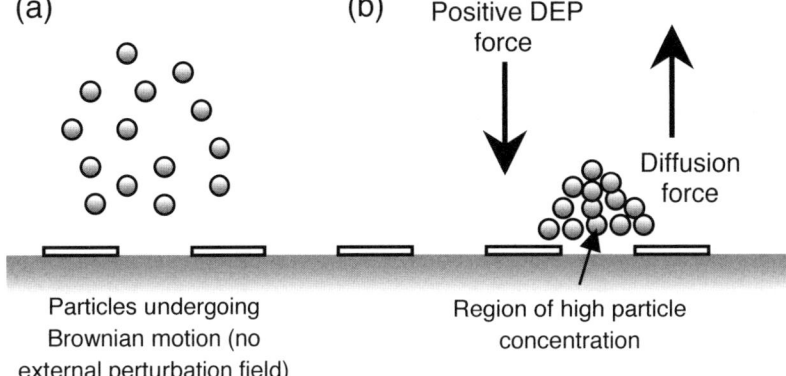

Fig. 12.11 Schematic diagram illustrating the forces and distribution of sub-micron particles in a DEP-FFF system. In (a) there is no deterministic force and particles assume a random distribution. When subjected to positive DEP, particles are pulled towards the electrodes. In the steady-state, the DEP flux and diffusion fluxes exactly balance leading to the skewed particle distribution, shown in (b).

disadvantage of E-FFF is that the absolute value of the force in the channel is low. As discussed in Chapter Five, when a DC voltage is applied to an electrode, a percentage of the potential is dropped at the electrode-electrolyte interface. In addition the voltage must be kept very low to avoid electrolytic decomposition of the water, so that this sets an upper limit of between 2 and 3 Volts. In a DC field, fractionation depends on the electrophoretic mobility of particles, *i.e.* the zeta potential. As a result, and because the device is symmetrical, a particle with the same magnitude but opposite sign of charge will fractionate at the same time. One will be attracted to the positive electrode and the other to the negative electrodes, but both will experience exactly the same fluid velocity.

By contrast, DEP-FFF systems can be "fine tuned" by adjusting the frequency of the driving field. For example, particles experiencing positive DEP of different magnitudes could be fractionated and at the same time populations experiencing negative DEP would be directed into the central part of the fluid stream and elute first.

Washizu *et al.* (1994) demonstrated that a micro-DEP-FFF system could be used to separate macromolecules and showed how separation of proteins and DNA could be achieved using a microelectrode array; he called his device a "dielectrophoretic chromatograph". DNA and proteins could be trapped on an electrode array by DEP in a flow-through chamber, however, owing to technical difficulties, true fractionation of a mixture was not demonstrated.

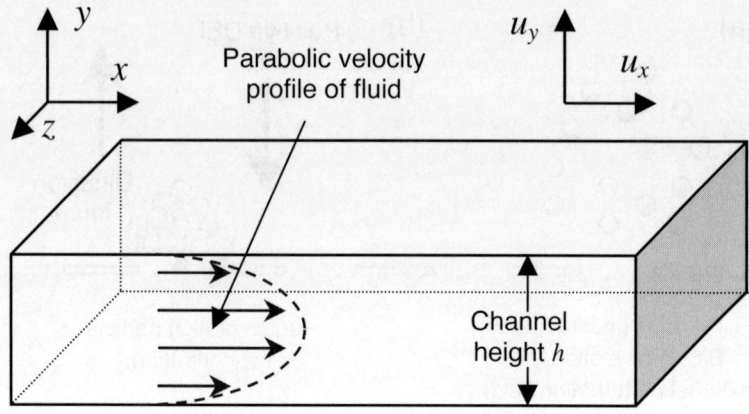

Fig. 12.12   Simplified model of a microchannel used to determine the
balance of fluxes in a sub-micrometre DEP-FFF system.

The theory of FFF for molecules in a DEP force field can be analysed in a
similar manner to that for particles in a DC electric field (see Giddings 1968 for an
explanation of the theory of electrophoretic FFF). Consider a distribution of
particles subjected to deterministic forces. The equilibrium concentration profile
occurs when there is a balance between the externally applied forces (such as DEP,
sedimentation, etc.) and the diffusion flux, *i.e.* when a fluid flows through the
system, the particle motion is governed by both drift and diffusion. With reference
to figure 12.12, a simplified analysis can be performed by treating the system in
two dimensions only. The drift velocity is the sum of velocities induced by the
external forces and the fluid flow

$$\mathbf{v}_{Drift} = \mathbf{v}_{Det} + \mathbf{u} \tag{12.10}$$

where **u** is the fluid velocity.

The flux is the sum of the diffusion flux (*i.e.* the concentration gradient) and
the flux due to the external forces, the deterministic and fluid velocities.

$$\mathbf{J} = -D\nabla n + (\mathbf{v}_{Det} + \mathbf{u})n \tag{12.11}$$

The rate of accumulation of solute at a point is given by the particle continuity
equation

$$\frac{\partial n}{\partial t} = -\nabla \cdot \mathbf{J} \tag{12.12}$$

Equations (12.11) and (12.12) are the governing equations which describe the evolution of the particle concentration profile in FFF. These equations are difficult to solve, however we can obtain a steady-state solution for the particle concentration profile in a channel without a fluid flow using the following assumptions:

- The only external force acting on the particles is the DEP force, balanced by Stokes force.
- The DEP force acts in one dimension (*i.e.* the far field approximation applies so that the x-component of the DEP force is ignored).

Application of a perturbation force gives a steady-state particle concentration that can be readily calculated from the balance of the DEP and diffusion fluxes, so that in terms of fluxes

$$\mathbf{J}_{Diff} = \mathbf{J}_{DEP} + \mathbf{J}_{Buoyancy} \tag{12.13}$$

i.e.

$$-D\left(\frac{\partial n}{\partial y}\right) + cv_{Det} = 0 \Rightarrow -D\left(\frac{\partial n}{\partial y}\right) + n\frac{|\mathbf{F}_{DEP}|}{f} = 0 \tag{12.14}$$

where f is the friction factor and the effects of gravity have been ignored. This equation can be integrated to give the concentration profile of particles as a function of space and time

$$n(x, y, t) = C_1(x, t) \exp\left(\int_0^h \frac{|\mathbf{F}_{DEP}|}{Df} dy\right) \tag{12.15}$$

where C_1 is an integration constant. This equation is solved to give the steady-state particle distribution. Using the analytical expression for the electric field gradient, an example of a typical concentration profile in a channel is shown in figure 12.13. The exponential shape of the particle distribution profile can be clearly seen. This figure also illustrates the dramatic effect of changing the sign of the DEP force (at 50 MHz).

A complete analysis of the time and spatial dependence of the particle distribution profile for a DEP-FFF device is an involved task. However, it is clear from equation (12.15) and figure 12.13 that the spatial dependence of the solute concentration profile is a function of several parameters, particularly frequency. Order of magnitude estimates of flux and particle separation can be made, and these point to the fact that DEP-FFF could be at least a factor of 10 more sensitive than electrophoretic FFF for separation of macromolecules. However, this is yet to be proven experimentally.

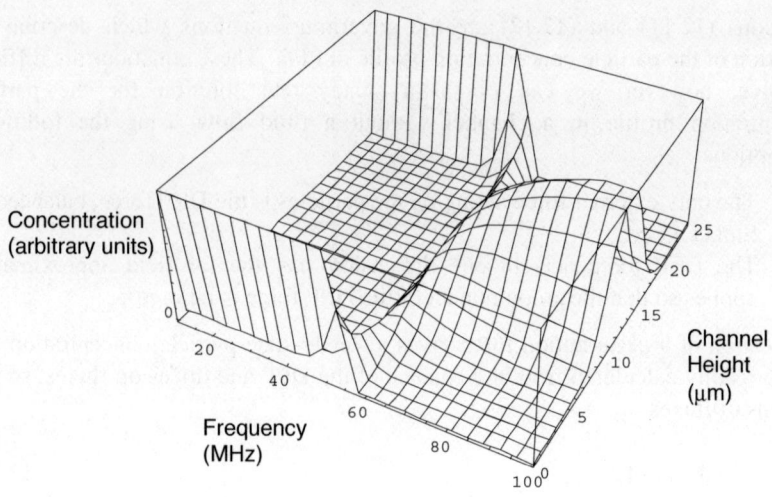

Fig. 12.13 Concentration profile for 60 nm diameter latex particles (with a surface conductance of 5 nS), plotted for frequencies from 10 MHz to 100 MHz in a solution of conductivity 140 mS m^{-1}. The DEP force is to zero at 50 MHz (the cross-over frequency). At frequencies below 50 MHz particles are concentrated at the channel wall, and above the concentration of particles is highest in the middle of the channel owing to negative DEP.

12.4   Single particle sorting and analysis systems

A major development in particle handling has occurred through the integration of several types of electrokinetic particle trapping and manipulation devices onto one microchip. This means that it will become possible to develop miniaturised devices incorporating much of the functionality seen in flow cytometer or fluorescently activated cell sorters (FACS) (Melamed *et al.* 1991). The group of Fuhr and co-workers has made remarkable progress in the development of a fully integrated particle processing chip (Fiedler *et al.* 1998; Müller *et al.* 1999). Particles are processed using sequential electrode arrays designed to focus, serially separate, trap and finally sort particles. The chip consists of several sets of micro-electrodes, as shown schematically in figure 12.14. Each electrode is fabricated on the bottom of a flow-through channel, with its mirror fabricated on the top of the channel (only the bottom is shown in the figure). The spacer between the two sets of electrodes is typically 30 – 50 μm, and each set of electrodes can be separately addressed with suitable AC fields and frequencies. Particles are suspended in an electrolyte of high conductivity such that the system operates under negative dielectrophoresis. With reference to this figure, the functioning of each of the different types of electrodes can be described as follows:

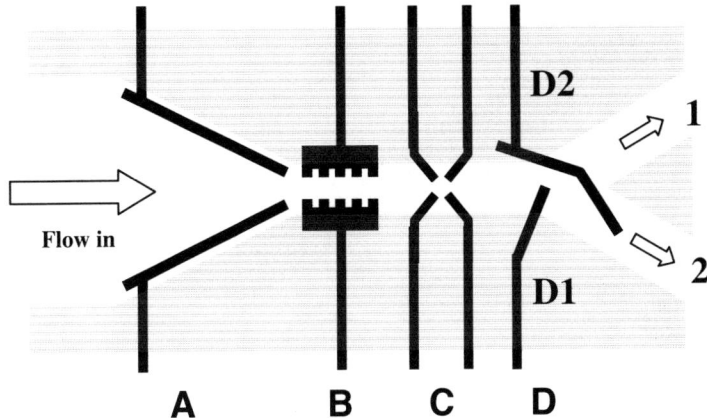

Fig. 12.14 Schematic illustration of a particle sorting and analysis system. AC electrokinetics is used to achieve a number of separate functions: particle focusing (**A**), alignment (**B**), analysis (**C**) and sorting (**D**). (Redrawn from Müller *et al.* 1999.)

(A) *Particle focusing.*

The purpose of this electrode array is to focus a wide spatial distribution of particles into a narrow beam. Particles enter the device from the left (figure 12.14) and are pushed hydrodynamically into the planar field funnel, marked A in the figure. These electrodes operate by providing a negative DEP force along their length, gradually pushing particles closer together to form a narrow stream. A photograph showing this effect is reproduced in figure 12.15 (a).

(B) *Alignment electrodes.*

Particle exiting the focussing electrodes need to be separated spatially so that they pass through further electrodes one at a time. This is achieved using the interdigitated fingers or castellated electrode array, labelled B in figure 12.14. The bottom electrode and its diagonal opposite electrode are connected to the same phase giving a series of spatially separated negative DEP traps. This electrode array both focuses the particles further (into the centre of the fluid stream) and also breaks up aggregates, so that they the emerge from the section as a single stream well separated from each other. A photograph showing this effect is reproduced in figure 12.15 (b). By applying a high enough voltage, particles can even be stably trapped against the Stokes force of the flowing liquid. For example, cells suspended in high conductivity phosphate buffered saline (PBS) can be held against a flow rate of 300 $\mu m\ s^{-1}$ (Müller *et al.* 1999).

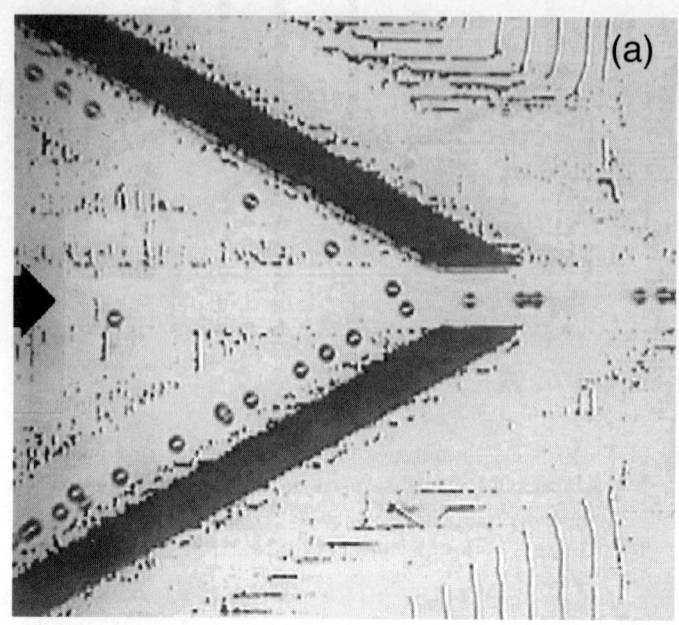

Fig. 12.15 Experimental photograph showing the functioning of the focussing device (A) and alignment electrodes (B), shown schematically in figure 12.14. (Reprinted with permission from Müller *et al.* 1999.)

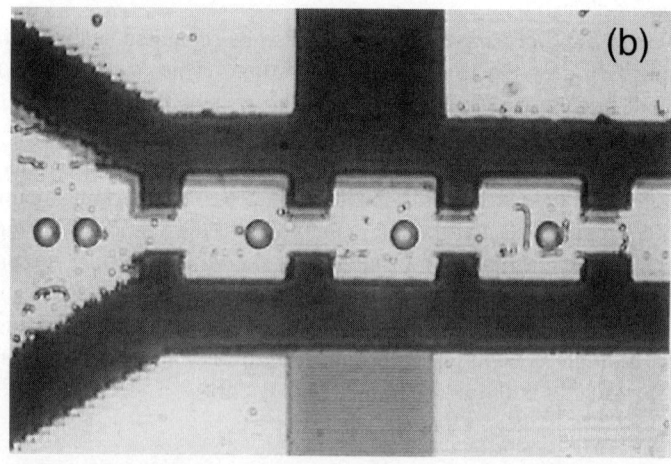

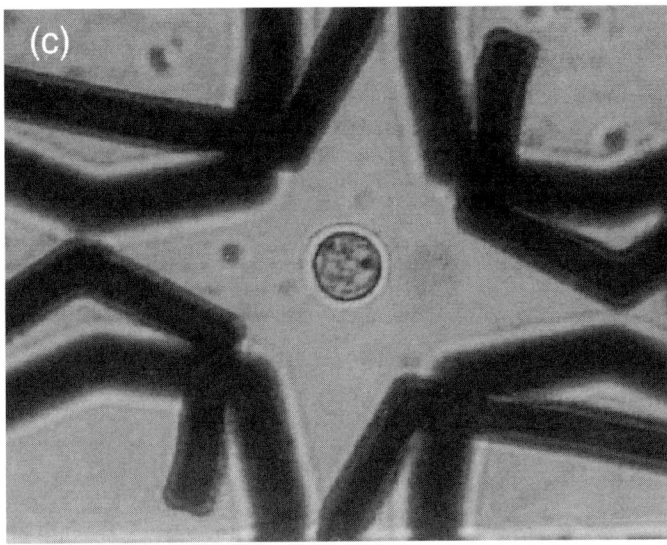

Fig. 12.15   Experimental photograph showing the functioning of the field
              cage (C) and switch (D), shown schematically in figure 12.14.
              (Reprinted with permission from Müller *et al.* 1999.)

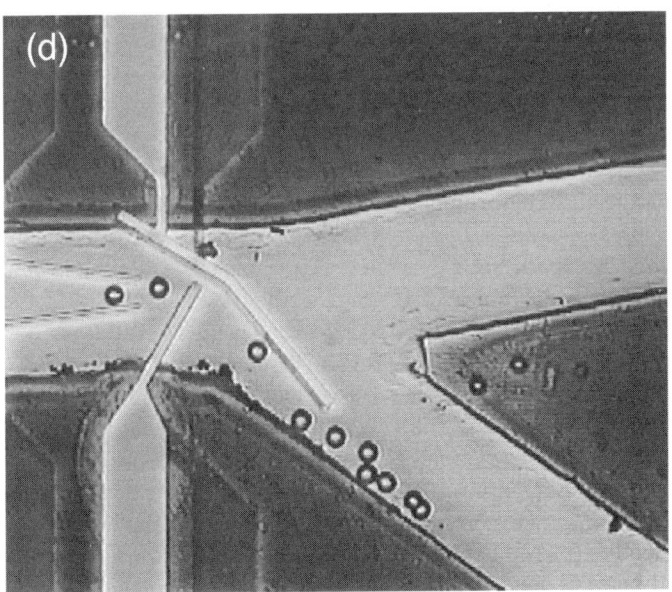

(C) *Field cage*

An octopole field cage, or particle trap, is created by arranging two opposing sets of quadrupole electrodes, as described in Chapter Eleven. Each half of the electrode set consists of four separate electrodes arranged as shown in figure 12.14, electrode set C. Single particles can be trapped here against the fluid flow. In addition, suitable design means that the cage can be arranged to hold only one particle at a time. Using rotating fields the dielectric properties of single particles can be analysed in such a cage. A photograph showing a particle held in a quadrupole trap is reproduced in figure 12.15 (c).

(D) *Switch*

The final component of the device is the switch, which is able to deflect particles into one of two outlet ports, thus sorting them sequentially. At point D in figure 12.14, the flow stream is split into two. Depending on which pair of energising electrodes is activated, particles are deflected to the upper or lower fluid stream. If the pair of electrodes marked D1 (top and bottom of device) is energised, then particles are pushed in to the lower fluid stream and emerge from outlet 1. If the other pair of electrodes marked D2 is energised, then particles are pushed into the lower fluid stream and emerge from outlet 2. This is shown in the photograph reproduced in figure 12.15 (d). This clearly shows how particles are being pushed towards the wall of the lower exit channel by the negative DEP forces produced by the upper pair of electrodes.

The 3-D micro-electrode system outlined above has been shown to be capable of processing cells and latex particles. The critical dimensions of the system must be chosen to allow sufficient trapping forces to be generated. Thus, for processing sub-micrometre particles, the system should be scaled appropriately, but this remains a technical challenge. These devices can also be combined with other particle handling or characterisation methods, such as ultrasonic trapping or optical trapping.

In this context a microsystem that combines laser traps with electrokinetic deflection has been developed for the screening of micro-organisms (Morishima *et al.* 1998; Arai *et al.* 2001). In this system, the microfluidic device consists of a main input and exit port, together with a side branch injection and drain port, as shown schematically in figure 12.16. Surrounding the injection port is a pair of micro-electrodes which are energised to produce a positive DEP force on particles. The design is such that these electrodes form a gate or filter at the point where the injection port meets the side branch. A suspension of bacteria is introduced into the injection port through a pipe and a single organism is selected and trapped in the centre of the injection port (point A), using a laser trap. Subsequently, all other bacteria are collected under positive DEP forces by the two electrodes at the mouth of the injection port, or are carried away to the drain by the fluid flow. The selected particle is then transported by manipulating the laser beam from point A to point B and released into the main fluid stream, thence to exit the device at the exit port.

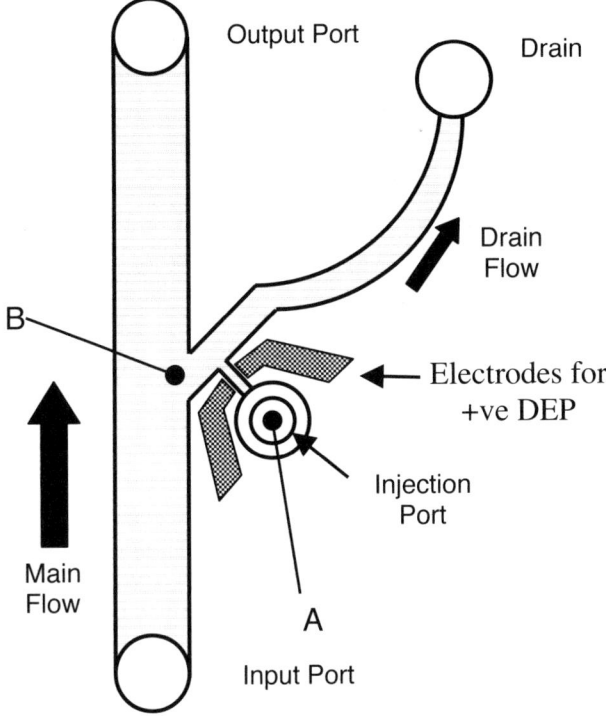

Fig. 12.16   A schematic diagram of a microfluidic device, which combines optical trapping with AC electrokinetic deflection. (Redrawn from Morishima *et al.* 1998.)

Particles can therefore be selected and manipulated into a fluid stream on a one-by-one basis. The system has been automated using image recognition software and feedback control to optimise the transportation speed (Arai *et al.* 2001).

12.5   Other separation methods

12.5.1   Combining DEP and EHD forces
Localised separation and positioning of particles can be achieved using a combination of electrohydrodynamic and dielectrophoretic forces. For example, using AC electroosmosis to drive fluid, it is possible to partition particles according to the balance between a Stokes force, caused by the localised fluid movement, and a DEP force. A photograph and diagram illustrating this effect is shown in figure 12.17(a) and (b). In figure 12.17(a) the separation of two different sized particles (93 nm and 216 nm diameter) on a castellated electrode array is illustrated diagrammatically (Green and Morgan 1998). At certain frequencies both

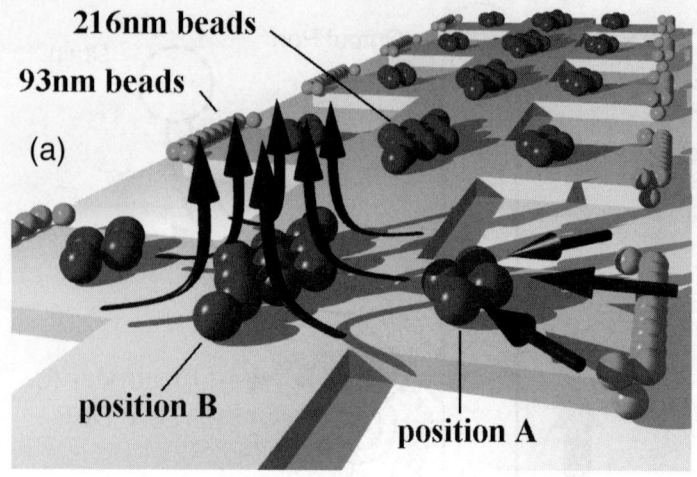

Fig. 12.17　Schematic diagram (a) and experimental image (b) showing how two different sized particles separate under the influence of a combination of DEP and EHD forces on a castellated microelectrode. The 93 nm beads remain held on the electrode edges by positive DEP and the 216 nm beads are pushed into the centre of the electrode by EHD induced flow. (Reprinted with permission from Green and Morgan (1998), copyright IOP Publishing Ltd – original in colour, www.iop.org)

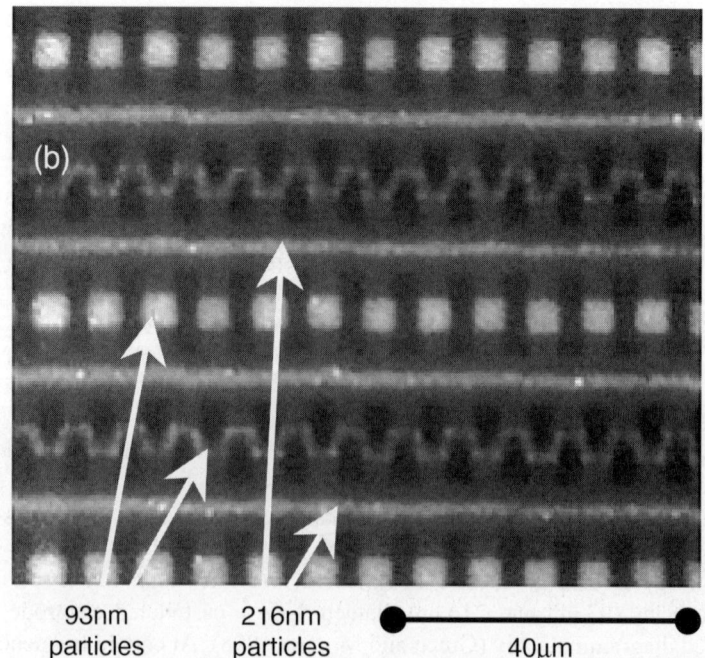

DEP forces and electroosmotic forces act on the particles. In this example, all the particles experience positive DEP, but owing to the local fluid motion the larger particles are pushed onto the electrode surface. They move to a stable region where the positive dielectrophoretic force and the viscous drag force balance each other, represented by positions A or B in the figure, depending on the applied voltage.

At the same time the smaller particles remain trapped at the electrode tips under positive DEP forces, leading to separation as shown in the photograph of figure 12.17(b). Under the experimental conditions used to take this photograph, the value of the real part of the Clausius-Mossotti factor for the 93 nm particle was approximately 20 times larger than for the 216 nm particle. Allowing for the difference in particle volumes, this means that the DEP force on the 93 nm particle is 1.6 times greater than on the 216 nm particle (for frequencies up to 1MHz). However, to a first approximation and for a constant fluid velocity, the viscous drag force on the 93 nm particles is 2.3 times smaller than on the 216 nm. The net balance of forces, therefore, is such that the larger particles move away from the electrode edges under the influence of EHD forces and are separated from the smaller particles. These remain trapped on the electrodes under dominant positive DEP forces. Such results demonstrate the potential for developing integrated microsystems that combine EHD forces with DEP forces for precise control of particle position, for aligning particles or for localised separation of particles.

12.5.2   Brownian ratchets

The concept of producing unidirectional motion of particles by exploiting diffusion and asymmetry in a potential function has been discussed in many papers. For example, Magnasco (1993) showed that a spatially asymmetric potential could generate particle motion. Curie (1894) suggested that spatially asymmetric structures could act as ratchets for colloidal particles. This was confirmed by Prost *et al.* (1994) who showed how microfabricated structures, providing spatially asymmetric potentials that could be switched on and off periodically in time, would cause particle transport. They constructed a microdevice that used DEP to generate asymmetric trapping potentials and demonstrated unidirectional transport of particles (Rousselet *et al.* 1994). DEP-based Brownian ratchets have also been reported by Faucheux and Libchaber (1995) and Gorre-Talini *et al.* (1997).

Brownian ratchets have attracted attention for their application to a range of separation science. For example, a device has been fabricated and used to separate short DNA fragments (Bader *et al.* 1999; Hammond *et al.* 2000). The DNA in this case was separated using switched DC voltages (at rates of up to 10 Hz) and the separation was caused by electrostatic charge-charge interaction rather than polarisability. The authors estimate that the device could be used to detect single nucleotide polymorphisms (SNPs), a position on a genome where multiple nucleotides can occur. They estimate that the separation of a 24mer from a 25mer oligonucleotide should be possible in a few seconds on a 1.25 cm long device.

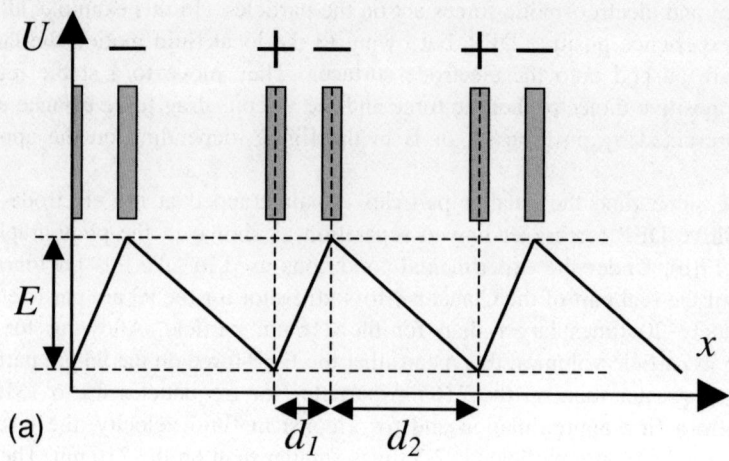

Fig. 12.18 Schematic diagram of the potential in a simple Brownian ratchet (a) and the corresponding motion of colloidal particles as the potential is periodically switched on and off (b).

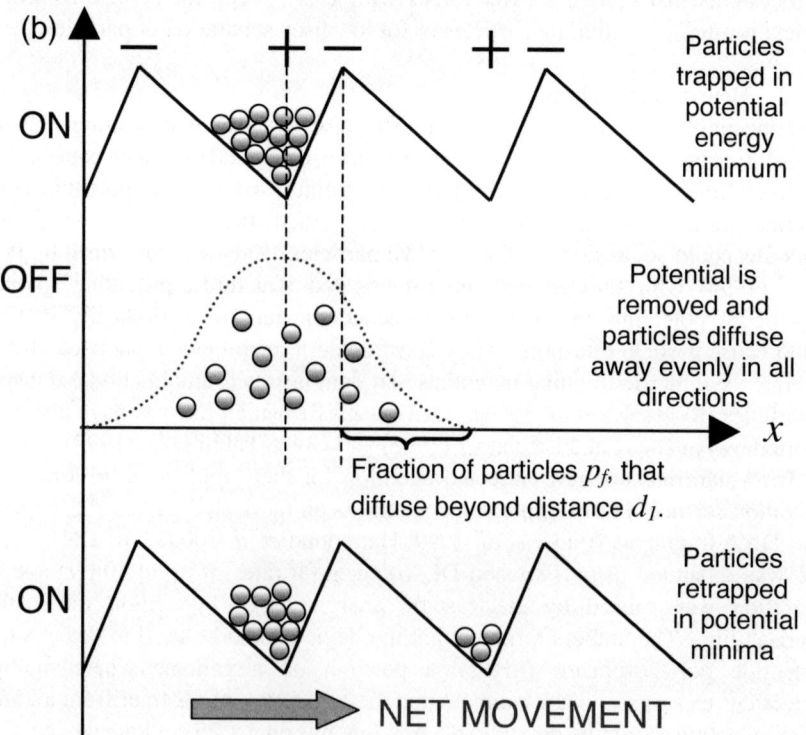

The functioning of a Brownian ratchet can be described as follows. Assume that a collection of particles is subjected to an asymmetric potential energy profile as shown in figure 12.18(a). The potential is switched on and off with times t_{ON} and t_{OFF}. The spatial periods of the potential profiles are d_1 and d_2 and the potential amplitude E is much greater than $k_B T$. The important feature of a Brownian ratchet is that the potential energy function is **periodic** (it repeats) and is also **asymmetric**. For the sake of analysis, it is convenient to think of a ratchet driven simply by charge-charge interactions (electrophoresis). In this case implementation of the device is trivial, requiring only a repeating electrode with gaps d_1 and d_2, and with alternating DC potentials of + and – as shown in the figure.

Imagine now using the device to pump a suspension of negatively charged particles. At time zero, the particles are all placed near the central electrode pair and the potential applied. During the period of the *ON* time, the negatively charged particles are attracted onto the positive electrode, and end up in the potential energy minimum indicated in figure 12.18(b). When the voltage is switched off, the particles begin to diffuse away. The profile of the particle distribution is Gaussian and is given by equation (9.5) of Chapter Nine

$$P(x, t_{OFF}) = \frac{1}{\sqrt{4\pi D}} e^{-\frac{x^2}{4D(t_{OFF})^{3/2}}} \tag{12.16}$$

where D is the diffusion coefficient of the particles.

After a given time some of the particles will have diffused further than the distance d_1 between the electrodes. When the voltage is turned back on, these particles are pulled towards the next potential energy minimum and are re-trapped. In other words, since the system of traps is asymmetric, more particles will on average move to the right than to the left, *i.e.* there is a net particle movement, as shown in figure 12.18(b). Repeating this process over and over again leads to unidirectional transport. The rate of drift depends very simply on the exponential of the diffusion coefficient of the particles during the time when the trap is switched off. Hence, particles of different sizes and diffusion coefficients move at different rates and not only particle transport but also fractionation is possible.

It can be shown that the probability of particles moving forward (P_f) or backward (P_b) is given by the following expressions

$$P_f = \frac{1}{2} \text{erfc} \left(\frac{d_1}{\sqrt{4Dt_{OFF}}} \right) \tag{12.17}$$

$$P_b = \frac{1}{2} \text{erfc} \left(\frac{d_2}{\sqrt{4Dt_{OFF}}} \right) \tag{12.18}$$

where erfc(…) is the complementary error function. With potential energy profiles

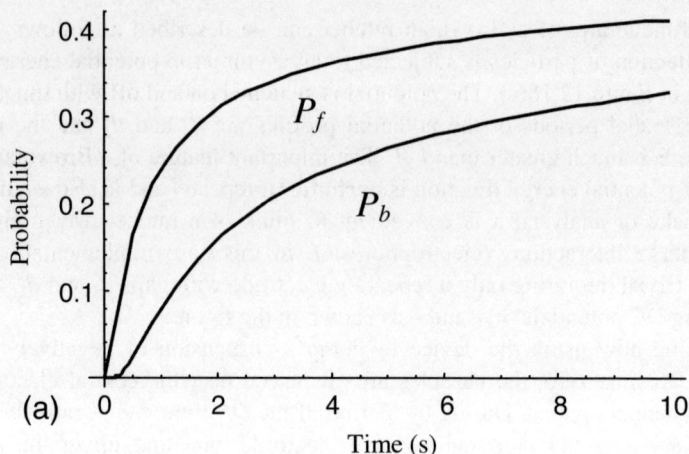

(a)

Fig. 12.19   (a) The forwards and backwards probabilities for a suspension of
colloidal particles in a Brownian ratchet as a function of OFF time.
(b) The difference in the forwards and backwards potential as a
function of OFF time.

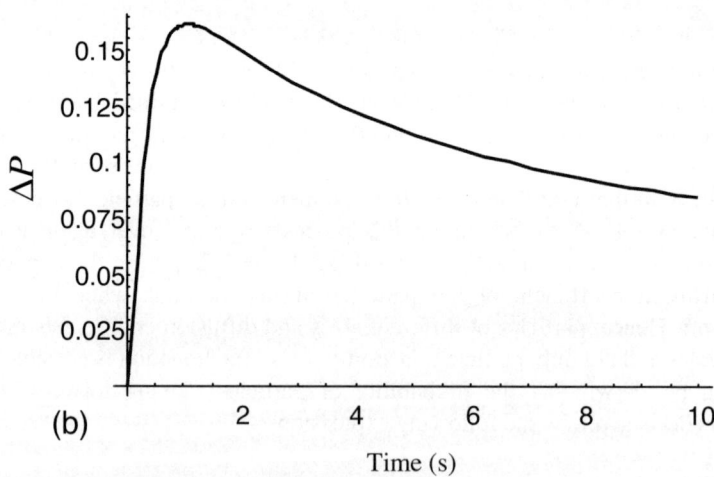

(b)

of the order of $\sqrt{D}$, the difference in probability between forward and backward
transitions can be quite large. For example, figure 12.19 shows a plot of the
forward and backward probability as a function of off time, t_{OFF}, for a particle of
100 nm diameter, with $d_1 = 2d_2 = \sqrt{D}$. Also shown is the difference between the
probabilities plotted as a function of t_{OFF}, which reaches a maximum at
approximately 1 second. This figure clearly shows how adjusting the ratio of *ON* to

Fig. 12.20   Schematic diagram of the "Christmas tree" electrode design used
as a dielectrophoretic Brownian ratchet. (Redrawn from Rousselet
et al. 1994.)

OFF times can be used to optimise the separation.

The fabrication of asymmetric structures for trapping and pumping particles by DEP has been reported in the literature and an example of a periodic asymmetric potential device is shown in figure 12.20. This electrode array is planar and has been called the "Christmas tree" electrode (Rousselet *et al.* 1994, Faucheux and Libchaber 1995). It produces an asymmetric DEP potential and has been shown to pump latex particles in the range 0.25 µm to 0.4 µm. The transport of the particles was found to be in semi-quantitative agreement with theory.

In figure 12.21 we have shown a different design of planar electrodes that behaves as a DEP-based Brownian ratchet. These electrodes work by simultaneously generating positive and negative DEP forces to trap and repel particles; analogous to the electrophoresis based ratchet for moving DNA shown in figure 12.18. Figure 12.21(a) shows the electrode array, which consists of a unit set of four electrodes, repeated along the length of the structure. These four electrodes are configured to give a simultaneous positive and negative DEP force during the *ON* time of the device. Thus, particles are collected at the tips of electrode set 1 and repelled from electrode set 2. A simpler, and perhaps more elegant device is shown in figure 12.22(a). In this case only three electrodes are used, producing the same simultaneous movement of particles by positive and negative DEP as the former electrode array.

Figure 12.21(b) and 12.22(b) show the corresponding potential energy for these electrodes plotted as a function of distance along the array. The plots were calculated from equation (2.13), with a particle size of 1 µm diameter and an

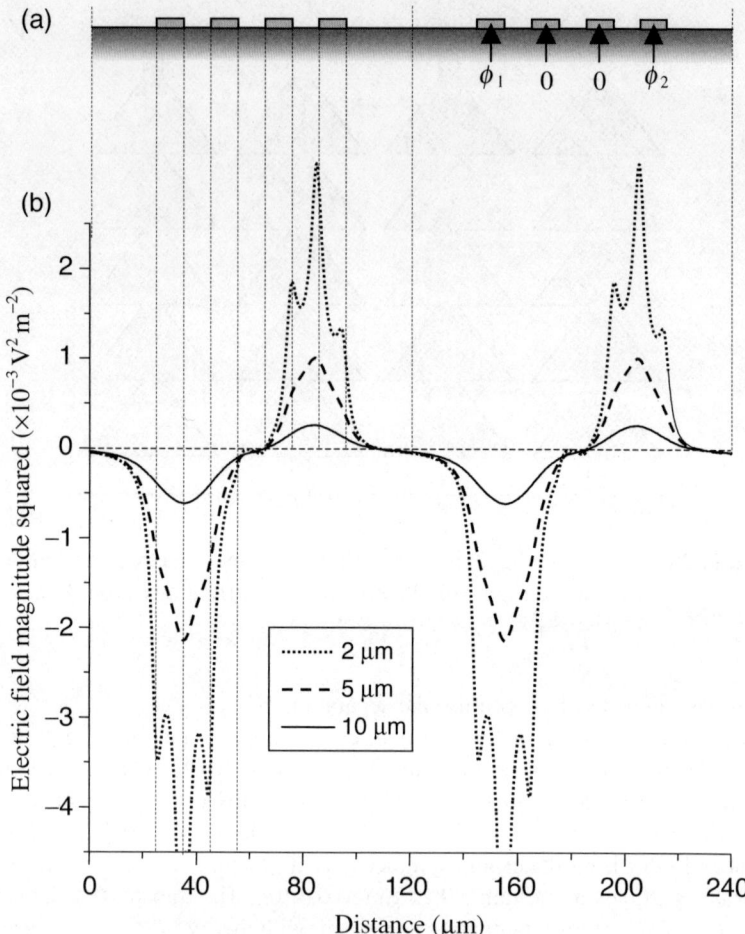

Fig. 12.21 Schematic diagram (a) and plots of the electric field magnitude squared (b) at different heights for a four-electrode dielectrophoretic Brownian ratchet. The electrodes are arranged in two pairs, with two signals of different frequency applied to each pair. One frequency is for positive dielectrophoresis and the other for negative. The four electrodes are repeated as shown and the DEP potential (described by the field magnitude squared) is asymmetric.

applied voltage of 1 Volt. The electrode spacing was 10 μm and the three plots shown were calculated for different heights above the electrode. As can be seen, there is a clear asymmetry in the potential energy profile, with local variations becoming more pronounced as the particle height above the electrode decreases. Experiments show that such devices function as DEP-based potential ratchets (unpublished data).

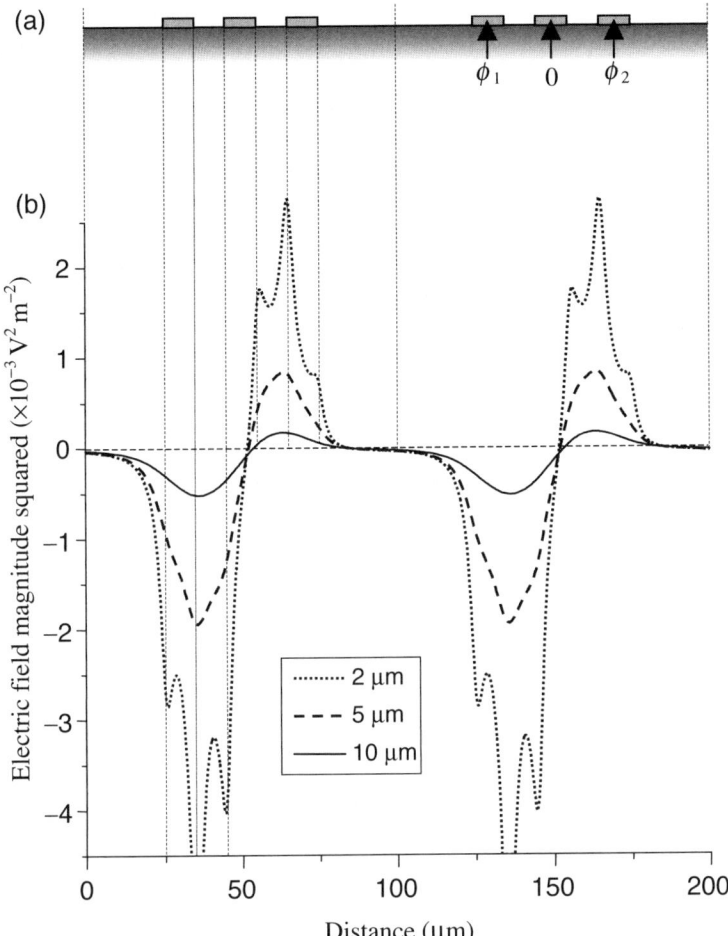

Fig. 12.22 Schematic diagram (a) and plots of the electric field magnitude squared (b) at different heights for a three-electrode dielectrophoretic Brownian ratchet. The electrodes are arranged with two signals of different frequency applied to the end electrodes, sharing a common ground electrode in the centre. One frequency is for positive dielectrophoresis and the other for negative. When the three electrodes are repeated as shown, the DEP potential (described by the field magnitude squared) is asymmetric. This arrangement produces a consistently more asymmetric potential than the four-electrode system shown in figure 12.21.

Gorre-Talini *et al.* (1997) used pseudo 3-D electrode arrays based on the shape of optical diffraction gratings. They measured different velocities for 1 μm and 0.5 μm diameter latex particles. The maximum velocity difference occurred with a t_{OFF} = 3 seconds and with $d_2/d_1 \approx 3$, where the ratio of velocities was approximately 10 with the smaller beads travelling at 0.2 μm s^{-1}. From calculations, the authors showed that these two different sized latex beads could be separated, albeit with a long processing time of eight hours for a 1 mm long device. This compares with the few minutes required to separate particles in a μDEP-FFF device. However, separation times are likely to be much shorter for smaller particles with bigger diffusion constants. Brownian ratchet separators may have advantages for use in pumping or separating small quantities of material such as macromolecules, DNA or viruses since the average drift velocity is a function of the diffusion constant.

12.6   Summary

In this chapter we have shown how dielectrophoresis can be used in particle separation science and technology. Recent technological advances now means that this technique has great potential as a method for separating micro- and nano-scale particles, particularly when used as part of an integrated microsystem. Further research will undoubtedly lead to the adoption of AC electrokinetic techniques in a range of new Lab-on-a-chip technologies.

12.7   References

Arai F., Ichikawa A., Ogawa M., Fukuda T., Horio K. and Itoigawa K. *High speed separation system of randomly suspended single living cells by laser trap and dielectrophoresis* Electrophoresis **22** 283-288 (2001).

Archer G.P., Render M.C., Betts W.B. and Sancho M. *Dielectrophoretic concentration of micro-organisms using grid electrodes* Microbios **76** 237-244 (1993).

Bader J.S., Hammond R.W., Henck S.A., Deem M.W., McDermott A., Bustillo J.M. Simpson J.W., Mulhern G.T. and Rothberg J.M. *DNA transport by micromachined Brownian ratchet device* Proc. Natl. Acad. Sci. USA **96** 13165-13169 (1999).

Becker F.F., Wang X-B., Huang Y., Pethig R., Vykoukal J. and Gascoyne P.R.C. *The removal of human leukaemia cells from blood using interdigitated microelectrodes* J. Phys. D: Appl. Phys. **27** 2659-2662 (1994).

Becker F.F., Wang X-B., Huang Y., Pethig R., Vykoukal J. and Gascoyne P.R.C. *Separation of human breast cancer cells from blood by differential dielectric affinity* Proc. Natl. Acad. Sci. USA **92** 860-864 (1995).

Berg H.C. and Purcell E.M. *A method for separating according to mass a mixture of macromolecules or small particle suspended in fluid I. Theory* Proc. Natl. Acad. Sci. USA **58** 862-869 (1967a).

Berg H.C. and Purcell E.M. *A method for separating according to mass a mixture of macromolecules or small particle suspended in fluid III. Experiments in a centrifugal field* Proc. Natl. Acad. Sci. USA **58** 1821-1828 (1967b).

Berg H.C., Stewart W.W. and Purcell E.M. *A method for separating according to mass a mixture of macromolecules or small particle suspended in fluid II. Experiments in a gravitational field* Proc. Natl. Acad. Sci. USA **58** 1286-1291 (1967).

Bike S.G. and Prieve D.C. *Electrohydrodynamics of thin double-layers - a model for the streaming potential profile* J. Colloid Int. Sci. **154** 87-96 (1992).

Black B.C. and Hammond   E.G.   *Separation by dielectric distribution: theory* J. Amer. Oil. Chem. Soc. **42** 931-935 (1965a).

Black B.C. and Hammond E.G. *Separation by dielectric distribution: application to the isolation and purification of soyabean phosphatides and bacterial spores* J. Amer. Oil. Chem. Soc. **42** 936 (1965b).

Caldwell K.D., Kesner L.F., Meyers M.N. and Giddings J.C.   *Electrical field flow fractionation of proteins* Science **176** 296-298 (1972).

Cheng J., Sheldon E.L., Wu L., Uribe A., Gerrue L.O., Carrino J., Heller M.J. and O'Connell J.P. *Preparation and hybridisation analysis of DNA/RNA from E.Coli on microfabricated bioelectronic chips* Nature Biotech. **16** 546-554 (1998).

Cui L. and Morgan H. *Design and fabrication of travelling wave dielectrophoresis structures* J. Micromech. Microeng. **10** 72-79 (2000).

Curie M.P. *Sur la symétrie dans les phénomènes physiques, symétrie d'un champ électrique et d'un champ magnétique* J. Physique **III** 393-414 (1894).

Crane J.S. and Pohl H.A. *The study of living and dead yeast cells by dielectrophoresis* J. Electrochem. Soc. **115** 584-586 (1968).

Docoslis A., Kalogerakis N., Behie L.A. and Kaler K.V.I.S. *A novel dielectrophoresis base device for the selective retention of viable cells in cell culture media* Biotechnology Bioengineering **54** 238-250 (1997).

Faucheux L.P. and Libchaber A.   *Selection of Brownian Particles*   J. Chem. Soc. Faraday Trans. **91** 3163-3166 (1995).

Fiedler S., Shirley S.G., Schnelle Th. and Fuhr G. *Dielectrophoretic sorting of particles and cells in a microsystem* Anal. Chem. **70** 1909-1915 (1998).

Gale B.K., Caldwell K.D. and Frazier A.B. *A micromachined electrical field flow fractionation (μ-FFF) system* IEEE Trans. Biomed. Eng. **45** 1459-1469 (1998).

Gascoyne P.R.C., Huang Y., Pethig R., Vykoukal J. and Becker F.F. *Dielectrophoretic separation of mammalian cells studied by computerized image analysis* Meas. Sci. Technol. **3** 439-445 (1992).

Giddings J.C. *Nonequilibrium theory of field-flow fractionation* J. Chem. Phys. **49** 81-84 (1968).

Giddings J.C. *Field flow fractionation of macromolecules* J. Chromatography **470** 327-335 (1989).

Giddings J.C. *Field flow fractionation: analysis of macromolecular colloidal and particulate materials* Science **260** 1456-1465 (1993).

Gorre-Talini L., Jeanjean S. and Silberzan P.   *Sorting of Brownian particles by the pulsed application of an asymmetric potential* Phys. Rev. E **56** 2025-2034 (1997).

Green N.G. and Morgan H.   *Dielectrophoretic separation of nano-particles* J. Phys. D: Appl. Phys. **30** L41-L44 (1997).

Green N.G., and Morgan H.   *Separation of sub-micrometre particles using a combination of Dielectrophoretic and Electrohydrodynamic forces* J. Phys. D: Appl. Phys. **31** L25-30 (1998).

Hammond R.W., Bader J.S., Henck S.A., Deem M.W., McDermott A., Bustillo J.M. and Rothberg J.M. *Differential transport of DNA by a rectified Brownian motion device* Electrophoresis **21** 74-80 (2000).

Happel J. and Brenner H. *Low Reynolds number hydrodynamics* Kluwer Academic Publ., Dordrecht (1983).

Hermanson G.T. *Bioconjugate Techniques* Academic Press, London (1996).

Huang Y., Wang X-B., Becker F.F. and Gascoyne P.R.C. *Introducing dielectrophoresis as a new force field for Field-Flow-Fractionation* Biophysical J. **73** 1118-1129 (1997).

Hughes M.P. and Morgan H. *Dielectrophoretic characterisation and separation of antibody coated sub-micron latex spheres* Anal. Chem. **71** 3441-3445 (1999).

Hughes M.P., Flynn M.F. and Morgan H. *Dielectrophoretic measurements of sub-micrometre latex particles following surface modification* Institute of Physics Conference Series **163** 81-84 (1999a).

Hughes M.P., Morgan H. and Flynn M.F. *The dielectrophoretic behaviour of sub-micron latex spheres: influence of surface conductance* J. Coll. Int. Sci. **220** 454-457 (1999b).

Lorz B., Simson R., Nardi J. and Sackman E. *Weakly adhering vesicles in shear flow: Tanktreading and anomalous lift force* Europhys. Lett. **51** 468-474 (2000).

Magnasco M.O. *Forced thermal ratchets* Phys. Rev. Lett. **71** 1477-1481 (1993).

Markx G.H., Talary M.S and Pethig R. *Separation of viable and non-viable yeast using dielectrophoresis* J. Biotechnology **32** 29-37 (1994a).

Markx G.H., Huang Y., Zhou X-F. and Pethig R. *Dielectrophoretic characterisation and separation of micro-organisms* Microbiology **140** 585-591 (1994b).

Markx G.H. and Pethig R. *Dielectrophoretic separation of cells: continuous separation* Biotechnology and Bioengineering **45** 337-343 (1995).

Markx G.H., Dyda P.A. and Pethig R. *Dielectrophoretic separation of bacteria using a conductivity gradient* J. Biotechnology **51** 175-180 (1996).

Markx G.H., Pethig R. and Rousselet J. *The dielectrophoretic levitation of latex beads, with reference to field flow fractionation* J. Phys. D: Appl. Phys. **30** 2470-2477 (1997).

Mason B.D. and Hammond E.G. *Dielectrophoretic separation of living cells* Can. J. Microbiol. **17** 879-888 (1971).

Melamed M.R., Lindmo T. and Mendelson M.L. *Flow Cytometry and Sorting* Wiley-Liss, New York (1991).

Morgan H., Hughes M.P. and Green N.G. *Separation of sub-micron bio-particles by dielectrophoresis* Biophys. J. **77** 516-525 (1999).

Morishima K., Arai F., Fukuda T., Matsuura H. and Yoshikawa K. *Screening of single Escherichia coli in a microchannel system by electric field and laser tweezers* Anal. Chem. Acta **365** 273-278 (1998).

Müller T., Gradl G., Howitz S., Shirley S., Schnelle T. and Fuhr G. *A 3-D microelectrode system for handling and caging single cells and particles* Biosensors & Bioelectronics **14** 247-256 (1999).

Pethig R., Huang Y., Wang X-B and Burt J.P.H.   *Positive and negative dielectrophoretic collection of colloidal particles using interdigitated castellated microelectrodes* J. Phys. D: Appl. Phys. **25** 881-888 (1992).

Pethig R. *Dielectrophoresis: using inhomogeneous A.C. electric fields to separate and manipulate cells* Crit. Rev. Biotechnol. **16** 331-348 (1996).

Pohl H.A. and Schwar J.P.   *Factors affecting separations of suspensions in non-uniform electric fields* J. Appl. Phys. **30** 69-73 (1959).

Pohl H.A. and Plymal C.E.   *Continuous separation of suspensions by non-uniform electric field in liquid dielectrics*   J. Electrochem. Soc. **107** 390-396 (1960).

Pohl H.A. and Hawk I.   *Separation of living and dead cells by dielectrophoresis.* Science **152** 647-649 (1966)

Pohl H.A. and Crane J.S. *Dielectrophoresis of cells* Biophys. J. **11** 711-727 (1971).

Pohl H.A. *Dielectrophoresis: application to the characterisation and separation of cells.* in Catsimpoolas N. (Ed) *Methods of Cell Separation Vol 1* Plenum Press, New York 67-169 (1977).

Pohl H.A. *Dielectrophoresis* Cambridge Uni. Press, Cambridge (1978).

Pohl H.A., Kaler K., and Pollock K.   *The continuous positive and negative dielectrophoresis of micro-organisms* J. Biol. Phys. **9** 67-86 (1981).

Prost J., Chauwin J-F., Peliti L. and Ajdari A.   *Asymmetric Pumping of Particles* Phys. Rev. Lett. **72** 2652-2655 (1994).

Rousselet J., Salome L., Ajdari A. and Prost J.   *Directional motion of Brownian particles induced by a periodic asymmetric potential* Nature **370** 446-448 (1994).

Sukumaran S. and Seifert U.   *Influence of shear flow on vesicles near a wall: a numerical study* Phys. Rev. E. **64** 011916 (2001).

van de Ven T.G.M, Warszynski P. and Dukhin S.S.   *Electrokinetic lift of small particles* J. Coll. Int. Sci. **157** 328-331 (1993).

Verschure R.H. and Ijlst L. *Apparatus for continuous separation of mineral grains* Nature **211** 619-200 (1966).

Wang X-B., Huang Y., Burt J.P.H., Markx G.H. and Pethig R.   *Selective dielectrophoretic confinement of bioparticles in potential energy wells* J. Phys. D: Appl. Phys. **26** 1278-1285 (1993).

Wang X-B., Vykoukal J., Becker F.F. and Gascoyne P.R.C. *Separation of polystyrene microbeads using dielectrophoretic gravitational field-flow-fractionation* Biophys. J. **74** 2689-2701 (1998).

Wang X-B., Yang J., Huang Y., Vykoukal J., Becker F.F. and Gascoyne P.R.C. *Cell separation by dielectrophoretic field-flow-fractionation* Anal. Chem. **72** 832-839 (2000).

Washizu M., Suzuki S., Kurosawa O., Nishizaka T. and Shinohara T. *Molecular dielectrophoresis of biopolymers* IEEE Trans. Ind. Appls. 30 835-843 (1994).

Williams P.S., Koch T. and Giddings J.C.   *Characterisation of near-wall hydrodynamic lift forces using sedimentation field-flow fractionation* Chem. Eng. Comm. **111** 121-147 (1992).

Williams P.S., Moon M.H. and Giddings J.C. *Influence of accumulation wall and carrier solution composition on lift force in sedimentation/steric field-flow fractionation* Colloids & Surfaces A **113** 215-228 (1996).

Yang J., Huang Y., Wang X-B., Becker F.F. and Gascoyne P.R.C. *Cell separation on microfabricated electrodes using dielectrophoretic/gravitational field-flow-fractionation* Anal. Chem. **5** 911-918 (1999).

Appendix A

Mathematical tools

Throughout the book, there are several instances where we have used mathematical programs to solve mathematical equations and to plot functions. To this end, many different programs are available, some of which are specifically aimed at solving electrostatics or electromagnetic problems. In this book we have used commercially available symbolic packages such as Matlab® or Mathematica®. In Chapter Ten, we calculated and subsequently plotted the analytical solution for the dielectrophoretic force. These plots were made using Mathematica®. Other plots in the first chapters of the book were made using Matlab®. There are also software packages dedicated to numerical calculations and throughout the book such calculations have been made using FlexPDE®.

In this Appendix we will examine in more detail these three software packages, showing how they can be used to solve different problems with an example in each case. The first example is using Matlab® to calculate and then plot the Clausius-Mossotti factor for the polarisation of multi-shelled spherical particle, as detailed in Chapter Three. The second example concerns the use of FlexPDE® to numerically simulate electric fields. In this context we also discuss the wider issues of the use of appropriate boundary conditions. Finally, we show a simple example of the use of Mathematica® to calculate the force from the analytical expression for the electric field, as detailed in Chapter Eight.

The inclusion of the examples programs should encourage the reader to develop further programs so that he or she may explore some of the more involved aspects of AC electrokinetic particle manipulation and characterisation.

A.1   Matlab®
In Chapter Three, we discussed the mechanism of interfacial polarisation and the creation of induced dipole moments in particles. In particular, section 3.2.1 showed that for a solid sphere the effective polarisability is given by equation (3.9) with the frequency dependence of the polarisability (or DEP force) described by the Clausius-Mossotti factor (equation (3.10))

$$\tilde{f}_{CM} = \frac{\tilde{\varepsilon}_p - \tilde{\varepsilon}_m}{\tilde{\varepsilon}_p + 2\tilde{\varepsilon}_m} \tag{A.1}$$

Plots of this factor as a function of frequency were shown in figures 3.4 and 3.5. In section 3.2.3, we elaborated on the solid sphere and showed how the shell model could be used to calculate the polarisability of, for example, a biological cell. In this case the particle complex permittivity is replaced by an expression representing the stratified nature of the cell (equations (3.21) and (3.22)). Plots of the factor f_{CM} for a particle with one and two shell were shown in figures 3.7 and 3.8. The following Matlab® script was used to generate these figures.

A.1.1   Matlab® script

The following script file can be copied and run from the Matlab® control window to produce the figures shown in Chapter Three. Alternatively, it can be saved as a plain text file with the suffix ".m" and called in the program as a function. The diameter, permittivity and conductivity of the different parts of the three models can be varied

```
% This is a script to model and plot the Clausius-Mossotti Factor
% script contains expressions for 0, 1, 2 shells with properties defined at the start.

clear all

eo = 8.854e-12;                              % permittivity (free)
em = 80*eo;        sm = 1e-4;                 % medium properties (general)

% permittivity    conductivity   radius
ep = 2.5*eo;       sp = 1e-2;     d = 2e-6;   % particle (zero shells)

% permittivity    conductivity   radius
em1 = em;          sm1 = sm;
ep2 = 10*eo;       sp2 = 1e-8;    d1 = 2.01e-6;   % particle (1 shell) membrane
ep3 = 60*eo;       sp3 = 0.5;     d2 = 2e-6;      % particle (1 shell) cytoplasm

% permittivity    conductivity   radius
em1 = em;          sm1 = sm;
e2 = 10*eo;        s2 = 1e-8;     de1 = 2.01e-6;  % particle (2 shells) membrane
e3 = 60*eo;        s3 = 0.5;      de2 = 2e-6;     % particle (2 shells) cytoplasm
e4 = 30*eo;        s4 = 0.1;      de3 = 0.5e-6;   % particle (2 shells) nucleus

% the frequency
f = logspace(1,9,100);              omega = 2*pi*f;

%  The calculation - zero shells
eph0 = ep - (i.*sp./omega);         emh0 = em - (i.*sm./omega);
```

```
fCM0 = (eph0 - emh0)./(eph0 + (2*emh0));          % Clausius-Mossotti factor

%  The calculation - 1 shell
gamma12 = d1/d2;                    extra1 = gamma12^3;
emh1 = em1 - (i.*sm1./omega);
eph2 = ep2 - (i.*sp2./omega);       eph3 = ep3 - (i.*sp3./omega);
fCM_23 = (eph3 - eph2)./( eph3 + (2*eph2));
ep23 = eph2.*(extra1 + (2.*fCM_23))./(extra1 - fCM_23);
fCM1 = (ep23 - emh1)./(ep23 + (2*emh1));          % Clausius Mossotti factor

%  The calculation - 2 shells
gamma12 = de1/de2;                  extra2 = gamma12^3;
gamma23 = de2/de3;                  extra3 = gamma12^3;
eh1 = em1 - (i.*sm1./omega);        eh2 = e2 - (i.*s2./omega);
eh3 = e3 - (i.*s3./omega);          eh4 = e4 - (i.*s4./omega);
fCM_34 = (eh4 - eh3)./(eh4 + (2*eh3));
ep34 = eh3.*(extra3 + (2.*fCM_34))./(extra3 - fCM_34);
fCMe_23 = (ep34 - eh2)./(ep34 + (2*eh2));
epe23 = eh2.*(extra2 + (2.*fCMe_23))./(extra2 - fCMe_23);
fCM2 = (epe23 - eh1)./(epe23 + (2*eh1));          % Clausius Mossotti factor

% Then the plotting part.
% The real parts are plotted as solid lines, the imaginary parts as dotted lines

semilogx(f,real(fCM0));             hold on
semilogx(f,imag(fCM0),':');         hold off
axis([1e1 1e9 -1 1]);               grid

figure
semilogx(f,real(fCM1));             hold on
semilogx(f,imag(fCM1),':');         hold off
axis([1e1 1e9 -1 1]);               grid

figure
semilogx(f,real(fCM2));             hold on
semilogx(f,imag(fCM2),':');         hold off
axis([1e1 1e9 -1 1]);               grid
```

A.2   FlexPDE®

FlexPDE® is a program designed to solve partial differential equations in two or three dimensions using the Finite Element Method (FEM) (Zienkiewicz and Taylor 1989, Backstrom 1999). The method works by dividing the problem space into a mesh of triangles in 2D, or tetrahedra in 3D. The solution in each element is represented by a second or third order polynomial function.

The program performs an adaptive refinement of the mesh by placing more elements where the error is greatest, in order to minimise the error in the solution.

In solving the electrostatic field, we begin with Poisson's equation for the electrical potential

$$\nabla \cdot (\varepsilon \nabla \phi) = 0 \qquad \qquad (A.2)$$

The boundaries of the system, *i.e.* the electrode array, the water and the glass surrounding the electrodes are then defined in terms of boundary conditions.

A.2.1   Boundary conditions

The electrodes are defined to be equipotentials, with a value equal to the applied voltage. Since the thickness of the electrode (~100 nm) is small compared with the size of the electrodes and the enclosing chamber (~100 μm), they can be represented by sections of the boundary between the glass and the water.

At interfaces between the electrolyte and the glass, the total normal current, free plus displacement, must be continuous. The jump condition at the interface is

$$(i\omega\varepsilon + \sigma)\frac{\partial \phi}{\partial y} = (i\omega\varepsilon_G + \sigma_G)\frac{\partial \phi_G}{\partial y} \qquad \qquad (A.3)$$

where ε_G, σ_G and ϕ_G are the electrical permittivity, conductivity and potential in the glass. The conductivity and permittivity of the glass are significantly smaller than those of the electrolyte and the boundary condition simplifies to the Neumann condition

$$\frac{\partial \phi}{\partial y} = 0 \qquad \qquad (A.4)$$

The boundary condition at any electrolyte/glass interface can be represented by this boundary condition to a reasonable degree of accuracy. It is important to note that if there is an upper chamber interface close to the electrodes, this must be included in the problem space.

Using the Finite Element Method, the solution is more accurate when a greater number of elements are used. However, this is at the expense of increased computation time. If the electrode array contains natural symmetry, odd and even mirror boundaries can be used to reduce the size of the problem space and increase accuracy of the final solution. The Neumann boundary condition (equation A.4) is the boundary condition of even symmetry; the boundaries are mirrored but with the same charges or potentials on either side. The Dirichlet boundary condition is $\phi = 0$ and this gives odd symmetry with boundaries mirrored but with opposite charges or potentials on either side. Combinations of boundary conditions can be used to represent an entire electrode array using the bare minimum of problem space.

In the following section, a FlexPDE® descriptor is reproduced. The electrode is represented by a section of the lower boundary with a constant value. The lower and upper glass surfaces of the chamber are described by the Neumann condition. Only one electrode is included in the problem space, with the edges of the problem space described by Dirichlet conditions at a distance from the electrode edge. This boundary condition creates electrodes of opposite potential twice as far away. The presence of two mirror boundaries repeats the array to infinite distance, an accurate representation of the potential in the middle of the array but not near the ends.

A.2.2   FlexPDE® descriptor

This descriptor file should be saved as a text filed with the suffix ".pde" and run in FlexPDE®. It will produce numerically calculated plots of the two-dimensional electrical potential, field and dielectrophoretic force for one electrode in an infinite interdigitated array. The size and scale of the electrode can be altered, as well as the applied potential. Compare these plots with those obtained using Mathematica® (next section) which plots the analytical solution for the field and DEP force.

```
title    'bar.pde'
         { solves the Laplace for interdigitated bar electrodes: solves over the whole
         of one electrode and plots the potential, field and grad(E^2) for the
         dielectrophoretic force }

select
         errlim = 1e-5            { sets the max value of error in convergence }
         ngrid = 100             { sets the starting node density in the mesh }

variables                         { sets the variables to be solved }
         phi                       { electrical potential }

definitions                       { defines problem parameters and values to be
                                   calculated from the resulting solution of phi }
         scale = 1e-6             { scale }
         L1 = 100*scale          { height of problem space }
         L2 = 5*scale            { half electrode width }
         L3 = 10*scale           { spacing d }
         L4 = 20*scale           { plot size }
```

```
        Vo = 1                      { voltage applied to electrode. NOTE value of 1 on
                                    one electrode means total potential difference
                                    across two electrodes of amplitude 2 and peak to
                                    peak value 4 }
        E = -grad(phi)              { electric field }
        magE = magnitude(E)         { electric field magnitude }
        gradEsqu = grad(magE^2)     { gradient of field mag squared for DEP force }
        magDEP = magnitude(gradEsqu)   { magnitude }

equations                          { defines the equation to be solved }
        div(grad(phi)) = 0          { Laplace's equation }

boundaries                         { defines the problem space as a boundary and the
                                    boundary conditions that define the problem }
        region 1
        start(-L3,0)
        natural(phi) = 0  line to (-L2,0)    { Neumann condition at glass }
        value(phi) = Vo   line to (L2,0)     { the electrode: equipotential value 1 }
        natural(phi) = 0  line to (L3,0)     { Neumann condition at glass }
        value(phi) = 0    line to (L3,L1)    { mirror boundary: Dirichlet condition }
        natural(phi) = 0  line to (-L3,L1)   { Neumann condition }
        value(phi) = 0    line to finish     { mirror boundary: Dirichlet condition }

plots                              { defines the plots to be shown of the solved
                                    potential and the parameters calculated from it }
        contour(phi)      zoom(-L3,0,L4,L4)
                                    { contour plot of equipotentials }
        vector(E)         norm zoom(-L3,0,L4,L4)
                                    { arrow plot of electric field }
        contour(magE)     log painted zoom(-L3,0,L4,L4) fixed range(5000,5e6)
                                    { colour plot of field magnitude, log scale }
        vector(gradEsqu) norm zoom(-L3,0,L4,L4)
                                    { arrow plot of grad(E^2), direction of positive DEP }
        contour(magDEP) log painted zoom(-L3,0,L4,L4) fixed range(1e13,1e20)
                                    { colour plot of grad(E^2) magnitude, log scale }
end
```

A.3   Mathematica®

The following script is a simple file that sums the Fourier series to give the analytical expression for the electric field above and interdigitated electrode array. The electrode gap and spacing is d and the applied voltage is Vo. The variable M tells the program how many terms to sum the series. If $M = 1$, the only the first term is calculated. M can also be set to infinity; to sum all terms in the series. The final line in the program plots the field gradient as a vector plot.

A.3.1   Mathematica® script

```
(************** Content-type: application/mathematica **************

            Mathematica-Compatible Notebook

This notebook can be used with any Mathematica-compatible application, such as
Mathematica, MathReader or Publicon. The data for the notebook starts with the
line containing stars above.

To get the notebook into a Mathematica-compatible application, do one of the
following:

* Save the data starting with the line of stars above into a file with a name ending in
.nb, then open the file inside the application;

* Copy the data starting with the line of stars above to the clipboard, then use the
Paste menu command inside the application.

Data for notebooks contains only printable 7-bit ASCII and can be sent directly in
email or through ftp in text mode.   Newlines can be CR, LF or CRLF (Unix,
Macintosh or MS-DOS style).

NOTE: If you modify the data for this notebook not in a Mathematica-compatible
application, you must delete the line below containing the word CacheID, otherwise
Mathematica-compatible applications may try to use invalid cache data.

For more information on notebooks and Mathematica-compatible applications,
contact Wolfram Research:
  web: http://www.wolfram.com
  email: info@wolfram.com
  phone: +1-217-398-0700 (U.S.)

Notebook reader applications are available free of charge from Wolfram Research.
**************************************************************)
```

```
(*CacheID: 232*)

(*NotebookFileLineBreakTest
NotebookFileLineBreakTest*)
(*NotebookOptionsPosition[    4120,      119]*)
(*NotebookOutlinePosition[    4765,      141]*)
(* CellTagsIndexPosition[    4721,      137]*)
(*WindowFrame->Normal*)

Notebook[{Cell[BoxData[\(\(\(Vi =
    Table[Vo*\(-Cos[wt + \((2  i - 1)\)*Pi/2]\), {i, 1,
      8}]\)\(;\)\)\(\ \ \)\( (*\ Sets\ up\ a\ table\ of\ voltages, \
    scaled\ by\ Vo\ *) \)\)\)\)], "Input"],

Cell[BoxData[\(\(\(M = 1\)\(;\)\)\(\ \ \)\( (*\ The\ number\ in\ the\ series\ *) \)\)\)\)], "Input"],

Cell[BoxData[\(\(\(V = \(-\((8  Vo/\((Pi^2)\))\)\)\)*
        Sum[\((Exp[\(-\((\((2*n)\) - 1)\)\)*Pi*y/\((2*d)\))]/\((\(\((\((2*n)\) - 1)\))\)\)^2)\)\n\t\t\t\t*\
\((Sin[\(\(\((\((2*n)\) - 1)\ \)\) \[Pi]\)\)/2]\ *\ Sin[\(3\ \(\((\((2*n)\) - 1)\) \[Pi]\)\)/2]*
Cos[\(\(\(\((\((2*n)\) - 1)\) \[Pi]\ x\)\)/\(2\ d)\)] + 4*Cos[\(\(\(\((\((2*n)\) - 1)\)\ \[Pi]\)\)/4]\ \
Sin[\(\(\(\((\((2*n)\) - 1)\)\ \[Pi]\)\)/4]\^2\ Sin[\(5\ \(\((\((2*n)\) - 1)\)\ \
\[Pi]\)\)/4]\ \ Sin[\(\(\(\ \)\)\(\(\(\((\((2*n)\) - 1)\)\ \[Pi]\ x\)\)/\(2\ d)\)])\), \
{n, 1, M}];\)\(\[IndentingNewLine]\) \( (*\ sums\ the\ series\ to\ M\ *) \)\)\)\)], "Input"],

Cell[BoxData[\(<< Calculus`VectorAnalysis`\)], "Input"],

Cell[BoxData[\(\(SetCoordinates[Cartesian[x, y, z]];\)\)], "Input"],

Cell[BoxData[\(\(\(Field = Evaluate[Grad[\(-V\)]]\)\(;\)\)\(\ \ \)\( (*\
    calculate\ the\ electric\ field\ *) \)\)\)\)], "Input"],

Cell[BoxData[\(\(\(Force = Evaluate[2*Pi*Grad[\((Field[\([1]\))]^2\ + \
        Field[\([2]\)]^2)\)]]\)\(;\)\)\(\ \ \ \ \)\( (*\
    calculate\ the\ relative\ DEP\ Force\ *) \)\)\)\)], "Input"],

Cell[BoxData[RowBox[{"<<", StyleBox["Graphics`PlotField`", "MR"]}]], "Input"],

Cell[BoxData[RowBox[{"<<", StyleBox["Graphics`Graphics`", "MR"]}]], "Input"],

Cell[BoxData[\(\(\(Force = FullSimplify[Grad[\((Field[\([1]\))]^2\ + \
        Field[\([2]\)]^2)\)]]\)\(\[IndentingNewLine]\) \( (*\
    gives\ the\ analytical\ solution\ for\ the\ first\ term\ in\ the\ \
series, \ i . e . \ when\ M\  = \ 1\ *) \)\)\)\)], "Input"],
```

Cell[BoxData[{\(d\ = \ 10\ 10\^\(-6\); \ Vo = 1;\ \ (*\ assigne\ values\ to\ variables, \
    d\ being\ the\ electrode\ gap\ and\ width\ *) \), \
"\[IndentingNewLine]", \(PlotVectorField[{Field[\([1]\)], Field[\([2]\)]}, {x, 0,
    40*10\^\(-6\)}, {y, 0, 20*10\^\(-6\)}]\ \ \ (*\
    plots\ the\ vector\ field\ for\ 0\ to\ 40 um\ and\ 0\ to\ 20 um\ *) \)}], "Input"]
},
FrontEndVersion->"4.1 for Microsoft Windows",
ScreenRectangle->{{0, 1024}, {0, 695}},
WindowSize->{819, 494},
WindowMargins->{{34, Automatic}, {Automatic, 45}}
]

(***
Cached data follows. If you edit this Notebook file directly, not using Mathematica,
you must remove the line containing CacheID at the top of the file. The cache data
will then be recreated when you save this file from within Mathematica.
***)

(*CellTagsOutline
CellTagsIndex->{}
*)

(*CellTagsIndex
CellTagsIndex->{}
*)

(*NotebookFileOutline
Notebook[{Cell[1705, 50, 203, 4, 50, "Input"],
Cell[1911, 56, 108, 2, 30, "Input"],
Cell[2022, 60, 684, 11, 122, "Input"],
Cell[2709, 73, 60, 1, 30, "Input"],
Cell[2772, 76, 72, 1, 30, "Input"],
Cell[2847, 79, 136, 2, 30, "Input"],
Cell[2986, 83, 224, 5, 50, "Input"],
Cell[3213, 90, 97, 3, 30, "Input"],
Cell[3313, 95, 96, 3, 30, "Input"],
Cell[3412, 100, 297, 7, 50, "Input"],
Cell[3712, 109, 404, 8, 70, "Input"]}
]
*)

(***
End of Mathematica Notebook file.
***)

A.4　　References

Zienkiewicz O.C. and Taylor R.L.　*The Finite Element Method (4th ed.)*　McGraw-Hill Book Co., London　(1989).

Backstrom G.　*Fields of physics by finite element analysis, an introduction* Studentlitteratur, Lund, Sweden　(1998).

Appendix B

Units, constants and notation

The information in this Appendix was compiled from a number of textbooks but mainly from the Bureau de Poids et Mesures, the agency responsible for maintaining the International System (SI) of Units (www.bipm.fr) and the National Institute of Standards and Technology (NIST: www.nist.gov).

The Appendix is divided into several sections, giving information on SI units and the values of physical constants from the above sources. A complete list of these would be lengthy and we have restricted ourselves to those relevant to this book. The third section contains a list of definitions of the symbols used in the book and the relevant units. The last section gives the definitions of some mathematical operators.

B.1   Units

The International System is based on seven well-defined units, referred to as *base units*. The units for other quantities are derived from these base units and are referred to as *derived units*, which we have separated into those with and without special symbols. In addition, there are some non-SI units commonly used in some fields which we have included along with the SI conversion factor.

B.1.1   Base units

quantity	SI base unit	symbol
electrical current	Ampere	A
luminous intensity	candela	cd
temperature	Kelvin	K
mass	kilogram	kg
length	metre	m
amount of substance	mole	mol
time	second	s

B.1.2   SI-derived units with special symbols

quantity	SI derived unit	symbol	expressed in SI units	expressed in SI base units
electric charge	Coulomb	C		$A\,s$
electric potential difference	Volt	V	$W\,A^{-1}$	$A^{-1}\,kg\,m^2\,s^{-3}$
capacitance	Farad	F	$C\,V^{-1}$	$A^2\,kg^{-1}\,m^{-2}\,s^4$
electric resistance	Ohm	Ω	$V\,A^{-1}$	$A^{-2}\,kg\,m^2\,s^{-3}$
electric conductance	Siemens	S	Ω^{-1}	$A^2\,kg^{-1}\,m^{-2}\,s^3$
magnetic flux	Weber	Wb	$V\,s$	$A^{-1}\,kg\,m^2\,s^{-2}$
inductance	Henry	H	$Wb\,A^{-1}$	$A^{-2}\,kg\,m^2\,s^{-2}$
force	Newton	N	$J\,m^{-1}$	$kg\,m\,s^{-2}$
pressure, stress	Pascal	Pa	$N\,m^{-2}$	$kg\,m^{-1}\,s^{-2}$
energy, work done	Joule	J	$N\,m$	$kg\,m^2\,s^{-2}$
power	Watt	W	$J\,s^{-1}$	$kg\,m^2\,s^{-3}$
plane angle	radian	rad	1	$m\,m^{-1}$
molar concentration	Molar	M		$mol\,l^{-1}$

B.1.3   Other SI-derived units

derived quantity	SI derived unit	symbol
electric field	Volts per metre	$V\,m^{-1}$
area	square metre	m^2
volume	cubic metre	m^3
velocity	metre per second	$m\,s^{-1}$
acceleration	metre per second squared	$m\,s^{-2}$
mass density	kilogram per cubic metre	$kg\,m^{-3}$
current density	Ampere per square metre	$A\,m^{-2}$
concentration	mole per cubic metre	$mol\,m^{-3}$

B.1.4   Non-SI units

quantity	value in SI units	symbol
atomic mass unit	$1.6605402 \times 10^{-19}\,kg$	au
atomic dipole moment	$3.33564 \times 10^{-30}\,A\,m\,s$	debye

B.2   Constants

constant	symbol	symbolic value	numerical value	units
pi	π		3.14159265	–
base of natural logs	e		2.718281828	–
Avogadro's number	N_A		$6.02214199 \times 10^{23}$	mol^{-1}
Boltzman constant	k_B		$1.3806503 \times 10^{-23}$	$J\ K^{-1}$
Gas constant	R	$N_A k_B$	8.314472	$J\ mol^{-1}\ K^{-1}$
elementary (electron) charge	q		$1.602176462 \times 10^{-19}$	C
Faraday	F	$N_A q$	96,485.3415	$C\ mol^{-1}$
Planck constant	h		$6.62606876 \times 10^{-34}$	J s
speed of light	c		2.99792458×10^{8}	$m\ s^{-1}$
Bohr radius	a_0		$0.5291772083 \times 10^{-10}$	m
permeability of free space	μ_o		$4\pi \times 10^{-7}$	$N\ A^{-2}$
permittivity of free space	ε_o	$\sqrt{c^2 / \mu_o}$	$8.854187817 \times 10^{-12}$	$F\ m^{-1}$

B.3   Symbols

This is a list of the symbols used in the book. If the symbol you are looking for does not appear in the list, other symbols can be found in the preceding tables for constants and units.

symbol	notes	units	base units
a	Particle radius		m
a_n	Half length of ellipsoid along axis n		
b	Average distance between charge sites		m
c	Concentration	$mol\ l^{-1}$	$10^3\ mol\ m^{-3}$
c_j	Concentration of ion j		
c_p	Concentration of phosphate groups: DNA		
c_p	Specific heat at constant pressure	$J\ kg^{-1}\ K^{-1}$	$K^{-1}\ m^2\ s^{-2}$
$\mathbf{d}$	Displacement vector		m
$\mathbf{d}$	Dipole charge separation vector		
d	Magnitude of $\mathbf{d}$		
d	Distance		

d_s	Thickness of Stern layer		m
$\tilde{d}_e$	Clausius-Mossotti factor equivalent		–
f	Frequency	Hz	s^{-1}
f_{MW}	Maxwell-Wagner relaxation frequency		
f_c	Charge relaxation frequency		
f_o	Zero force frequency		
f	Friction factor	$kg\,s^{-1}$	$kg\,s^{-1}$
f_θ	Rotational friction factor	$N\,m\,s\,rad^{-1}$	$kg\,m^2\,s^{-1}\,rad^{-1}$
$\tilde{f}_{CM}$	Clausius-Mossotti factor		–
f	Body force	$N\,m^{-3}$	$kg\,m^{-2}\,s^{-2}$
f_g	Gravitational body force		
f_E	Electrical body force		
f	Function relating electrophoretic mobility and zeta potential		–
g	Gravitational acceleration		$m\,s^{-2}$
g	Magnitude of g		
h	Unit (hour)	h	$\times 3600s$
h	Height		m
i	Complex unit $i^2 = -1$		–
k	Thermal conductivity	$J\,s^{-1}\,K^{-1}\,m^{-1}$	$K^{-1}\,kg\,m\,s^{-3}$
k_{DEP}	Exponential factor: far field DEP force		–
l_o	Typical length		m
l_{entry}	Entry length for Poiseuille flow		m
m	Mass		kg
m	Dimensionless electroosmotic contribution factor		–
m	Number of particles in a pearl chain		–
m	(subscript) medium		–
n	Number density		m^{-3}
n_o	Bulk number density		
p	Pressure	Pa	$kg\,m^{-1}\,s^{-2}$
p_o	Typical pressure		
p'	Dimensionless pressure		–
p	(subscript) particle		–

p	Dipole moment	C m	A m s
p$_{av}$	Average dipole moment		
p$_n$	Dipole moment along axis n	Debye	
$\tilde{\mathbf{p}}$	Dipole moment phasor		
r	Radial direction in (spherical) polar co-ordinates		–
r	Position vector in spherical polar co-ordinates		m
$\hat{\mathbf{r}}$	Unit vector in radial direction		–
s	Arbitrary distance of integration		m
t	Time		s
t_o	Typical time/time of observation		
t'	Dimensionless time		–
u	Fluid velocity		m s^{-1}
u_o	Typical fluid velocity		
u_s	Velocity of sound		
u'	Dimensionless fluid velocity		–
v	Particle velocity		m s^{-1}
v$_t$	Terminal velocity		
v$_c$	Mean charge drift velocity		
v$_{DEP}$	Dielectrophoretic velocity		
x	x co-ordinate		–
x	(subscript) x component of vector		–
x	Position vector		m
$\hat{\mathbf{x}}$	Unit vector in x-direction		–
x'	Dimensionless position		–
y	y co-ordinate		–
y	(subscript) y component of vector		–
$\hat{\mathbf{y}}$	Unit vector in y-direction		–
z	z co-ordinate		–
z	(subscript) z component of vector		–
$\hat{\mathbf{z}}$	Unit vector in z-direction		–
z_j	Valency of ion j		–
A	Hamaker constant		–
A	Area		m^2
A	Constant in DC impedance		–

A	Stability factor (Chapter Eleven)		–
A_n	Depolarising factor along axis n		–
A_n	Fourier coefficients in field expansion		–
A_{DEP}	Magnitude coefficient in far field DEP force		–
B	Factor in calculation of depolarising factor		–
C	Capacitance	F	A s
C_{DL}	Double layer capacitance		
C'	Specific capacitance	F m^{-2}	A m^{-2} s
C'_{DL}	Specific double layer capacitance		
C'_d	Specific diffuse layer capacitance		
C'_s	Specific Stern layer capacitance		
D	Diffusion coefficient		m^2 s^{-1}
$\mathbf{D}$	Electric flux density	C m^{-2}	A m^{-2} s
Du^d	Dukhin number		–
$\mathbf{E}$	Electric field	V m^{-1}	A^{-1} kg m s^{-3}
$\tilde{\mathbf{E}}$	Electric field phasor		
$\mathbf{E}'$	Local electric field		
$\mathbf{E}_n$	Electric field along axis n		
$\mathbf{E}_{rms}$	Root-mean-square electric field		
$\mathbf{E}_0$	Applied electric field		
$\mathbf{E}_1$	Perturbation electric field		
$\mathbf{F}$	Force	N	kg m s^{-2}
$\mathbf{F}_{arb}$	Arbitrary force		
$\mathbf{F}_{EP}$	Electrophoretic force		
$\mathbf{F}_{DEP}$	Dielectrophoretic force		
$\mathbf{F}_\eta$	Viscous drag force		
$\mathbf{F}_q$	Force on a charge		
$\mathbf{F}_g$	Gravitational force		
$\mathbf{F}_{Bouyancy}$	Buoyancy force		
$\mathbf{F}_{Rand}$	Random force (Brownian motion)		
$\mathbf{F}_{thr}$	Threshold force		
I	First ionisation potential	J	kg m^2 s^{-2}
$\mathbf{J}$	Flux		m^{-2} s^{-1}
$\mathbf{J}_{Diff}$	Diffusion flux		

Symbol	Description	Unit	SI base
J	Electrical current density	$C\,m^{-2}\,s^{-1}$	$A\,m^{-2}$
$\mathbf{J}_c$	Conduction current density		
$\mathbf{J}_d$	Displacement current density		
$\mathbf{J}'$	Surface current density	$C\,m^{-1}\,s^{-1}$	$A\,m^{-1}$
K_s	Surface conductance	S	$A^2\,kg^{-1}\,m^{-2}\,s^3$
$\tilde{K}_n$	Clausius-Mossotti factor equivalent along axis n		$-$
M	M-factor from diffuse layer polarisation		$-$
N	Total number of particles		$-$
P	Probability		$-$
P	Polarisation	$C\,m^{-2}$	$A\,m^{-2}\,s$
$\mathbf{P}_{ae}$	Atomic and electronic polarisation		
$\mathbf{P}_{or}$	Orientational polarisation		
Q	Charge	C	A s
Q	Volume flow rate		$m^3\,s^{-1}$
R	Angular velocity/Rate of rotation	$rad\,s^{-1}$	$rad\,s^{-1}$
R	Resistance	Ω	$A^{-2}\,kg\,m^2\,s^{-3}$
R_{DL}	Double layer resistance		
R_P	Particle resistance		
R_T	Total resistance		
S	Surface of integration		$-$
T	Temperature		K
U	Potential energy	J	$kg\,m^2\,s^{-2}$
U	Debye interaction energy		
V, V_o	Applied voltage	V	$A^{-1}\,kg\,m^2\,s^{-3}$
Z	Impedance	Ω	$A^{-2}\,kg\,m^2\,s^{-3}$
Z_{DL}	Double layer impedance		
Z_p	Particle impedance		
Z_{ch}	Pearl chain impedance		
Z_E	Electrode suspending medium		

Symbol	Description	Units	Units
α	Polarisability		
$\tilde{\alpha}$	Complex polarisability		
α_e	Electronic polarisability		
α_a	Atomic polarisability		
α_{ae}	Atomic & electronic polarisability	F m^2	$\text{A}^2\,\text{kg}^{-1}\,\text{s}^4$
α_d	Orientational polarisability		
α_i	Counterion/interfacial polarisability		
α_T	Total polarisability		
$\tilde{\alpha}_n$	Complex polarisability along axis n		
β	Power in spread of relaxation times		–
β	Power in complex phase impedance		–
χ	Electric susceptibility		
χ_{ae}	Atomic/electronic susceptibility		–
χ_{or}	Orientational susceptibility		
δ	Loss angle		–
δ	Dielectric increment/decrement		–
ε	Permittivity		
ε_o	Permittivity of free space		
$\tilde{\varepsilon}$	Complex permittivity		
ε'	Real part of permittivity		
ε''	Imaginary part of permittivity	F m^{-1}	$\text{A}^2\,\text{kg}^{-1}\,\text{m}^{-3}\,\text{s}^4$
ε_{Soltn}	Solution permittivity		
$\varepsilon_{StaticW}$	Static permittivity		
ε_{hf}	High frequency permittivity		
ε_{lf}	Low frequency permittivity		
ε_r	Relative permittivity		
$\tilde{\varepsilon}_{r,d}$	Complex relative permittivity		
$\varepsilon_{r,s}$	Static relative permittivity		–
$\varepsilon_{r,\infty}$	High frequency relative permittivity		
ϕ	Electrical potential		
ϕ_o	Surface potential		
ϕ_d	Potential Stern/diffuse layer interface		
$\tilde{\phi}$	Complex potential phasor	V	$\text{A}^{-1}\,\text{kg m}^2\,\text{s}^{-3}$
ϕ_R	Real part of potential phasor		
ϕ_I	Imaginary part of potential phasor		

ϕ	2nd angle spherical polar co-ordinates	rad	–
γ	Radius ratio in multishell model		–
γ	Factor in Gouy-Chapman theory		–
γ	Factor in Complex dipole expression		–
η	Viscosity	Poise (P) ×0.1 Pa s	$kg\ m^{-1}\ s^{-1}$
φ	Volume concentration factor		–
κ	Reciprocal Debye length		m^{-1}
λ	Ionic conductivity	$S\ m^2\ mol^{-1}$	$A^2\ kg^{-1}\ s^3\ mol^{-1}$
λ_D	Debye length		m
μ	Mobility	$m^2\ V^{-1}\ s^{-1}$	$A\ kg^{-1}\ s^2$
μ_{EP}	Electrophoretic mobility	$m^2\ V^{-1}\ s^{-1}$	$A\ kg^{-1}\ s^2$
μ_{DEP}	Dielectrophoretic mobility	$m^4\ V^{-2}\ s^{-1}$	$A^2\ kg^{-2}\ s^5$
ν	Orbiting frequency	Hz	s^{-1}
θ	Angle	rad	–
θ	1st angle in polar coordinates	rad	–
ρ	Charge density	$C\ m^{-3}$	$A\ m^{-3}\ s$
ρ_b	Bound volume charge density		
ρ_f	Free volume charge density		
ρ_b	Charge density in double layer		
ρ_m	Mass density		$kg\ m^{-3}$
ρ_p	Particle mass density		
σ	Conductivity	$S\ m^{-1}$	$A^2\ kg^{-1}\ m^{-3}\ s^3$
$\tilde{\sigma}$	Complex conductivity		
σ_q	Surface charge density	$C\ m^{-2}$	$A\ m^{-2}\ s$
σ_{qo}	Surface charge density		
σ_{qs}	Diffuse layer surface charge density		
σ_{qd}	Stern layer surface charge density		
τ	Relaxation time		s
τ_{or}	Orientational relaxation time		
τ_{MW}	Maxwell-Wagner relaxation time		
τ_n	Relaxation time along axis n		
τ_A	Acceleration time constant (momentum relaxation time)		
τ_q	Charge relaxation time		
τ_α	Alpha relaxation time		
τ_o	Observation time		

υ	Volume		m^3
ϖ	Dummy integration frequency variable		$rad\ s^{-1}$
ω	Angular frequency		$rad\ s^{-1}$
ω_{MW}	Maxwell-Wagner relaxation frequency		
ω_q	Charge relaxation frequency		
ξ	Charge density parameter		–
ψ	Dimensionless potential		–
ζ	Zeta potential	V	$A^{-1}\ kg\ m^2\ s^{-3}$
Δ	r.m.s. displacement		m
Γ	Torque	N m	$kg\ m^2\ s^{-2}$
Γ_{ROT}	Electrorotational torque		
Λ	Molar conductivity	$S\ m^2\ mol^{-1}$	$A^2\ kg^{-1}\ s^3\ mol^{-1}$
$\Lambda_{c=0}$	Limiting molar conductivity		
Π	Frequency dependent factor in electrothermal force		–
Ω	Dimensionless frequency		–
Re	Reynold's number		–
eRe	Electrical Reynold's number		–
$\Im$	Constant of relation between Λ and $\Lambda_{c=0}$		–
Υ	Retardation coefficient		–

B.4   Operators and notation

symbol	mathematical significance
Δ ...	Change in ...
<...>	Time average operator
|...|	Magnitude operator
∇...	Del operator
.	Dot product
$\times$	Vector cross product
Re[...]	Real part of ...
Im[...]	Imaginary part of ...
*	Complex conjugate
<<	Much less than
>>	Much greater than

Index